$2340

S0-EKW-398

PAID AUG 2 7 1984

systems analysis and design:
a case study approach

systems analysis and design:
a case study approach

Robert J. Thierauf

Chairperson, Department of Management and Information Systems
D. J. O'Conor Memorial Professor of Business Administration
Xavier University

George W. Reynolds

Adjunct Professor, Xavier University
Systems Project Manager

CHARLES E. MERRILL PUBLISHING COMPANY
A Bell & Howell Company
Columbus   Toronto   London   Sydney

Published by
Charles E. Merrill Publishing Company
A Bell & Howell Company
Columbus, Ohio 43216

This book was set in Quorum and Bauhaus.
Production Editor: Cherlyn B. Paul
Cover Design Coordination: Will Chenoweth
Cover photo: Larry Hamill

Copyright © 1980, by Bell & Howell Company. All rights reserved. No part of this book may be reproduced in any form, electronic or mechanical, including photocopy, recording, or any information storage and retrieval system, without permission in writing from the publisher.

Library of Congress Catalog Card Number: 79-90796
International Standard Book Number: 0-675-08172-6

Printed in the United States of America

5 6 7 8 9 10—85 84 83

dedicated to
the D. J. O'Conor Family

preface

This book proposes a new thrust in teaching the traditional systems analysis and design course by stressing an interactive and distributed data processing approach to systems. The concepts presented here enable a systems analyst to operate successfully in the 80s and beyond: a comprehensive treatment of design theory and principles, and great emphasis on the application of systems design to the real world. More specifically, several chapters of the text are devoted to a typical systems design case study—namely, the ABC Company—whereby students are shown how to design an order-entry system. Based upon this illustrated case, they are then required to design a finished product inventory system or an accounts receivable system that complements the one presented. Thus, students are exposed to the underlying principles and practices of systems analysis and design, as well as given an opportunity to design an interactive computerized (management information) system in a distributed processing environment, currently operated in a batch processing mode.

Additionally, emphasis is placed on relating the human factor to the new systems design. Such a presentation helps alleviate students' fear that computer systems are impersonal and insensitive to the human element in business, government, and education today. Consideration for the human element goes a long way toward resolving the problem that people are just numbers; the focus in this book is on utilizing the vast capabilities of personnel at all levels in an organization.

Within the foregoing framework of principles and practices in systems analysis and design, the book contains certain student-oriented learning features. These include
> **chapter objectives and outlines** to help students place each chapter in its proper perspective.

listing of design principles to give students a handle on the important design principles in a typical system project.
master case study to bridge the gap between principles and practices.
summaries of important subjects to help students bring together the important material in the chapter.
self-study exercises consisting of true-false questions, fill-ins, and problems, to help students evaluate their comprehension of the chapter.
answers to self-study exercises in the Appendix for students' convenience.

The structure of the book follows a logical sequence to insure a comprehensive treatment of systems analysis and design. The major areas covered are described below.

Part I: An overview of systems analysis and design Chapter 1 provides an introduction to systems analysis and design, and chapter 2 discusses the standard tools employed by system personnel. In chapter 3, an overview of the major subsystems of the ABC Company (the text's master case study) is presented.

Part II: Systems analysis of present business information system The introductory investigation of a system project is the subject matter of chapter 4. In chapter 5, the detailed investigation of the present system, or systems analysis, is examined, along with the concluding investigation phase of a system project, resulting in an exploratory survey report to top management. Chapter 6 focuses on analyzing the present order-entry system of the ABC Company.

Part III: Systems design of new management information system An overview of systems design is treated in chapter 7. In contrast, chapters 8 and 9 center on the design of system output and data files, respectively. This material provides the background necessary to design the required output reports and data base (chapter 10) for the order-entry system of the ABC Company. Similarly, chapters 11 and 12 set forth the important design criteria for input and procedures, respectively, and chapter 13 applies these criteria to the ABC Company. Lastly, chapter 14 examines the design of system controls, and chapter 15 relates this area to the ABC Company. Hence, all important design criteria from theoretical and practical viewpoints have been presented in sufficient detail to give the student a link between system theory and practice.

Part IV: Beyond systems analysis and design In chapters 16 and 17, equipment selection and system implementation are presented, respectively. These areas complete the essential parts of a typical system project.

The text is appropriate for any course covering the fundamentals of systems analysis and design, whether taught over a quarter or semester. It can be helpful in company training programs where an understanding of the design efforts involved in a system project is desired. Also, it can be supplemented with other material as deemed important by the instructor. Thus, the text is flexible enough to accommodate most learning experiences related to systems analysis and design.

In an undertaking of such magnitude, we are deeply indebted to the many people who have contributed their time and effort to make this publication possible. We are grateful to Richard Bialac and William Kramer, both of Xavier University, as well as Ronald Black, Rhode Island Junior College, and H. C. Munns, Daytona Beach Community College, for their helpful reviews in earlier stages of this project.

August, 1979

Robert J. Thierauf
George W. Reynolds

contents

1   **part one: an overview of systems analysis and design**

3   **chapter 1: introduction to systems analysis and design**

Introduction to Systems Analysis and Design, 6; Current Management Information Systems, 9; Underlying Concepts of Management Information Systems, 18

29   **chapter 2: standard tools of systems analysis and design**

Overview of Standard Tools of Systems Analysis and Design, 31; Flowcharts, 32; Flowchart Types, 34; Decision Tables, 43; Sampling Techniques, 45; Time-Charting Techniques in a System Project, 48

61   **chapter 3: major subsystems—abc company**

Overview of Business Organizations, 63; ABC Company—Master Case Study, 64; Major Subsystems in Industrial Organizations—ABC Company, 70; Major Subsystems in Service Organizations, 75; Relationship of Systems Analysis and Design to Major Subsystems, 75

xiii

part two: systems analysis of present business information system
p. 81

chapter 4: initiation of a system project
p. 83

System Project Affects the Entire Organization, 85; Major Steps of a System Project, 87; Introductory Investigation, 88; Organization of System Project, 89; Definition of the Scope of System Project, 92; Determination of the Schedule of System Project, 95

chapter 5: systems analysis of present system
p. 99

Objectives of Systems Analysis, 101; Methods of Systems Analysis, 102; Systems Analysis—Detailed Investigation of Present System, 103; Systems Analysis—Concluding Investigation, 110; Systems Analysis—Exploratory Survey Report to Top Management, 119

chapter 6: systems analysis of the batch order-entry system—abc company
p. 125

Introductory Investigation of New Order-Entry System, 127; Systems Analysis—Detailed Investigation of Present Batch Order-Entry System, 134; Systems Analysis—Concluding Investigation of New Order-Entry System, 146; Systems Analysis—Exploratory Survey Report to Top Management, 153; Case Study 1: Finished Project Inventory, 158; Case Study 2: Accounts Receivable, 162

part three: systems design of new management information system
p. 167

chapter 7: overview of systems design
p. 169

Approaches to Systems Design, 171; Imaginative Systems Design, 172; Characteristics of a Well-Designed System, 173; Steps in Systems Design, 178

chapter 8: design of system output
p. 195

Common Types of System Output, 198; Incorporate System Design Principles—Output, 203; Design of System Output, 205

chapter 9: design of system data files
p. 225

Common Types of File Storage, 228; Common Methods of File Organization, 239; Incorporate System Design Principles—Data Files, 243; Design of System Data Files, 246; Data Base Management Systems, 257

265   **chapter 10: systems design of order-entry output and data base—abc company**

Design of Order-Entry Output—ABC Company, 268; Design of Order-Entry Data Base—ABC Company, 280; Case Study 1: Finished Product Inventory, 292; Case Study 2: Accounts Receivable, 299

303   **chapter 11: design of system input**

Common Methods of System Input, 306; Incorporate System Design Principles—Input, 316; Design of System Input, 318

333   **chapter 12: design of system methods and procedures**

Basic DP Functions that Comprise System Methods and Procedures, 336; Incorporate System Design Principles—Methods and Procedures, 343; Design of System Methods and Procedures, 345

357   **chapter 13: systems design of order-entry input and procedures—abc company**

Design of Order-Entry Input—ABC Company, 359; Design of Order-Entry Procedures—ABC Company, 374; Case Study 1: Finished Product Inventory, 385; Case Study 2: Accounts Receivable, 389

391   **chapter 14: design of system controls**

Introduction to Design of System Controls, 393; Internal Control, 396; Incorporate System Design Principles—Controls, 400; Design of System Controls, 402; Questionnaire for Evaluating System Controls, 413

419   **chapter 15: systems design of order-entry controls—abc company**

Design of Order-Entry Controls—ABC Company, 421; Case Study 1: Finished Product Inventory, 435; Case Study 2: Accounts Receivable, 436

437   **part four: beyond systems analysis and design**

439   **chapter 16: equipment selection**

Equipment Selection Process, 442

455   **chapter 17: system implementation**

System Implementation, 457; Operational Review of System, 469

475	appendix: answers to self-study questions
503	bibliography
505	index

part one

an overview of systems analysis and design

chapter 1

introduction to systems analysis and design

OBJECTIVES:

- To distinguish between systems analysis and design—the subject matter of this text.
- To demonstrate the relationship of systems analysis to systems design in a typical system project.
- To examine the basic types of management information systems that can be designed.
- To set forth the underlying concepts of current management information systems.

IN THIS CHAPTER:

    The Challenge of the Computer World for Systems Analysts
INTRODUCTION TO SYSTEMS ANALYSIS AND DESIGN
    What Is Systems Analysis?
    What Is Systems Design?
    Relationship of Systems Analysis to Systems Design
CURRENT MANAGEMENT INFORMATION SYSTEMS
    Batch MIS
    Batch Processing Mode
    Interactive MIS
    Interactive Processing Mode
    Relationship of Systems Design to Management Information Systems
UNDERLYING CONCEPTS OF MANAGEMENT INFORMATION SYSTEMS
    The Management Information System Concept
    The Distributed Processing System Concept
    The Modular System Concept
CHAPTER SUMMARY
QUESTIONS
SELF-STUDY EXERCISE

The analysis, design, and implementation of a new system, no matter what type, is not a simple task. In order to install the new system effectively, a detailed and carefully devised plan must be initiated by management and followed by all personnel. The large outlays for equipment, programming, conversion, and related activities require systematic procedures for implementation. Otherwise, vast sums of money can be wasted. From this perspective, the purpose of this and succeeding chapters is to explore the steps from a feasibility study through system implementation for a typical system project. The end result of these systematic steps is an effective system that both conforms with organization objectives and lies within the time and budget constraints of the system project.

Initially, the chapter examines the challenging world of computers and defines the terms **systems analysis** and **systems design.** Because the text will center on the analysis and design of management information systems (MIS), current computer approaches within an MIS environment will be investigated. The main emphasis today is on computer systems having the capability to interact with the user, that is, an interactive MIS mode. Additionally, emphasis in the chapter is placed on the underlying concepts of management information systems. These provide a basis for placing current and future MIS trends in their proper perspective. As in all chapters of the text, important material is summarized periodically throughout the chapter.

The Challenge of the Computer World for Systems Analysts

Behind the flashing lights, spinning tape reels, and subdued hum of the modern computer lies the story of one of the world's most dynamic fields. Despite its fantastic growth, the computer is still in its infancy; many new and exciting applications are being developed every day. The increasing number of computer applications will result in daily encounters between the computer and the average person. As a computerized approach to daily living continues to develop, there will be an increasing demand for systems analysts. Based upon the user's requirements, **systems analysts** devise efficient patterns of information flow from source to computer. They also define the computer process necessary to turn raw data into useful information, plan the distribution and use of results, and test the working system in operation.

To get a glimpse of the challenging future of computers, consider the following events for one household. It is 8:00 A.M. When the alarm clock goes off, the bedroom curtains swing apart and the thermostat raises the heat to the desired level. The percolator in the kitchen starts. Even the back door opens to let out the family pets. The television set comes on with the news, including a selective rundown (ordered the night before) of the latest events affecting the economy. After the news, the TV displays the mail from correspondents who have dictated their messages into the computer network. A button on a bedside genie box is pressed that issues a string of personal and business memos. After a shower (which turned itself on at exactly the right temperature), one household member dresses, eats, and goes out to the car—whose engine, naturally, is running.

Meanwhile, another member of the household concentrates on the TV screen for a readout of comparative prices at the local stores. Following visual consultations with the baker and the grocer, he or she presses a button on the kitchen terminal to order supplies for tonight's dinner party. Other keys on the kitchen terminal order favorite recipes from the memory bank, tell the computer to calculate the ingredients for eight servings, and direct the oven to preheat to the correct temperature for each dish, starting at 5:30 P.M.

Overall, the computer has been an integral part of the daily functioning of this family. Although the foregoing scenario may be years away, the basic technology exists to accomplish all of the tasks described above.

For all its amazing capabilities today and tomorrow, the computer still owes its essential power to **people**. These include individuals who design and build these machines (computer scientists), people who analyze problems and devise solutions using computer technology (systems analysts), those who program instructions and change data into a form usable by the computer (computer programmers), and other individuals who provide service in engineering, sales and marketing, customer service, manufacturing, operation of the computer, and in data entry and related clerical jobs.

Of all the foregoing individuals, systems analysts are foremost in computer application since they constitute the human interface between the computer and the application. Hence, they must be able to think logically; to organize, analyze, and handle data and information systematically; to attend closely to detail and accuracy; and to be imaginative in devising new solutions to existing problems. In essence, systems analysts develop the methods and procedures necessary to implement specific computer applications. Since this text provides information on the analysis and design of computer systems, the authors welcome the reader to this challenging world and hope that its rewards are gratifying.

INTRODUCTION TO SYSTEMS ANALYSIS AND DESIGN

Because analyzing, designing, and implementing a new system is a difficult and costly undertaking, there is great need for a highly systematic and analytic approach in a system project, namely, **systems planning.** Systems planning encompasses three objectives: **what** must be done, **how** to do it, and **when** to do it. There are important relationships among these three objectives. For example, we are generally constrained in what we decide to do by limitations in capabilities and/or resources. Often, we simply do those things that we know how to do and ignore other important areas without weighing the alternatives for best results.

In planning for an MIS environment, a patchwork approach to systems development must be avoided. The patchwork approach can result in the development of unrelated and incompatible subsystems. Thoughtful systems planning centers around a total system that provides coherence of architecture design, methods, standards, operating procedures, and other commonalities important for economy in development, implementation, and operation. The overall system should be structured as a set of integrated subsystems and component parts that are flexible enough to accommodate changes of organizational activities.

Inasmuch as systems planning provides an overall framework for implementing any type of MIS project, it is necessary to examine its two essential elements, the **feasibility study** and **system implementation.** Throughout this text, the focus is on the detailed components of the feasibility study, namely,

- introductory investigation
- systems analysis
- systems design
- equipment selection

In part two, an introductory investigation and systems analysis of the present system are presented, followed by part three—systems design of management information systems. Finally in part four, equipment selection is examined. Thus, the main thrust of this text is on systems analysis and design.

Systems implementation, the second phase in a system project, is also covered in the final part (chapter 17). It should be noted that systems planning is not restricted to implementation of the current project; it also includes establishing long-range plans that must be updated continuously. Periodic review of the installed system determines its continuing relevance to the business environment. This area is also covered briefly in the final chapter. An overview of these parts is found in figure 1.1, p. 8.

What Is Systems Analysis?

Systems analysis centers on analyzing the present system and includes two basic steps: detailed investigation, and a concluding investigation. These steps constitute a logical framework for the systems analysis phase of the feasibility study and serve as a basis for preparing an exploratory survey report to top management.

After an **introductory investigation** or orientation where selected committees are formed, the scope of the project is defined, and comparable items are set forth (as in chapter 4), a **detailed investigation** (first step) of the current system in operation is pursued. This allows the study group to comprehend the full scope of their undertaking. Chapter 5 will outline this phase in greater detail, highlighting the areas to be investigated. In a similar manner, systems analysis of the ABC Company (the text's case study) is the subject matter of chapter 6. Thus, the detailed investigation of the present system is the focal point of systems analysis.

In the **concluding investigation** (second step), the study group reaches one of two pivotal conclusions. Either the present system appears better than any one of the new systems conceived, or one or more new systems appear to be superior to the present system. In the second case, the feasibility study proceeds to the next step—systems

8 Introduction to Systems Analysis and Design

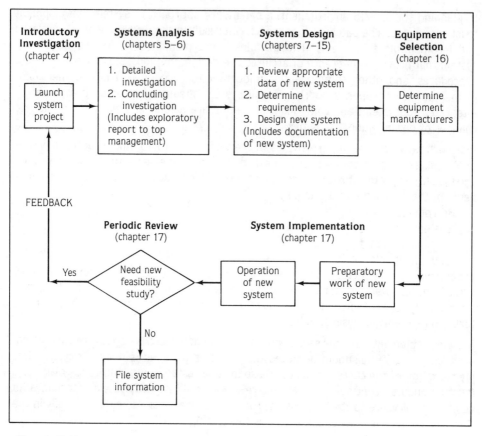

figure 1.1 An overview of the important parts of a system project, specifically, a management information system (MIS) project.

design. In the first case, the study group may go back and examine additional systems or study new equipment developments since the analysis was initiated. If the exploratory survey still indicates a negative answer, the feasibility study will cease, awaiting new developments.

What Is Systems Design?

Like systems analysis, systems design (see chapters 7–15) encompasses certain steps: review appropriate data, determine requirements for a new system, and design the new system. Within this framework, the creative talents of the systems analyst can be employed to the fullest.

Once the data from the exploratory survey report—in particular, the new system recommendation (first step)—have been reviewed, its design requirements must be thoroughly analyzed (second step). Specifically, the following areas are examined: policies consistent with organization objectives, outputs, data base (on-line and off-line files), methods, procedures, inputs, system controls, and system performance. Likewise, consideration at all times must be given to the **human factor.**

The real essence of systems design is devising a system that meets the preceding requirements; this is the third step. At this stage, the creative talents of systems analysts are necessary to develop an efficient and economical system. In essence, **systems design** involves the creation and development of new inputs, a data base, methods, procedures, and outputs for processing business data and producing meaningful operational and managerial reports that conform to organization objectives. And last, the system designed must incorporate good design principles. Documentation is needed during the entire system design process.

A logical starting point for systems design may be an activity that dominates the entire system, an area that will be the most costly for the new system, or an area that will reap many tangible and intangible benefits. Based upon the area or activity selected, each important element must be analyzed and alternative ways of data handling must be developed. As indicated previously, considerable creativity is involved in devising subsystems and their component parts within the framework of the new system recommendation.

Relationship of Systems Analysis to Systems Design

After the system project is launched, systems analysis is an intensive study of the present facts about a system while systems design, the next step, is the creation of a new system. Although these steps chronologically follow one another (figure 1.1), the imagination and creativity involved is a very important difference. On the one hand, these qualities are not necessary to analyze the present system; on the other hand, they figure heavily in imaginative new systems design. Mainly for this reason, systems analysis is generally relegated to junior systems analysts who are capable of systems analysis and straightforward design work, and systems design is given over to the senior systems analysts who are capable of complex design work. In this manner, the special design skills and talents of systems personnel can be employed to their fullest. This is particularly true when innumerable system approaches are available within the framework of the selected system alternative. The almost infinite variety of alternatives make the design task a much more challenging one than systems analysis.

CURRENT MANAGEMENT INFORMATION SYSTEMS

Management information systems can be characterized in two basic ways, batch processing and interactive processing. Under the batch processing approach, transactions are accumulated into batches which are processed periodically. The transactions in a batch may be in sequential or random order. Using the interactive (i.e., on-line real-time) processing approach, records are stored on line and updated as the transactions occur. The term "on line" refers to the fact that input/output devices, data files, and comparable equipment are connected to a computer such that a transaction may be entered at

10 Introduction to Systems Analysis and Design

once or information may be retrieved relatively immediately at any time. The concept of "real time" means that information is entered or retrieved relatively quickly (usually, via a terminal of some type) in order to control the operating environment. When comparing these two in an MIS environment, the requirements of interactive processing for on-line data in real time are considerably different from those of a batch processing mode.

Peripheral file storage devices attached to computer systems can be of two types, sequential access or direct (random) access. In sequential access files, the data are stored in some predetermined order; before a record can be read all preceding file records must be read, as in magnetic tape files. When operating with direct or random access files, the unit is capable of locating and reading any record without having to read other records. Typical on-line computer devices that have this ability are magnetic disks, magnetic drums, and mass storage.

Batch MIS

The batch processing approach to MIS is characterized by the periodic processing of batches of accumulated transactions. Normally, the accumulated transactions must be sorted into a desired order before the transactions are processed further, and a sufficient number of transactions must have accumulated before it is economically feasible to

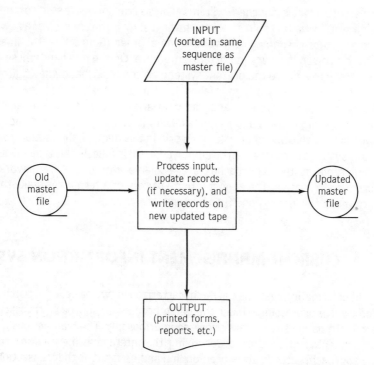

figure 1.2 Batch processing with sequential access file storage on magnetic tape.

process the data further. Batch processing is fundamentally associated with records that are maintained on punched cards, magnetic tapes, and magnetic disks.

Sequential access files   In a sequential access system, the entire master file is read and written each time transactions are processed against the master file. This updating procedure requires sorting all input data (i.e., transactions) in the same sequence as the master file (figure 1.2). Batch processing with sequential access file storage is ideally suited for payroll and accounts payable applications. It usually costs less per transaction to process batches than to process each transaction, as it occurs, immediately on line. However, for other processing runs that require timely information for immediate decisions, the interactive processing approach may be justified.

Direct (random) access files   Batch processing is not limited to sequential access files. In an MIS environment, the batch approach can also use direct, or random, access file devices depicted in figure 1.3. The most common form of random access storage is the magnetic disk which allows the direct updating of the desired record. For applications involving few transactions relative to the size of the master file, the direct access file approach is widely used. An added advantage of direct access over the sequential access file approach is that there is no need to sort the input data into a specific sequence. Also, sequential access updating means new records are created; after updating, both old and new records exist. In direct access, the old record is updated and rewritten to disk. Thus, certain MIS applications are logical candidates for direct access batch processing versus sequential access batch processing.

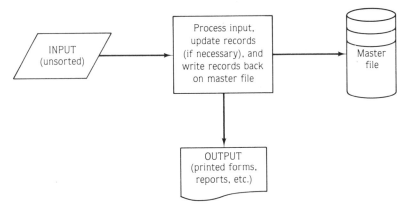

figure 1.3   Batch processing with direct (random) access file storage on magnetic disk.

Batch Processing Mode

The batch processing mode for management information systems may consist of local batch processing or remote job entry processing. Local batch processing means that data, accumulated into batches on the premises of an organization, are sent directly to the computer center for processing, and reports or other types of printed documents are returned. The opposite of on-site batch processing is remote job entry whereby data are sent over a data transmission system to the data center for processing. Reports are returned to the remote site via the same data transmission system.

Local batch   Local batch processing is illustrated in figures 1.2 and 1.3. The outstanding feature of sequential batch processing (figure 1.2) is that a new updated master file is written each time data must be changed. In contrast, direct (random) access batch processing (figure 1.3) does not require the computer to write an entire new file; rather, only those records affected by the current transaction being processed are updated while all others are ignored. Neither of these local batch processing methods utilize data communications equipment since input data are processed on site.

Remote job entry   Remote job entry combines data transmission with batch processing and is used to process data created at points distant from the main computer installation. Based upon a predetermined schedule, data are transferred over the data communications channel to the main computer site in one of two ways: either it is written onto a machine-processable medium, such as magnetic tape or disk, for future processing (off-line) or it is fed into the computer itself for immediate processing (on-line). The transfer of output information involves basically the same steps, but in the opposite direction; that is, from the main computer complex to the outlying data processing locations.

When remote job entry is used in an **off-line mode**, it is because there is not a continuing need for an immediate response. All data are first captured on some machine-processable medium before computer processing at the central facility is initiated. The chief advantage of this off-line approach is the comparatively low data communications cost. The flowchart for off-line remote job entry is depicted in figure 1.4.

The other approach to remote job entry is an **on-line mode** whereby accumulated data are transferred to and from the computer with little manual intervention. It is also possible to maintain a permanent communication link with the central computer, which eliminates the need for establishing a manual connection (possibly via telephone, depending on the mode of operation) when a group of data is to be transferred. This type of remote job entry is utilized where there are a large number of transactions to be transferred and/or a large number of locations for transferring data. The flowchart for on-line remote job entry processing is illustrated in figure 1.5, p. 14. Transferred information is stored on secondary storage devices, such as magnetic disk, drum, or tape, for subsequent processing.

Interactive MIS

Of the two basic types of management information systems, the trend is toward more interactive processing. Transactions and inquiries into the system are processed as they occur. Data stored on-line in direct access files are always up-to-date to reflect the current operating status, whether it be data related to marketing, manufacturing, finance, accounting, personnel, engineering, or some other functional area of an organization. However, it should be noted that magnetic tapes in a batch processing mode are

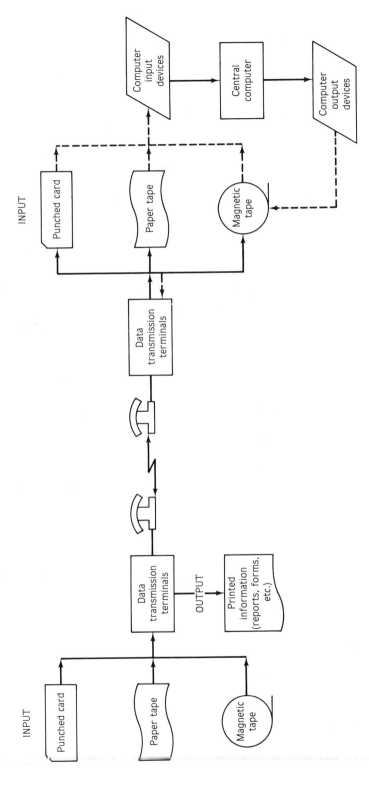

figure 1.4 The transfer of data (input) to and the receipt of information (output) from the central computer facility in an off-line remote job entry mode.

13

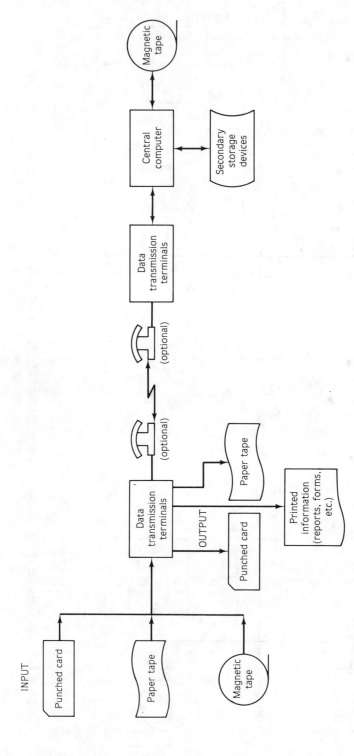

figure 1.5 The transfer of data (input) to and the receipt of information (output) from the central computer facility in an on-line remote job entry mode.

essential to log messages, take snapshots of before-and-after master file records in case of data base recovery, and back up programs and files periodically in a typical interactive system.

Direct (random) access files In an interactive system, transactions and inquiries are processed as they occur. Each transaction originating point has an input/output (I/O) terminal connected to the computer. The terminal is used to send transactions to and receive responses from the central computer complex. All information is stored in direct access file devices that are available at all times for immediate interrogation (figure 1.6). Typical applications include sales order processing, accounts receivable, reservation systems, bank deposit and withdrawal accounting, law enforcement intelligence systems, patient hospital records, savings and loan deposit accounting, and stock market information.

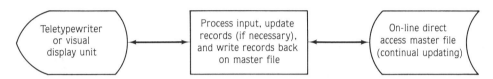

figure 1.6 Interactive processing with direct access file storage.

Interactive Processing Mode

The interactive processing mode found in management information systems may consist of a time-sharing system or a real-time system. A time-sharing system is an on-line system designed to serve the problem-solving needs of many users. On the other hand, a real-time management information system is designed to fulfill the data processing requirements of only one versus many organizations "sharing" a time-sharing system. The operating environmental characteristics of both systems are quite similar.

Time sharing Time-sharing systems utilize remote keyboard terminals for program development and testing, as well as for entering data into and retrieving information from the system. Figure 1.7 (next page), shows secondary storage devices of varying types (namely, direct and sequential) are employed. To service a large number of users, the amount of time available to each incoming program is allocated and controlled by the central processor. The computer controls the input terminals in such a manner that each user appears to have sole access to the computer. In reality, the computer is servicing many different programs at the same time. With this time-sharing approach, one computer is capable of handling a large number of users, each unaware of the others.

An important advantage of time sharing, like real time, is its "conversational mode," that is, its ability to analyze each line of the program for syntax as the program is being entered. As errors are detected by the computer, they are typed or displayed on the user's terminal for corrective action. In the same manner, data entered via input/output terminals are tested for accuracy or reasonableness before further processing occurs. If

16 Introduction to Systems Analysis and Design

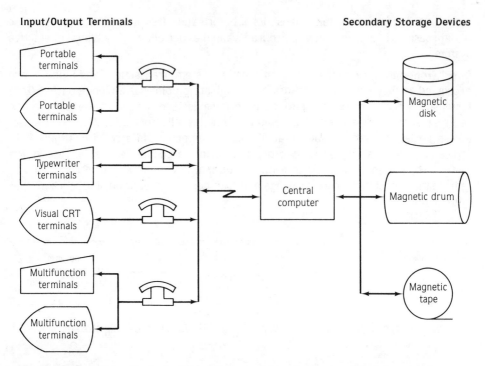

figure 1.7 Time sharing with input/output terminals plus direct (random) access and sequential access file storage.

errors are detected, they are typed or displayed for appropriate action before processing proceeds.

Real Time Real-time systems, especially when they are an integral part of advanced management information systems, must be capable of processing large volumes of information in relatively short intervals. In addition, there must be a fast conversion (turnaround) of input to output in order to affect the functioning of the MIS environment at that time.

Real-time systems vary in their capabilities. Most maintain a continuous connection between the many terminals (commonly called **dedicated terminals**) geographically dispersed and the central computer facility (figure 1.8). Generally, an input transaction from an on-site or an outlying location triggers the immediate processing of the data with the "answer" being returned in seconds. Not only is the central computer capable of updating data files (on secondary devices, such as magnetic disks) from input transactions, but also it can route output to another terminal location either immediately or at a near-future time.*

---

*For a comprehensive understanding of real-time management information systems, see Robert J. Thierauf, **Systems Analysis and Design of Real-Time Management Information Systems** (Englewood Cliffs, N. J.: Prentice-Hall, Inc., 1975).

Current Management Information Systems   **17**

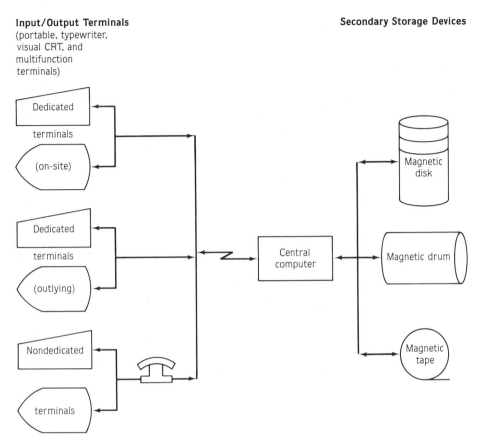

figure 1.8   Real-time processing with input/output terminals plus direct (random) access and sequential access file storage.

SUMMARY: Current Management Information Systems

Batch Processing   Transactions are accumulated and processed periodically; items in the batch may be in sequential or random order. Sequential or direct (random) access files are used.

Interactive Processing   Transactions and inquiries into the system are processed in order of occurrence. Direct access files are employed. A **real-time** system is an interactive processing system designed to fulfill the data processing requirements for one organization; a **time-sharing** system (also interactive) is designed to serve the problem-solving needs of many users, i.e., many organizations.

Relationship of Systems Design to Management Information Systems

The preceding differentiation between batch and interactive processing modes serves to place past and present system designs in their proper perspective. With the arrival of the computer, organizations found that they could produce many more accounting reports utilizing a batch processing mode than some type of manual or punched card procedures. This led ultimately to the development of **responsibility reporting systems** that were capable of producing accounting-oriented reports at all levels of management. In essence, actual performance could be measured against flexible budget amounts, thereby holding managers responsible and accountable for operating results.

These accounting-oriented approaches satisfied management in the 1950s and 1960s—the initial period of computers. The dynamics of the ever-changing business world, the complexity of producing needed managerial information, and the recognition of the computer's potential by personnel in the data processing field provided the initial thrust for better systems design. Simultaneously, management began to realize that the information potential of the computer had not been fully exploited. Based on these initial developments, **integrated data processing systems** were developed and further refined into **integrated management information systems**. These systems, like responsibility reporting systems, were concerned with reporting historical results (backward-looking) and generally batch-oriented. The next system development, currently in vogue, is on interactive systems, i.e., **real-time management information systems**. Because of their importance, the focus of this text will be on interactive systems concerned with reporting current and future information (forward-looking) as exemplified in the ABC Company—the text's master case study. As will be seen in the next section, the focus of management information systems is away from past accounting-oriented information and toward current and future management information for doing a more effective job of planning, organizing, directing, and controlling an organization's activities.

UNDERLYING CONCEPTS OF MANAGEMENT INFORMATION SYSTEMS

Now that the basic types of management information systems have been explored, it will be helpful to investigate their underlying concepts. Although many basic concepts have been developed over the years, the discussion will center around the more important ones. These include
- the management information system concept
- the distributed processing system concept
- the modular system concept

As time passes, newer concepts will supplement the current listing.

The Management Information System Concept

A most important systems design concept currently is one referring to **management information systems,** sometimes called **management information and control systems** because their output is information reports and control reports. Even though it is possible to have noncomputerized management information systems, this book stresses the computer as an essential part. External and internal information is channeled into a computer system which generates output that is meaningful and effective for managerial decisions. Common data files and control reports are two key aspects of a management information system.

Common data files Fundamentally, a computerized management information system means capturing originating data as close to their source as possible, feeding the data directly or almost directly into the computer system, and permitting the system to utilize common data files (a data bank, such as magnetic tapes for a batch system, or a data base, such as magnetic disks for an interactive system) that can produce different outputs (figure 1.9). In this type of environment, a single piece of information is entered into the data processing system only once. From then on, it is available to serve all processing requirements until its usefulness is exhausted. Under these ideal conditions,

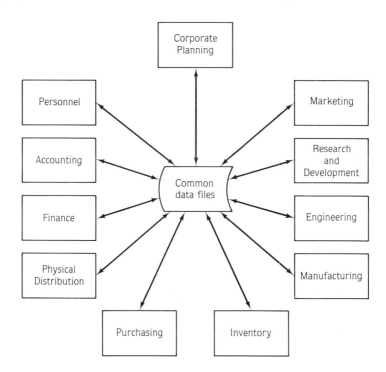

figure 1.9 An essential part of a management information system is common data files (a data bank or data base) that can serve different functional areas.

adequate and correct information can be presented to managers in a coordinated rather than in a segmented manner. Such a system provides the required information on a timely basis to help management plan, organize, direct, and control an organization's activities. Thus, the output of a management information system is report-oriented, being position- or department-oriented for meeting specific managerial requirements. In other words, a management information system is not an improved accounting reporting system, but rather an orderly flow of timely information for meeting specific management needs.

Control reports   Control reports produced by this system include those necessary to service the basic business functions of an organization. Among these are sales forecasting, shipping and warehousing, finished goods replenishment, production control, materials management, manufacturing cost control, personnel skills and control, and management incentives. Other reports include short- and long- range financial and operating budgets, monthly financial and operating statements, and various historical data for short- to long-range planning. Sales order-entry statistics, including sales quotas, salespeople's compensation, and shipping orders, are additional examples of output that provides input to many other subsystems.

An essential part of any management information system is **feedback** which shows what has happened to the financial plan in view of actual results to date or what would happen if hypothetical changes in the plan were made. Control reports, then, are concerned with monitoring actual results via feedback in order to determine whether organizational activities are proceeding in accordance with plans and standards.

Definition   The **management information system concept** can be defined as a collection of subsystems and related program parts or modules that are interconnected in a manner which fulfills the information requirements necessary to plan, organize, direct, and control business activities. It is a system for producing and delivering timely information that will support management in accomplishing its specific tasks in an enterprise. Among the many benefits of such a system are

faster and better information
greater facility to carry out managerial functions
improved decision making
more effective use of employees and facilities
prompt communication
very little need to recreate data
elimination of peak period volume reports
prompt correction of out-of-control conditions

In addition, an advanced management information system should be computer based, integrated with all subsystems, accessible through a common data bank or data base, utilize I/O (input/output) terminals at on-site and remote locations, provide timely information through use of communications capability, and interactive between the user and information available.

For a management information system to qualify as excellent in terms of **speed**, the system should be capable of sending the information to the user within the operational time span in which it is needed to make the right decision or to take the appropriate action. To qualify as excellent in **content**, the data should be relevant to the user's needs in terms of the problem under study. Using the criterion of **cost**, the management information system should compare favorably with alternative methods available and with the benefits mentioned above. Additional criteria for measurement of MIS can be employed. However, the foregoing qualifications highlight the essential considerations for converting to such a system.

The Distributed Processing System Concept

One of the newer system concepts underlying business information systems is the **distributed processing system concept**. It is actually a spin-off of the small business computer concept. In many cases, the computer equipment used is the same; however, the accent is on who is using it and how it is programmed.

Focus on local processing An essential part of distributed data processing (DDP) is the increased responsiveness of the DP function to the user's needs by providing data entry/inquiry capabilities—i.e., processing power—at the lower levels, local and regional. Not only does this give local and regional managers more control over and involvement with the data and the system, but also takes a burden off the central computing facility. And because more processing is done locally and regionally, the overall system can be utilized to do what it does best—repetitive processing with improved responsiveness.

The distributed processing concept is an answer to difficulties experienced with earlier management information systems. The use of large centralized computers created large data input bottlenecks so that the feedback of the business data necessary to run the business occurred after long delays. In effect, distributed processing arose out of the need to channel data processing power where it is needed to handle DP operations that can be done more efficiently in the field (local and regional levels) than at the home office.

Relationship with centralized processing The distributed processing concept still allows the use of centralized computing facilities. Those DP operations that can be best handled by the local and regional on-site computers stay at that level. However, summary information that is needed for higher levels of management are forwarded from the lower levels to the centralized computer facility for appropriate processing. In this manner, detailed information needed for daily operations remains at that level; summary information needed to plan, organize, direct, and control overall operations is forwarded to the central computer.

Three-level system approach* Because a distributed data processing environment can be structured in many ways, it is helpful to think in terms of multilevel approaches for system implementation. Fundamentally, three different levels which make such an approach attractive for a wide range of applications include:

    first-level system—local or regional processing of source data entry with great emphasis on interactive DP equipment

---

*For a comprehensive study of this new thrust in MIS, see Robert J. Thierauf, **Distributed Processing Systems** (Englewood Cliffs, N. J.: Prentice-Hall, Inc., 1978).

second-level system—local or regional processing of source data entry, transaction processing, and preparation of operational management reports with great accent on interactive DP equipment

third-level system—network of dispersed systems at the various levels in an organization. A widely used approach is the **hierarchical** or **tree network** where processing at the lower organization levels is kept at that level, except for forwarding summary information to the higher levels (figure 1.10).

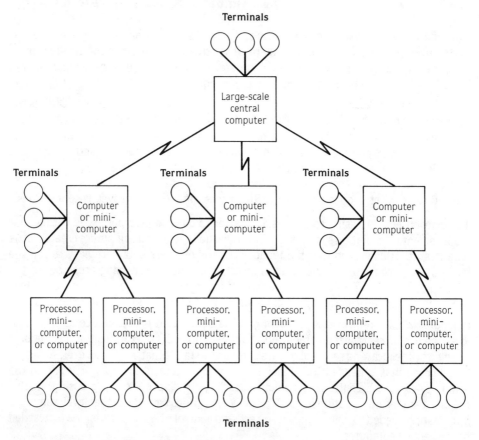

figure 1.10 A hierarchical or tree network in distributed processing system. Most distributed processing networks are currently of this type.

Definition The **distributed processing system concept** can be defined as an approach to placing low-cost computing power at the various points of data entry and linking these points, where necessary, with a centralized computer via a distributed communications network. In effect, it is an approach to placing computing power where it is needed in an organization for efficient and economical DP operations. Similarly, it is a viable alternative to centralized data processing because of the declining costs of programmable terminals, microprocessors, microcomputers, minicomputers, and small business computers. Because the focus of current distributed processing systems is placing computer power at lower levels in an organization, their output generally centers on assisting operational (lower) and functional (middle) management.

The Modular System Concept

As a means of obtaining systems design objectives, DP personnel have found it necessary to formulate the **modular system concept**. Under this concept, separate but detailed information system modules are identified. For example, the finance function (major module) can be subdivided into three intermediate modules (cash management, capital budgeting, and sources of funds), which can be further divided into minor and basic modules (figure 1.11, p. 24). Thus, with this approach, the system can be subdivided into its component parts, from the highest level to the lowest level.

Modular design Not only can the modular (or building block) approach be used for analyzing the present system, but also it can be employed for designing a new system. This approach results in developing a system, and applications within it, in an orderly and planned fashion. Inasmuch as the focus in this text is on utilizing the modular approach, further discussion of this important design concept will be left until part three, Systems Design of New Management Information Systems.

Modular programs Another important use of the modular system concept is its application to computer programs. The modular system concept refers to the capability to develop computer programs and to change individual program modules without the need to reconstruct or redo the entire computer program. Also, this modularity or building block approach allows programming personnnel to test separately each subsection (containing several modules) of the final program.

To utilize this concept effectively, computer applications must specify the interacting functions among modules; otherwise, many program modules will be developed and available but not used. Hence, an important characteristic for this system concept is the integration of the modules as they are developed. For example, it would be advisable to integrate the modules required for the various computer programs under corporate planning, especially budgets, with accounting and finance before developing other program modules for marketing, research and development, engineering, manufacturing, inventory, purchasing, physical distribution, and personnel. However, once program modules have been written, the task of making major changes to programs will be minimized by modular programming, because only the individual program modules requiring correction need be changed. New program modules can be tested before inserting them in the required program.

Definition The **modular system concept** can be defined as an efficient design method for breaking down a system and its computer programs into their lowest-level component parts so that modules can be logically grouped for implementation and ease of alteration. As will be demonstrated throughout the various chapters on the design of an order-entry system for the ABC Company, it is an effective way to "get a handle" on how to design a system no matter what area or level is being investigated. From this view, it matters little what type of management information system is being designed.

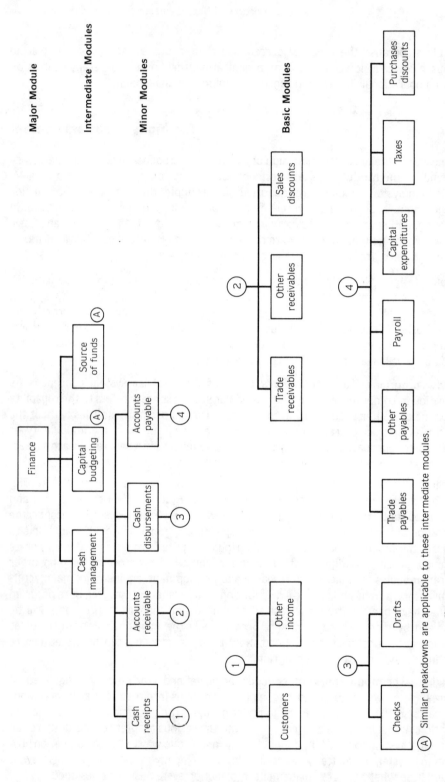

figure 1.11 The modular system concept applied to an organization's finance function.

> **SUMMARY: Underlying Concepts of Management Information Systems**
>
> **The Management Information System Concept**   A group of subsystems interconnected in such a manner that the managerial information requirements necessary to plan, organize, direct, and control business operations are met. It is a system approach for producing and forwarding timely information that will support management in fulfilling its specific tasks.
>
> **The Distributed Processing System Concept**   An approach to placing low-cost computer power at the various processing points in an organization and linking these points, if necessary, with a centralized computer. It is an alternative to centralized processing due to the declining costs of computing devices, thereby placing computer power where it is needed for efficient and economical DP operations.
>
> **The Modular or Building Block Concept**   An efficient design approach for breaking down a system and its related computer programs into their lowest-level component parts in order to group them logically for implementation and later alterations. It is an approach that can be employed to design any type of management information system.

> ## CHAPTER SUMMARY:
>
> This first chapter has introduced the remainder of the text by providing an overview of systems analysis and design and their relationship to a system project. Similarly, the current types of management information systems and their relationship to systems analysis and design were presented. To facilitate a better understanding of management information systems, which will be analyzed and designed in future chapters, their underlying concepts were discussed at some length. No matter what type of MIS is designed and implemented, an effective way of structuring the new system is to employ the modular or building block design approach. In this manner, the essential job of systems design will be made easier. Fundamentally, this initial chapter has set forth a conceptual framework for designing an interactive management information system, such as that found in the text's master case study of the ABC Company.

Questions

1. a. Distinguish between systems analysis and systems design.
   b. Which of the two requires the most creative talents of DP personnel?
2. What is the relationship of systems analysis and systems design to a system project? Explain thoroughly.

26 Introduction to Systems Analysis and Design

3. a. What are the two basic types of management information systems? Explain.
   b. What types of files are employed for each basic type?
4. a. Differentiate between interactive and real-time processing.
   b. Differentiate between real-time processing and time-sharing.
5. Contrast the similarities and differences between a management information system and a distributed processing system.
6. What is the importance of the modular system concept in designing any type of system? Explain thoroughly.

Self-Study Exercise

True-False:

1. _____ Preceding the systems analysis phase is the equipment selection phase of the feasibility study.
2. _____ Basically, systems analysis centers on system planning of the new system.
3. _____ An important consideration when designing a new system is the human factor.
4. _____ The batch processing approach allows data to be entered as they occur or information to be retrieved relatively quickly.
5. _____ The batch processing approach allows the use of direct or random access files.
6. _____ Remote job entry is used in an off-line mode when there are not enough transactions to warrant a permanent communication with the main computer.
7. _____ In an interactive processing system, transactions and inquiries are processed as they occur.
8. _____ Underlying many design approaches to computers is the management information system concept.
9. _____ The trend toward centralization is found in distributed processing systems.
10. _____ Underlying the building block approach in the design of management information systems is the modular system concept.

Fill-In:

1. In _____ _____, what must be done, how to do it, who must do it, and when to do it must be decided.
2. The _____ _____ of the present system is frequently referred to as systems analysis.
3. The real essence of _____ _____ is devising a system that meets specific requirements.
4. For an effective and economical system, good _____ _____ should be incorporated in the various systems design phases.
5. The concept of _____ _____ means that information is entered or retrieved relatively quickly via a terminal of some type in order to control the operating environment.
6. In _____ access files, the data are stored in some predetermined order.

Self-Study Exercise

7. The trend continues toward more _____ _____ for management information systems.
8. An essential part of _____ _____ is providing data entry/inquiry capabilities (i.e., processing power) at the lower levels of an organization.
9. Another name for the modular approach to systems design is the _____ _____ approach.
10. The modular system concept is applicable to systems design as well as _____ _____ .

chapter 2

standard tools of systems analysis and design

OBJECTIVES:

- To set forth the standard tools used in the analysis and design of a typical systems project.
- To present the standard flowchart symbols utilized by systems analysts in developing a new system.
- To examine those situations where flowcharts are preferred over decision tables.
- To present the standard time-charting techniques used in managing a systems project.

IN THIS CHAPTER:

OVERVIEW OF STANDARD TOOLS OF SYSTEMS ANALYSIS AND DESIGN
FLOWCHARTS
    Value of Flowcharts
    Standard Flowchart Symbols
    Flow of Symbols
FLOWCHART TYPES
    System Flowchart
    Flow Process Chart
    Document Flowchart
    Program Flowchart
    Computer Prepared Program Flowchart
DECISION TABLES
    Value of Decision Tables
    Decision Table Components
    Symbols and Rules
SAMPLING TECHNIQUES
    Statistical Sampling
    Document Sampling
    File Sampling
TIME-CHARTING TECHNIQUES IN A SYSTEMS PROJECT
    Gantt Charts
    PERT Charts
CHAPTER SUMMARY
QUESTIONS
SELF-STUDY EXERCISE

The preceding chapter introduced systems analysis and design. This background provides a basis for introducing the standard tools utilized by systems analysts. Great accent is placed on system flowcharts since they provide a means of seeing on paper the interworkings of not only the present system, but also the new one to be installed. Not only do these flowcharts serve as a basis for designing and reviewing the system under study, but they also serve as documentation. From this view, flowcharts serve several purposes for assisting designers in accomplishing the arduous task of detailed design for simple to complex systems.

Although many specialized tools of systems analysis and design have been developed over the years, only the more popular ones are presented in this chapter—namely, system flowcharts, flow process charts, document flowcharts, and program flowcharts. Complementary to these flowcharts is a discussion on decision tables and sampling techniques. Additionally, the chapter surveys time-charting techniques, which are a means of bringing together the many activities involved in a systems project. These include Gantt charts and PERT/time networks. Overall, the chapter's main thrust is directed toward exploring the standard "tools of the trade" that assist systems analysts in undertaking a system project in an efficient and economical manner.

OVERVIEW OF STANDARD TOOLS OF SYSTEMS ANALYSIS AND DESIGN

The systems analyst has a variety of standard tools to draw upon when undertaking a system project; they can be classified into two groups—**display tools** and **charting techniques.** Display tools are the most fundamental tools of systems personnel. They are an effective way of organizing data to facilitate systems analysis and design as well as provide a means for conveying ideas to the user. The more popular display tools include

    system flowcharts
    flow process charts
    document flowcharts
    program flowcharts
    decision tables

Because of their importance, display tools will be covered in subsequent sections of the chapter. In addition, they are compatible with the modular system concept presented in the prior chapter.

Fundamentally, charting techniques are designed to assist in the administration of the many steps involved in systems analysis, systems design, equipment selection, and systems implementation. Discussion in the chapter will center on the most widely used charting techniques, namely, Gantt charts and PERT charts.

Although other standard tools exist, most are essentially variations of the foregoing display tools and charting techniques. Instead of an explanation of all possible types, the ones set forth above are covered in sufficient detail within the chapter to explain their use in analyzing and designing a new system—in particular, an interactive management information system for the ABC Company.

FLOWCHARTS

A **flowchart** is widely used for a better understanding of existing or proposed methods, procedures, and systems. It is defined as a graphical representation of the definition, analysis, or solution of a problem using symbols to represent operations, data flow, equipment, and the like. In essence, it is a diagrammatic representation of a series of events.

For a computerized system, there is need for a visual display of accurate and detailed end-to-end activities. The flow of data and paper work from the input stages through the many intermediate stages (including complex computer programs) to the final outputs must be explicitly detailed for effective communication. Otherwise, it is difficult to follow the logical flow of data processing activities. Hence, the need for flowcharts is paramount in computer systems.

Value of Flowcharts

It has been said that "a picture is worth a thousand words." A flowchart is a picture of some part of a DP system.

Aids understanding A flowchart shows explicitly what is happening and in what order, and has the capability to detect gaps in procedures and overlaps in system activities. It is much easier to comprehend what is occurring with diagrams than with a written description. When one is forced to diagram on paper the steps involved in a DP procedure, errors and omissions stand out.

Effective communicator The flowchart can communicate the interworkings of a new method, procedure, or system to other interested parties. Likewise, it is a succinct presentation of data flow to management and operating personnel for controlling the organization activities. Thus, a second value of the flowchart is its use to communicate to personnel other than the originator.

Permanent record A third value of flowcharting is its existence as a permanent record which does not depend on oral communication. Since the chart is written, it is available for review in terms of accuracy and completeness. It also provides a basis for analyzing and comparing present and proposed systems so that efficiency, cost, timeliness, or other relevant factors may be improved. Regardless of the purpose in developing a flowchart, it serves as a basis for system documentation, now and in the future.

Standard Flowchart Symbols

Over the years, there has been a concerted attempt to standardize flowcharting symbols. The rationale for standardization is that anyone can interpret accurately the work of another. This is particularly important today because of the high job mobility of systems personnel; the person preparing the flowchart today may not be the one interpreting it tomorrow. If standard flowcharting symbols are used, the amount of confusion about the exact meaning intended is kept to a minimum. Thus, standardized flowchart symbols have been developed by the United States of America Standards Institute (USASI) and the International Standards Organization (the international counterpart of USASI).

Standard flowchart symbols that indicate the type of operations to be performed by computer and related equipment consist of general and specialized ones. A template, like the one pictured in figure 2.1, aids in drawing those symbols. Generally, these standard symbols can be classified into three main categories: basic, input/output (I/O) and file, and processing. This classification, along with typical payroll examples, is illustrated in figures 2.2 through 2.4, respectively (pp. 34–36).

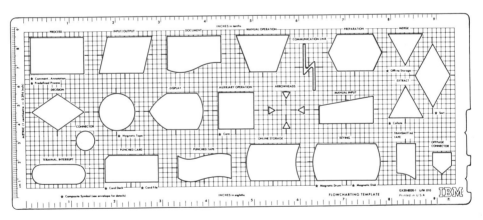

figure 2.1 A template for drawing standard flowchart symbols.

Flow of Symbols

No matter what combination of above flowchart symbols is used, flowcharts are constructed to follow our natural tendency to read from left to right and top to bottom. At times, it is desirable to deviate from this pattern in order to achieve symmetry and to emphasize important points. **Solid flowlines** are drawn to indicate the direction of the flow whereas **dotted flowlines** depict a transfer of information and annotated information. In either case, flowlines can be drawn horizontally, vertically, or diagonally, as needed, for a meaningful flowchart.

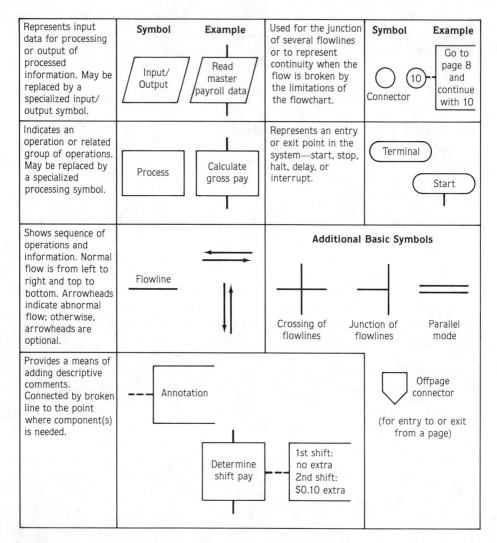

figure 2.2 Flowchart symbols and illustrated payroll examples for basic modules.

FLOWCHART TYPES

There are several types of flowcharts found in a DP installation. Among the most frequently used are system flowcharts and program flowcharts. A **system flowchart** depicts the flow of data through all parts of a system with a minimum of detail: generally, it shows where input enters the system, how it is processed and controlled, and how it leaves the system in the form of storage or output. On the other hand, a **program**

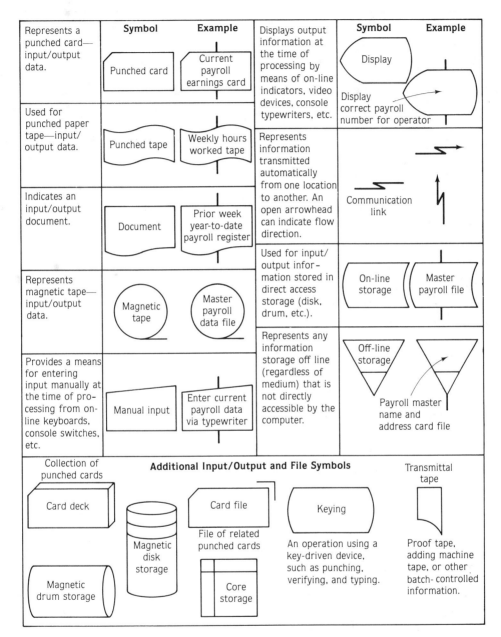

figure 2.3 Flowchart symbols and illustrated payroll examples for input/output files.

36 Standard Tools of Systems Analysis and Design

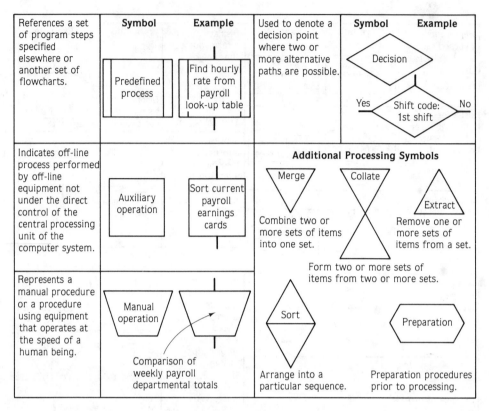

figure 2.4 Processing flowchart symbols and illustrated payroll examples.

flowchart is more detailed and supplements a system flowchart. Since a computer must be directed according to a detailed set of instructions called a **program** (stored internally in the computer), these flowcharts are a necessity for programming computer applications.

Within the framework of the various type flowcharts, there are no formal rules for determining the level of detail to include; it depends on the purpose for which the flowchart is used. However, for a successful installation of any DP equipment, there comes a point at which all methods and procedures must be flowcharted in detail in order for the system to be implemented. Also, the lowest level of detail not only is necessary for documentation but also forms the basis for issuing instructions to organization personnel.

System Flowchart

System flowcharts, sometimes referred to as **procedural flowcharts,** show the sequence of major activities that normally comprise a complete operation. They are generally prepared to assist all organization personnel—in particular, the systems analyst—in understanding some specific data processing operation as well as obtaining an overview of the operation itself. Before a system flowchart can be drawn, the area under study must be clearly defined. Questions relating to the type and number of inputs (source documents), exceptions, transactions, files, and reports must be answered. Similar questions refer to the relationship of the area under study to other functional parts of the system, the timeliness of data, and the source(s) of various data. Answers to these typical questions provide the necessary information for the initial system flowchart.

Raw material inventory illustration Figure 2.5, p. 38, shows a typical system flowchart for a computer raw material updating procedure. Inspection of this flowchart indicates that current transactions are processed as they occur. These include raw material receipts from vendors, in-plant transfers to finished goods inventory, physical inventory count changes (physical inventory counting is performed on a rotating basis for counting raw materials once a month), and miscellaneous adjustments based on spoilage, scrappage, obsolescence, shrinkage, and similar items. Also, automatic purchasing of raw materials is performed on line. Cards are punched to signal excess inventory and certain inventory errors that are the result of previously mentioned on-line activities. In essence, this approach to raw material inventory allows inquiry into the system at any time for updated information that is critical to maintaining continuous manufacturing operations.

Flow Process Chart

Flow process charts depict the sequence of operations by functions and individuals, as shown in figure 2.6, p. 39. Details are clearly shown so that they can be questioned and analyzed with the view of determining better methods of operation. Within the chart, various noted operations are represented symbolically, and a reminder of who, what, where, when, why, and how operations are listed in the order in which they can be considered during the investigation and analysis. Operational time, recommended action, and summary space are provided on the chart. In addition to documentation, it serves as a work simplification technique.

Document Flowchart

Document flowcharts, like system flowcharts, are suitable for tracing the origin of input data through each phase of processing and communication into files and, finally, out of files for desired outputs—many in the form of reports. They show the way various forms, documents, or reports move from person to person or from department to department; this is extremely helpful in understanding and obtaining an overview of paper flow within an organization for a specific function. Although no special flowcharting symbols are needed, normally the departments or individuals involved are identified on the top of the sheet. An example of this flowcharting technique for raw material purchases is found in figure 2.7, p. 40.

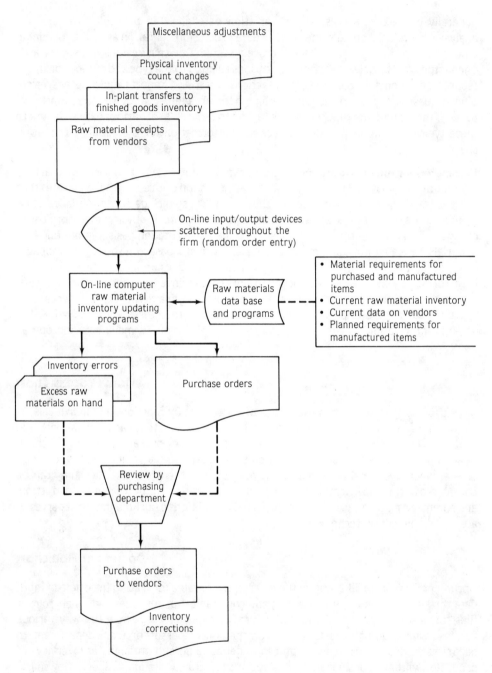

figure 2.5 System flowchart for a raw material inventory updating run.

		PAGE	OF

PRESENT METHOD ☐
PROPOSED METHOD ☒
MAN ☐ MATERIAL ☐

SUMMARY	PRESENT		PROPOSED		DIFFERENCE	
	NO.	TIME	NO.	TIME	NO.	TIME / PERCENT
○ OPERATIONS	5	65 hr	5	6.25 hr	0	.25 hr
⇨ TRANS.	2	—	—	—	2	—
☐ INSPECTIONS	1	.25	—	—	1	.25
D DELAYS	—	—	—	—	—	—
▽ STORAGES	4	1.25	3	.75	1	.50
DISTANCE		15 FT.		15 FT.		15 FT.

TASK: DOCUMENTING AND AUDITING ACCOUNTS PAYABLE #1
PROCESS BEGINS: RECEIPT OF INVOICES
PROCESS ENDS: PREPARATION OF CHECKS
DATE:
PREPARED BY: R.R.

#	DETAILS OF METHOD	SYMBOLS	DIST	TIME	NOTES
1	D - SORTS, DOCUMENTS & NUMBERS ALL INVOICES	● ⇨ ☐ D ▽	—	.50	
2	D - SORTS ALL M/R'S, FILES M/R'S AND INVOICES IN FOLDERS	● ⇨ ☐ D ▽		.25	
3	D - FILES FOLDERS	○ ⇨ ☐ D ▼	5	.25	
4	C - AUDITS - VERIFY RECEIVING AGAINST BILLING & CHECK MATH ACCURACY	○ ⇨ ■ D ▽	5	3.00	
5	C - WRITES DISTRIBUTION & SORTS BY DISCOUNT RATE	● ⇨ ☐ D ▽		.50	
6	C - FILES BY DUE DATE - GIVES CONTRACTS TO SUPV.	○ ⇨ ☐ D ▼	5	.25	
7	B - RE-AUDITS PRIOR TO DISCOUNT DATE	○ ⇨ ■ D ▽		2.00	
8	B - FILES BY CUSTOMER FOR CHECK PREPARATION	○ ⇨ ☐ D ▼		.25	
9		○ ⇨ ☐ D ▽			
10		○ ⇨ ☐ D ▽			
11		○ ⇨ ☐ D ▽			
12		○ ⇨ ☐ D ▽			
13		○ ⇨ ☐ D ▽			
14		○ ⇨ ☐ D ▽			
15		○ ⇨ ☐ D ▽			
16		○ ⇨ ☐ D ▽			
17		○ ⇨ ☐ D ▽			
18		○ ⇨ ☐ D ▽			
19		○ ⇨ ☐ D ▽			
20		○ ⇨ ☐ D ▽			
21	B = SR. AUDIT CHECK	○ ⇨ ☐ D ▽			
22	C = AUDIT CLERK	○ ⇨ ☐ D ▽			
23	D = CLERK	○ ⇨ ☐ D ▽			
24		○ ⇨ ☐ D ▽			

figure 2.6 A typical process flowchart. (Robert J. Thierauf, **Systems Analysis and Design of Real-Time Management Information Systems,** © 1975. Reprinted by permission of Prentice-Hall, Inc., Englewood Cliffs, New Jersey.)

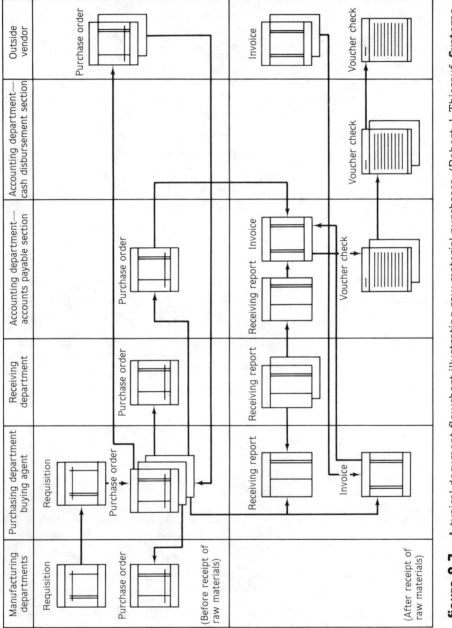

figure 2.7 A typical document flowchart illustrating raw material purchases. (Robert J. Thierauf, **Systems Analysis and Design of Real-Time Management Information Systems,** © 1975. Reprinted by permission of Prentice-Hall, Inc., Englewood Cliffs, New Jersey.)

Program Flowchart

Program flowcharts, sometimes referred to as **block diagrams,** describe the specific steps and their sequence for a particular computer program. When a program is extremely simple, a flowchart may not be necessary. However, for most programs, it is necessary to have a sequence of operations and decisions that detail the computer program steps. Otherwise, the programmer would have a difficult task in coding the program properly. Again, the program flowchart also provides an excellent means of

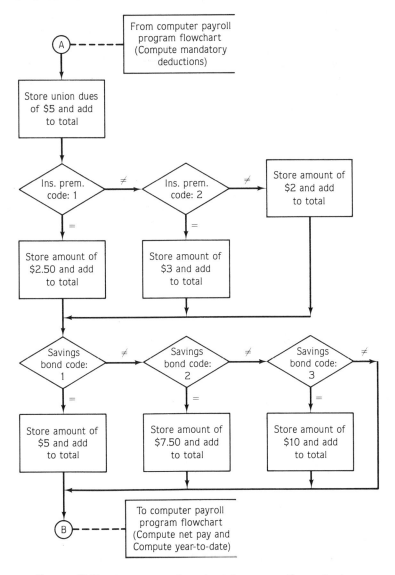

figure 2.8 A program flowchart for computing voluntary deductions.

documenting the program. The program flowchart, then, has three important uses: to aid program development, to serve as a guide for coding, and to document the program.

Compute voluntary deduction illustration A computer program flowchart for computing voluntary deductions is on p. 41, figure 2.8. It consists of a series of operations and decisions regarding the proper weekly voluntary payroll deductions. In the first processing symbol, union dues of $5.00 are deducted for each employee; likewise, this value is added for all employees. For the first set of decision symbols, if the insurance code is 1, then $2.50 will be deducted; similarly, if the insurance code is 2, $3.00 will be deducted; otherwise, the insurance deduction will be $2.00. In all cases, each insurance deduction is added for a grand total. In the last set of decisions, a comparison is made to determine whether the code is 1, 2, or 3. The amount deducted for these codes is $5.00, $7.50, and $10.00, respectively; like before, each amount is added for a grand total of all employees. However, if the code is not 1, 2, or 3, no deduction is made. Thus, within this program flowchart, comparisons are made to determine the deductions for insurance premiums and savings bonds while union dues of $5.00 are deducted for all employees.

Computer Prepared Program Flowchart

An extremely helpful flowcharting technique is utilization of the computer itself with a special flowchart writing program. Using asterisks and other special print characters, the printer plots the outlines of the symbols and converts them as if they were flowcharted manually. The advantage of such a method should be apparent—revisions can be made easily once the initial table is written and cards have been punched correctly. The updating process consists of repunching cards that represent program flowchart changes. This approach can keep the laborious task of updating complex program flowcharts to a minimum.

SUMMARY: Flowcharts and Flowchart Types

Flowcharts Pictorial representations of data flow. They aid in understanding what is happening, are an effective communicator to those using them, and serve as a permanent record. There are three major categories of **standard flowchart symbols**.

1. Basic symbols:
    - input/output
    - process
    - flowline
    - annotation
    - connector
    - terminal
2. Input/output and file symbols:
    - punched card
    - punched tape
    - document
    - magnetic tape
    - manual input
    - display
    - communication link
    - on-line storage
    - off-line storage
3. Processing symbols:
    - predefined process
    - auxiliary operation
    - manual operation
    - decision

Flowchart Types
1. **System flowchart or Procedural flowchart**   Depicts the flow of data through the major parts of a system with a minimum of detail. For the most part, if shows where input enters the system, how it is processed and controlled, and how it leaves the system to storage and/or output.
2. **Flow process chart**   Shows the sequence of operations by functions and individuals. Detailed operations can be analyzed to determine ways of improving them.
3. **Document flowchart**   Traces input data through each phase of processing and communication into files and, finally, out of files as output to satisfy managerial and operational personnel needs.
4. **Program flowchart or Block diagram**   Portrays the various arithmetic and logical operations as well as the sequence of those operations for a particular computer program.

DECISION TABLES

A **decision table** is similar to a flowchart in use and construction. It can be used independently of or to complement a flowchart. A decision table shows conditions and actions in a simplified and orderly manner. By presenting logical alternative courses of action under various operating conditions, a decision table enables one to think through a problem and present its solution in compact notation. It allows a computer problem to be divided into its logical segments that provide a multilevel structure in the problem's analysis. At the highest level, decision tables can be used for an overall system by referencing to lower-level tables, resulting in a modular or building block approach. The purpose of a decision table, then, is to assemble and present complex decision logic in such a way that its meaning can be readily understood.

Over the years, simplified forms of decision tables have been utilized. Tax rate and insurance rate tables are forms of decision tables. Likewise, the price list of various product lines, expressed in terms of quantity discounts, package sizes, and product specifications, is another example. These common examples have become the foundation of decision tables.

Value of Decision Tables

Like flowcharts, the value of decision tables derives from the table format; it is a complete statement of the problem from the outset. Several studies have shown that approximately 50 percent of total costs related to computer system development have been attributed to poor problem definition and ineffective documentation.

Compact notation   Many pages of a computer program flowchart can be condensed into a single logic table. Studies indicate that people with no previous computer experience have programmed applications in less time with decision tables than experienced programmers who fail to use them. Also, those who must review a program need not go through page after page of flowcharts to follow the logic of the program. Generally,

44 Standard Tools of Systems Analysis and Design

decision tables are easier for the nonsystems person to understand; they are easier to check by observation for consistency and completeness.

Easy to modify program Another important value of the decision table's compact notation is the ease of modifying and updating a program as the system changes. Adding new conditions or changing a given decision rule action will not require substantial reformulation, as can be the case with flowcharts. In addition, introduction of new decision tables into a system will not require substantial revision to system flowcharts.

Ability to produce a machine-language program Today, a most significant value of decision tables is their ability to produce a machine language computer program. In such cases, a translator converts the decision rules into a programming language without human intervention and can, in the ordinary course of operations, further convert this language into the appropriate machine language. Thus, this computerized process reduces the amount of time necessary to produce a completed program for testing sample input data.

Compatible with flowcharts Although decision tables have many advantages over program flowcharts, they are not complete substitutes in a business environment. Business processing is still primarily sequential, using input data to process controlled tasks, revise storage files, and prepare output. Flowcharts are still needed to show this sequential process and are useful when there are few conditions and only simple decisions or when presenting the combined system of decision logic, programs, and computers. Instead of showing each branch representing an individual decision, flowcharts can include a single block representing a complete decision table. From this viewpoint, decision tables and the various types of flowcharts complement and supplement each other.

Decision Table Components

A decision table is divided into four basic elements with provision for other information that may be helpful for interpreting the final results. This is shown below.

IF (condition statement)	Condition stub	Condition entry
THEN (action statement)	Action stub	Action entry
	Other information	

All conditions are listed in the upper part of the decision table and represent the contents of decision and branching symbols on the program flowchart. Since these represent the condition of the computer at a particular time, or an "if" situation, it may be necessary to perform a specific operation on data at this time. Actions taken correspond to the processing symbols on a flowchart, or a "then" situation, and are listed in the second half of the decision table. Thus, conditions and actions have an "if ... then" relationship; that is, if a set of conditions exist, then the indicated actions are performed. It should be noted that a condition cannot appear in the action area, nor can an action appear in the condition area.

Symbols and Rules

Various symbols can be used in a decision table for the condition entry, namely, yes (**Y**), no (**N**), greater than (>), equal to (=), less than (<), and blank (—). The action entry to be performed is an **X** or a blank (—). A blank in either case means that the condition or action is not applicable. Each column in the decision table makes up a rule that corresponds to one of the many possible paths of the program flowchart. Basically, a decision table relates given conditions to the corresponding actions, with a column of entries that forms a rule. Alternative conditions resulting in other actions that constitute other rules in the table are written side by side.

Checking inventory level illustration An example of a decision table, table 2.1, p. 46, checks the inventory level against incoming orders. The decision table was first developed by determining the possible conditions. After exhausting the condition list, the action entries are listed. For the first rule, if the inventory is greater than or equal to the order amount (**Yes** condition), the ordered item would be shipped, noted by the **X** in the appropriate space. Since the remaining actions are not relevant to this condition, the — appears in the first column. The next eight rules are interpreted in a similar manner. The decision table shows a total of nine separate decision rules.

An investigation of table 2.1 indicates little difference between rules 6 and 9 since both result in the same action, that is, holding an order for one day. In such cases, the two rules can be combined into one rule. The combined rule is:

Decision rule 6	Decision rule 9	Combined rule
N	N	N
Y	N	—
—	—	—
Y	Y	Y

SAMPLING TECHNIQUES

In addition to flowcharts and decision tables, sampling techniques can be used by the systems analyst to obtain information about an existing system. Likewise, they can be employed to determine requirements for a new system. Basically, there are three major categories—statistical sampling, document sampling, and file sampling.

Statistical Sampling

Statistical sampling involves making a set of observations to infer conclusions about a problem or an operation. It is most often applied to problems which otherwise would involve collecting and accumulating detailed data to provide information about an operation.

table 2.1 Decision table for checking inventory level.

DECISION TABLE

Table name: INVENTORY LEVEL CHECKING

Date: November 5, 1980 Preparer: Robert J. Thierauf

	Rule number											
	1	2	3	4	5	6	7	8	9	10	11	12
Condition												
Inventory available ≥ ordered amount	Y	N	N	N	N	N	N	N	N			
Inventory available < ordered amount	—	Y	Y	Y	Y	Y	N	N	N			
Partial shipment of goods	—	Y	Y	N	N	—	—	—	—			
Back order of goods	—	Y	N	Y	N	—	Y	N	—			
Additional goods due in next day	—	N	N	N	N	Y	N	N	Y			
Action												
Item shipped	X	—	—	—	—	—	—	—	—			
Partial shipment	—	X	X	—	—	—	—	—	—			
Back order unshipped balance	—	X	—	X	—	—	X	—	—			
Out-of-stock notice sent	—	—	X	—	X	—	—	X	—			
Order held for entire day	—	—	—	—	—	X	—	—	X			

Other Information:

For example, to estimate the annual clerical time associated with manual preparation of invoices, an analyst might measure the clerical time to be 250 minutes for preparing fifty invoices. Based on this sample, it would be possible to project the clerical effort required to prepare all invoices for the year. Using an estimated annual sales of $12 million and an average amount of $1000 per sales invoice, the calculation of annual clerical time is as follows:

Estimated number of annual invoices = $\dfrac{\$12{,}000{,}000}{\$1{,}000}$ = 12,000 invoices

Clerical effort per invoice = $\dfrac{250 \text{ minutes}}{50 \text{ invoices}}$ = 5 minutes per invoice

Estimated annual clerical time = 12,000 invoices × 5 minutes per invoice
= 60,000 minutes or 1,000 hours per year

Thus, 1,000 hours is the estimated annual time to prepare invoices manually.

Document Sampling

A portion of the time spent interviewing people associated with a system under study should be spent discussing forms or reports currently used. The system analyst should always attempt to obtain sample forms, source documents, worksheets, and reports associated with the system under study. Those forms processed by several people should be collected at various stages of completion to aid in understanding the activities involved. From these sample documents, the analyst can get a picture of the paper flow, learn what is presently being done and by whom, and identify sources of information used to complete the form or report. Just as with statistical sampling, document sampling can be of great assistance to the systems analyst in reviewing the present system and developing the new one.

File Sampling

File sampling refers to using extract programs to study the contents of computer files used in the system under study. The extracts provide accurate information about the volume, format, and content of records in the file, as well as the data integrity of specific data elements in the file. In studying an order-entry system, for example, an analyst could write a computerized extract program to do the following:

1. Sort the open-order file into customer number sequence.
2. Read the open-order file and, on a change in customer number, increment a counter by one so that the number of current customers can be determined.
3. Examine the order date and increment a counter for each order over five days.
4. Increment a counter for each open-order file record read.
5. Print every hundredth record.

This sampling process informs the systems analyst of the total number of customers and the number of open orders, and indicates the efficiency of current order-entry processing. The data elements of every hundredth record on the file could be studied intensively to yield such useful information as: which data fields, if any, are blank or contain meaningless data; what is the content of each data field, and is it consistent with the analyst's understanding of the data that should be in each field; and what are reasonable ranges of values for various numeric fields. Quite often, detailed analysis of records using file sampling will raise many important questions that otherwise would not have been identified until much later.

> **SUMMARY: Decision Tables and Sampling Techniques**
>
> **Decision Tables**   Combine "conditions" (IF) to be considered in the description of a problem along with the "actions" (THEN) to be undertaken. They allow a problem to be divided into its logical segments and represented in compact notation. Also, they can be used in the place of flowcharts.
>
> **Decision Table Components**   A decision table is divided into four basic elements with provision for other information that may be helpful in interpreting the results.
>
> - IF (condition statement):   Condition stub   Condition entry
> - THEN (action statement):   Action stub   Action entry
> - Other information
>
> Symbols used for the "condition" entry are:
> - **Y** (Yes)
> - **N** (No)
> - **>** (greater than)
> - **=** (equal to)
> - **<** (less than)
> - **—** (blank)
>
> Symbols used for the "action" entry are:
> - **X** (applicable)
> - **—** (blank)
>
> **Sampling Techniques**   There are three categories.
> 1. **Statistical sampling**   Involves making a set of observations to infer conclusions about a problem or an operation.
> 2. **Document sampling**   Focuses on reviewing sample documents so that information about their paper flow, their sources of information, and like items can be identified.
> 3. **File sampling**   Refers to using computerized extract programs to study the contents of computer files.

TIME-CHARTING TECHNIQUES IN A SYSTEM PROJECT

The preceding techniques are of great assistance to the systems analyst in understanding the present system and designing a new one. There is also a great need for bringing together the many activities involved in a system project—from systems analysis to system implementation. This involves the use of time-charting techniques, such as Gantt charts and PERT/time networks.

As will be demonstrated below, the activities of a system project should be scheduled in sufficient detail so that each important milestone can be planned and controlled. Even through uncertainty may exist for some activities, accurate times should be developed. Generally, past experience can be utilized as a starting point in scheduling.

The scheduler must determine appropriate starting dates for each activity. Analysis of these times generally indicates whether organization personnel have ample time to complete all the necessary tasks, especially in light of the equipment delivery date; overtime and additional personnel may be required at certain periods to meet the delivery date. The capability today to foresee a future problem is a great help to the DP group in planning and controlling a systems project.

Gantt Charts

The Gantt chart was first formulated by Henry L. Gantt in 1917 as a means of controlling the production of war materials. It is currently used in the scheduling of a systems project, starting from the first major phase to the last one. It can also be used for the various phases of a systems study, that is, for systems analysis, systems design, equipment selection, and/or system implementation. The complexity of the project determines what phase or phases should be charted.

Gantt chart components Within a Gantt bar chart, the horizontal axis is used to depict time, with activities listed vertically in the left-hand column. Ordinarily, the chart is used to compare planned performance against actual performance. Comparison of the planned and actual times indicates whether the system project is ahead, behind, or on schedule. As illustrated in figure 2.9, the solid line is the scheduled event and the dotted line is the completed event. Only the introductory investigation phase has been completed to date. As of March 1, the project is on schedule.

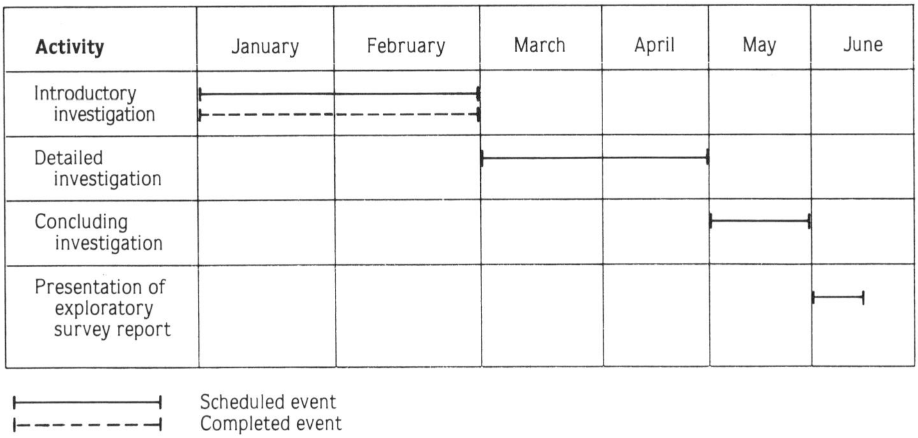

Activity	January	February	March	April	May	June
Introductory investigation	⊢―――	――――⊣				
Detailed investigation			⊢――	―――⊣		
Concluding investigation					⊢――⊣	
Presentation of exploratory survey report						⊢―⊣

⊢――――⊣ Scheduled event
⊢― ― ―⊣ Completed event

figure 2.9 Typical Gantt chart for scheduling the systems analysis phase of a system project.

Another way of depicting the various events is to utilize colored lines or bars. The scheduled events, for example, would appear in one color and the actual times achieved are added in another color as the various activities are accomplished. In this manner, a visual review of the chart will reveal the current status of the project.

An important advantage of this bar chart over other time-charting techniques is its simplicity. Managers are accustomed to working with charts; in relatively simple projects, they can comprehend a schedule and its status easily from a Gantt chart. However, the Gantt chart fails to show the relationship between one event and another. This shortcoming is not found in PERT/time networks discussed below.

PERT Charts

PERT, an acronym for Performance Evaluation and Review Technique, is illustrated in figure 2.10 for a system project. The circled numbers in the illustration are **events** which represent the completion of **activities**. No time is associated with events, unlike activities, depicted as arrows. All events are numbered from the lowest (first event) to the highest (last event).

PERT chart components The time within the triangles represents expected time (t_e) which is calculated by the formula,

$$t_e = \frac{a + 4m + b}{6}$$

where **a** equals the most optimistic time, **m** equals the most likely time, and **b** equals the most pessimistic time. The values developed from the formula have proven to be more

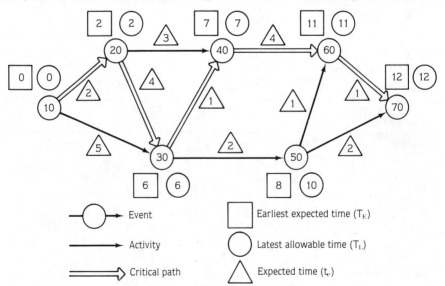

figure 2.10 PERT/time network depicting the critical path and the earliest expected and latest allowable times.

accurate than those from any other method. Those appearing in figure 2.10 were calculated on this basis.

Earliest expected times (T_E) and latest expected times (T_L) can be calculated, shown in the boxes and circles above the events, respectively. The difference between these two points represent **slack** (extra) time available for each activity within the project. In those cases where the earliest and latest times are equal, the event is on the critical path. In a PERT network, dummy arrows (dotted lines) do not represent activities, but rather connect related events in the network.

The double lines shown in the illustration represent the longest time path through the PERT network, the **critical path.** Any delays in starting or completing times on this path will delay the end completion date. Initial consideration should be given to working personnel overtime or reallocating qualified personnel to the critical activities. The critical path requires careful scrutiny as the project progresses. Those activities not on the critical path allow some slack time for completion; however, if they should exceed their expected time, they may be on the critical path. In essence, all noncritical activities have the potential of becoming critical ones if not carefully managed.

SUMMARY: Time-Charting Techniques in a Systems Project

Gantt Charts   Bar charts depict time on the horizontal axis and activities listed vertically in the left-hand column. Generally, they are used to compare planned performance against actual performance; comparison of these times indicates whether the project is ahead, behind, or on schedule as of any given date. Gantt charts are useful for managing straight-forward systems projects.

PERT Charts   Time charts relate activities of a systems project to one another. Events (shown as circles) which represent the completion of activities are not associated with time whereas activities (shown as arrows) are. The longest time through the numbered network represents the critical path (shown as a double line). Those activities which are not on the critical path have available slack time for completion. PERT charts are useful for managing complex systems projects.

CHAPTER SUMMARY:

To assist systems analysts in analyzing the present system and designing a new one, the fundamentals of flowcharts and decision tables were presented in this chapter. Although the focus of this text is on system flowcharts; flow process charts, document flowcharts, and decision tables along with statistical techniques also have their place in systems work. Flowcharts and decision tables can be used to assist systems personnel in understanding the level of programming for the new system. For simple and straight-forward computer programs, the use of program flowcharts might be the best method; however, for complex programs, decision tables might be preferred. Thus, the utilization of program flowcharts, decision tables, or a combination of the two depends on the complexity of programming activities.

In order to bring together the many activities associated with a typical systems project, there is need for some type of time-charting technique. For relatively small and straight-forward projects, the Gantt chart is preferred. Where there are many or complex activities, as found in an interactive management information system, it is advisable to employ PERT for planning and controlling purposes. This is particularly true when the systems project will take several years. Changes to computerized PERT programs can be made frequently, thereby resulting in new output for more effective management of the project.

Questions

1. What are the values of using system flowcharts in an MIS project? Explain.
2. Distinguish between a system flowchart and a program flowchart.
3. Why not save the expense of developing flowcharts for analyzing and designing a new system by using a narrative form? Criticize this viewpoint.
4. a. How does a program flowchart differ from a decision table?
   b. Which should be preferred when programming and why?
5. Distinguish between Gantt charts and PERT charts.
6. Distinguish between statistical sampling and file sampling.
7. Why is there a need for a time-charting technique when implementing a comprehensive system, such as an interactive management information system?

Self-Study Exercise

True-False:

1. _____ System flowcharts are equally applicable to the analysis and design of batch and interactive processing systems.
2. _____ A decision table is a pictorial representation of data flow.
3. _____ Generally speaking, there are no official flowcharting symbols.
4. _____ A system flowchart shows what operations are performed by a computer program.
5. _____ System flowcharts' only useful purpose is for documentation.
6. _____ Decision tables are capable of being converted directly to machine language.
7. _____ Gantt charts are basically bar charts that relate time to specified activities.
8. _____ The primary functions of PERT charts are for planning and controlling complex systems projects.
9. _____ In a PERT network, time is associated with events only.
10. _____ There is no provision for slack time in a PERT network.

Self-Study Exercise 53

Fill-in:

1. A flowchart has certain important values. It aids understanding, is an effective _____, and serves as a permanent _____.
2. A flowchart is a _____ representation of data flow within an organization.
3. A management report can be represented by a _____ symbol.
4. A diamond indicates a _____ symbol.
5. The basic _____ symbol is represented by a rectangle.
6. The basic _____ symbol is represented by a small circle.
7. The most significant feature of decision tables is their _____ notation.
8. Various types of flowcharts and decision tables complement and supplement each other in _____ and _____ a new system.
9. A Gantt chart is fundamentally a _____ chart.
10. A PERT network consists of numbered _____ and _____.

Problems:

1. The American Company is currently using a time-sharing service for order processing. The input/output devices used are teletypewriter units that are capable of interrogating on-line computer files, such as finished goods inventory balances and customer names and addresses. The order processing system using time sharing begins with the orders received from customers, whether via telephone or mail. The customer orders form the basis for interrogating the on-line computer files.

    The daily order processing system is illustrated in the system flowchart on p. 54. Fill in the blank flowchart symbols with the appropriate operation to be performed.

2. The production evaluation system using time sharing for The American Company begins with the production scheduling program. The input from order processing forms the basis for scheduling production to optimize the use of manufacturing facilities and manpower.

    The daily production evaluation system is illustrated in the system flowchart, p. 55. Fill in the blank flowchart symbols with the appropriate operation to be performed, punched card name, or computer program name.

3. To effect more efficient and economical order processing/billing, The American Company has decided to initiate a new computerized system. Basically, customer orders are received via mail and serve as input to the system. After orders are batched and control totals are established, they are forwarded to the key punch section for key punching and verifying. Next, the computer edits and sorts the transaction cards (items to be shipped) by customer number. Not only is an error listing produced along with control totals (for checking with the beginning control totals), but also a transaction file (by customer number sequence) is produced. In turn, this file provides the necessary input for the combined inventory updating and billing run. In addition to updating inventory files for items shipped, the customer master file is updated for items charged to the customer account. Likewise, customer invoices and shipping orders are printed along with control totals for comparison with the first totals established over customer orders.

    Based upon the foregoing narrative, prepare a system flowchart for depicting the flow of inputs and outputs from the first computer run to the second computer run. Also, show the balancing of control totals where necessary.

Standard Tools of Systems Analysis and Design

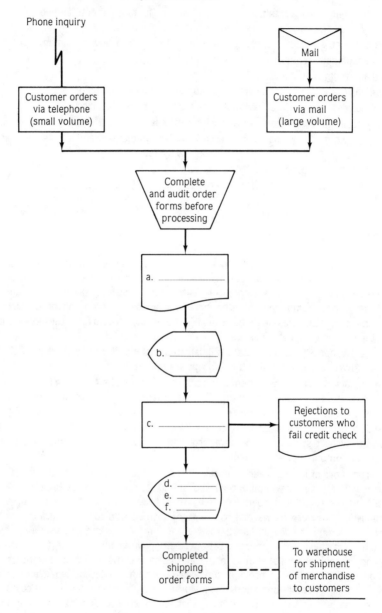

4. To assist marketing management in getting greater control over their salespeople, The American Company has decided to install a new computer system. Fundamentally, weekly master charge cards (punched cards) which have been previously sorted by customer account number, provide input for the computerized sales report run. The weekly and year-to-date

Daily Production Evaluation System Using Time Sharing System Flowchart

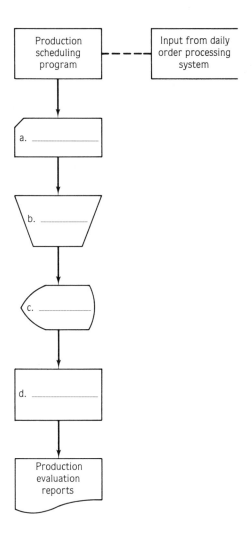

(YTD) customer sales master files are updated, and a weekly and year-to-date sales report by customers is printed. The aforementioned file is then sorted by salesperson number and written on a computerized file which, in turn, provides the necessary input for printing a weekly and year-to-date sales report by salesperson. The totals of this run are then compared to the weekly and year-to-date sales report by customer.

Based upon the foregoing narrative, prepare a system flowchart that depicts the flow of inputs and outputs for the three computer runs. Also, show the balancing of control totals where necessary.

5. The finished goods inventory program (batch processing) for The American Company begins with initializing the master file. Basically, the program makes the necessary additions (card

#1), deductions (card #2), and adjustments (card #3) to the current master finished goods inventory file before writing the updated master file. The processing procedures are illustrated in the program flowchart below.

Fill in the blank flowchart symbols with the appropriate operation(s) or logical decisions.

6. The customer billing program (batch processing) for The American Company is flowcharted on p. 58. However, for more concise notation, the decision table is preferred.

   Complete the decision table form on p. 59 after examining the program flowchart closely.

7. The American Company is currently developing a computerized payroll system which consists of many parts.

   a. For one segment of the new system, develop an appropriate **program flowchart** and the corresponding **decision table** for computing gross pay: total hours = regular hours + overtime hours (at time and a half).

**Weekly Finished Goods Inventory
Program Flowchart**

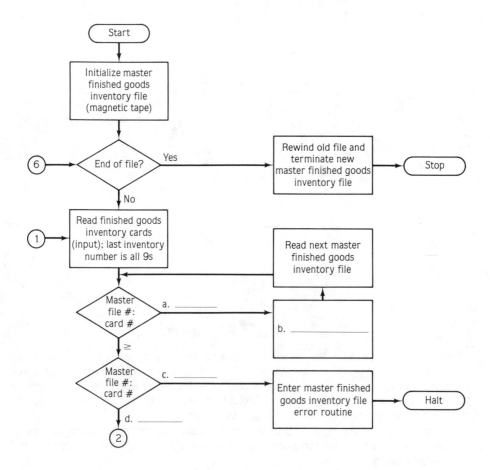

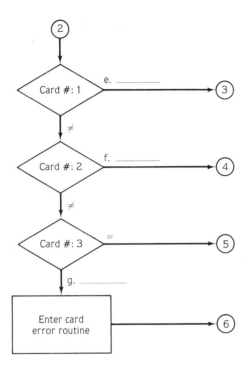
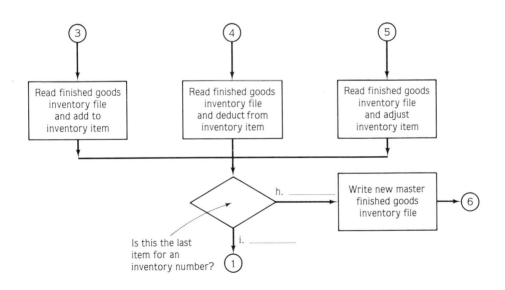

**Customer Billing
Program Flowchart**

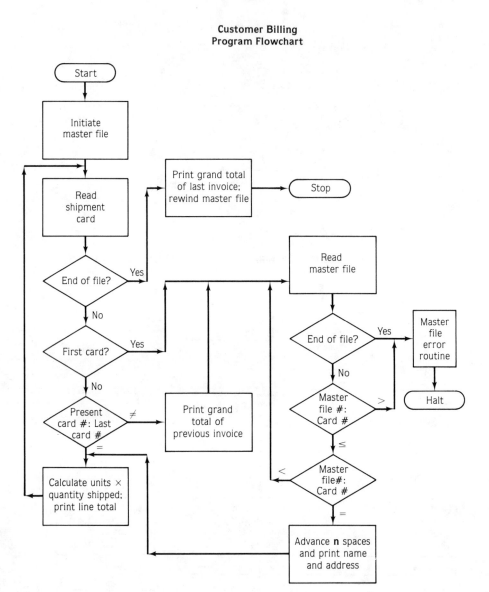

b. For another segment of the new system, develop an appropriate **program flowchart** and the corresponding **decision table** for computing the savings bond deduction: amount equal to 5% of the gross pay or $3.00 per week.

c. For a third segment of the new system, develop an appropriate **program flowchart** and the corresponding **decision table** for computing the credit union deduction: amount equal to 10% of the gross pay or $15.00 per week.

Self-Study Exercise

DECISION TABLE	Table name: CUSTOMER BILLING PROGRAM								Page 1 of 1			
	Chart no.: Billing–1				Prepared by:				Date:			
	Rule number											
	1	2	3	4	5	6	7	8	9	10	11	12
Condition												
Action												

Other Information:

chapter

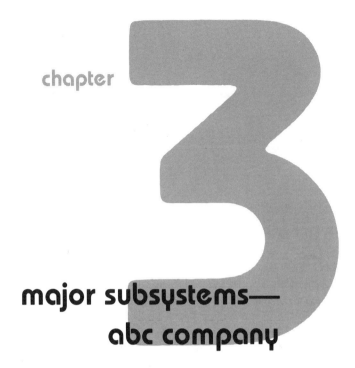

major subsystems—
abc company

OBJECTIVES:

- To distinguish between industrial and service organizations when undertaking a systems project.
- To set forth the necessary background for the ABC Company, the text's master case study.
- To examine the many subsystems found in a typical industrial organization, for example, the ABC Company.
- To explore the basic subsystems that are found in typical service organizations.

IN THIS CHAPTER:

OVERVIEW OF BUSINESS ORGANIZATIONS
    Industrial Organizations
    Service Organizations
ABC COMPANY—MASTER CASE STUDY
    Data Processing Organization Structure
    Overview of Major Subsystems
MAJOR SUBSYSTEMS IN INDUSTRIAL ORGANIZATIONS—ABC COMPANY
    Marketing Subsystem
    Research and Development, and Engineering Subsystems
    Manufacturing Subsystem
    Purchasing and Inventory Subsystems
    Physical Distribution Subsystem
    Accounting and Finance Subsystems
    Personnel Subsystem
MAJOR SUBSYSTEMS IN SERVICE ORGANIZATIONS
RELATIONSHIP OF SYSTEMS ANALYSIS AND DESIGN TO MAJOR SUBSYSTEMS
CHAPTER SUMMARY
QUESTIONS
SELF-STUDY EXERCISE

Now that an overview of systems analysis and design has been presented and the standard tools of a systems project have been discussed, we can concentrate on the basic types of business organizations and their subsystems. Specifically discussed are industrial and service organizations, which are prime candidates for undertaking systems analysis and design work. It is within these two broad categories of business organizations that the designing of new management information systems will be examined.

Following a discussion of business organizations, a brief background on the ABC Company is presented. This overview approach to the text's master case study is necessary before the many subsystems can be explored. Fundamentally, these subsystems, which operate in a **batch processing mode,** include the following:
- marketing
- research and development, and engineering
- manufacturing
- purchasing and inventory
- physical distribution
- accounting and finance
- personnel

This logical breakdown of subsystems, then, will form the basis for developing a detailed order-entry system in part three, Systems Design of New Management Information Systems that operates in an **interactive processing mode.**

OVERVIEW OF BUSINESS ORGANIZATIONS

Although many types of business organizations are currently found in our economy, most business can be segregated into two basic types, **industrial** and **service** organizations. The distinctive feature of the industrial organization is the manufacture of goods. On the other hand, the activities of service organizations center only on the sale and distribution of goods and services.

Industrial Organizations

Since industrial organizations focus on the manufacture of some type of product(s), they can be defined as organizations of **people, materials,** and **machinery** that are backed by **money** and under **management** control to produce, sell, and distribute goods of some type. Utilizing this definition (traditionally called "the 5 Ms" when the word "men" was used instead of "people"), the industrial organization is typically more complex in terms of organization structure than the service organization. This should be somewhat obvious from the prior listing of subsystems for the ABC Company, a typical industrial organization.

Service Organizations

While industrial organizations are oriented toward the production of goods, service-oriented business organizations are oriented toward the sale and distribution of goods and services. Typically, such organizations include retailers, wholesalers, distributors, utilities, and the like. In addition, another type of service organization is the "governmental organization" which includes federal, state, county, and city governments, as well as educational institutions. Both governmental and educational organizations are normally concerned with providing some type of services.

ABC COMPANY—MASTER CASE STUDY

The ABC Company, the text's master case study, is an organization which manufactures products for the consumer market. Its sales are currently $75 million per annum and are projected to be approximately $100 million in five years. Its product line consists of three hundred products which include variations within basic product groups. Variations of these basic products are for specific customers whose requirements differ because of the markets they serve. Certain products are shipped directly from the plants to large volume customers; however, customer orders are usually filled from one of the five field warehouses, strategically placed throughout the country.

Corporate headquarters are centrally located in St. Louis; the locations of the company's manufacturing plants are found in Minneapolis, Philadelphia, and Los Angeles. Each manufacturing plant has an attached warehouse where finished goods are held temporarily for direct shipment or shipment to field warehouses. Presently, the firm's entire operation includes approximately 2000 employees.

The ABC Company is organized as shown in figure 3.1. The president and chief executive officer reports to the board of directors and is assisted by the corporate planning group. The executive vice-president, in turn, reports to the president. In a similar manner, eight vice-presidents (marketing, research and development and engineering, manufacturing, purchasing and inventory, physical distribution, accounting and finance, personnel, and management information system) report to the executive vice-president. Various corporate headquarters managers, plant managers, and warehouse managers report to their respective vice-presidents.

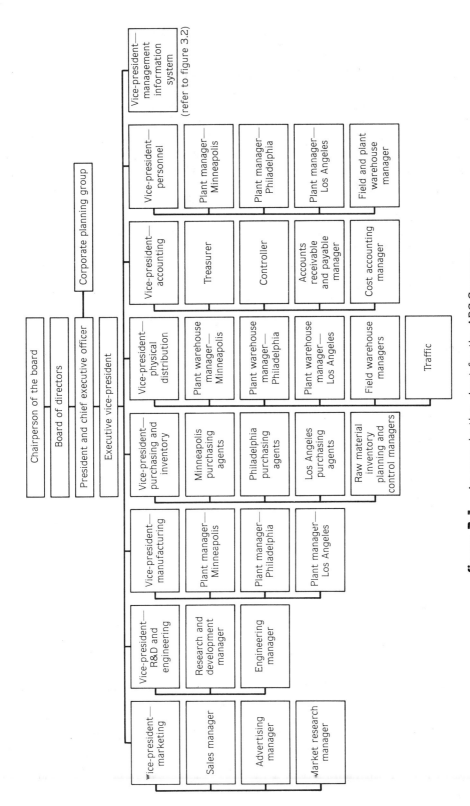

figure 3.1 An organization chart for the ABC Company.

The ABC Company has progressed through a series of data processing systems that have been guided by the constraints of available equipment, management acceptance of data processing, and an increasing base of experience in the technology. Many years ago, manual systems were augmented by adding machines and calculators. Punched card equipment and tabulating machines provided expanded system capabilities. As computers became available, the data processing system was adapted to utilize these machines, and translating data into information was accomplished much faster. Also, the computer provided capacity for the development of new applications. Presently, the company uses a batch processing information system. The problem encountered with the present system is that information (for management and nonmanagement purposes) is not sufficiently timely to effect changes in the current operating environment. Too often, information is received by corporate headquarters too late to effect the necessary control over manufacturing and warehousing operations. Thus, the batch processing mode, whereby data are collected for a period of time before being processed and summarized, does not facilitate day-to-day decision making.

In view of this problem, the MIS vice-president of the ABC Company initiated a business system study to improve the planning process for the development of new systems and the application of data processing technology. This study was conducted over a period of six months and involved members of top management—vice-presidential level and higher. The study identified the following key objectives to be met, in part, through future system work:
1. provide better customer service
2. improve selling efficiency
3. assist production in more closely meeting customer needs
4. provide more timely information for analysis of the company's present and future operations at both plant and corporate levels
5. improve coordinated control of the overall organization

As a result of this study, it is apparent that a major change was needed in the ABC Company's system development philosophy. The company-wide information system has to be upgraded to effect better decision making on a day-to-day basis. Specifically, what is needed is a forward-looking system that promotes control over present and future operations. Because an interactive system operating in a real-time mode is capable of meeting such a goal, the MIS vice president initiated a major program to meet the key corporate objectives for future system work, which included the use of interactive systems where feasible and appropriate.

Data Processing Organization Structure

The present data processing organization chart in a batch processing environment is illustrated in figure 3.2 for the ABC Company. The system and programming manager and the DP operations manager report directly to the vice-president of the management information system. On the systems side, the systems analysts and the programming supervisor report to the system and programming manager. Generally, the systems and programming positions will be unchanged in the new organization structure with the exception that program development and testing can take place in a user/machine mode (by utilizing one of the on-line I/O devices).

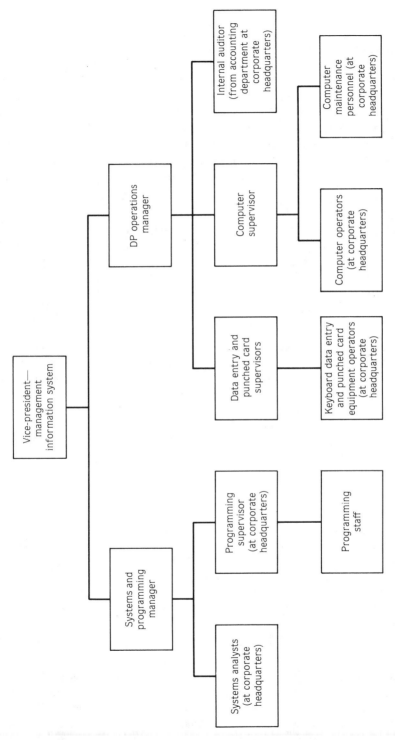

figure 3.2 An organization chart for data processing in a batch processing environment for the ABC Company.

67

From an operational viewpoint, the data-entry and punched card supervisors, the computer supervisor, and the internal auditor (from the accounting department) report to the DP operations manager. Unlike systems and programming, an important organizational change is contemplated with the interactive system. Data-entry terminal supervisors will manage on-line typewriter and CRT terminals at the corporate headquarters, plants, and warehouses (plant and field). In contrast, all data-entry and punched card equipment are presently under their control. Also, the computer supervisor currently manages the computer's operators and the computer maintenance servicepeople. Thus, the batch processing organization chart for current operations per figure 3.2 will be changed for the interactive capabilities of the new system.

Although specific positions and levels of system work are clearly set forth (figure 3.2), the system project cannot be delegated completely to analysts who are asked to find "the answer." The company's management must work closely with systems analysts to create an awareness of value systems and premises used in planning and decision making. **Top management must exert leadership and operating managers must actively cooperate with computer specialists in order to form an effective team to design and implement the new system.** Only when the company's managers get involved, cooperate with DP specialists, and relate their problems to the computer's capabilities can the feasibility study objectives be fully realized.

Similarly, systems analysts need to communicate and work with operating personnel on the detailed methods and procedures of the new system. This is necessary in order to determine how exceptions and problems are to be handled. Also, cooperation is needed to realistically evaluate the decision-making process at all levels and the necessary information flow for that process. Systems analysts and programmers must work with people above or below them, as well as on the same level, for effective results since the company's managers and operating personnel generally will accept only methods and procedures that they fully comprehend.

Overview of Major Subsystems

From an overview standpoint, the ABC Company can be described as a "material flow company." This concept for the three manufacturing plants is illustrated in figure 3.3, shown as a double line arrow on the outer rim of the system flowchart. Purchased materials and manufactured goods for stock flow at various stages of the production process. As they flow through the process, the materials take on a variety of forms and shapes until they become finished goods. Next, the finished goods flow through the physical distribution system either via direct shipments to customers or through company-owned warehouses until they reach the customers. Thus, in this **materials flow** concept several of the major subsystems are involved.

Coupled with the materials flow is a corresponding **information flow** (figure 3.3). Materials information flow is a most important factor in coordinating the diversified activities of the three manufacturing plants and warehouses (plant and field) with corporate headquarters. It must be comprehensive, thereby integrating decision making throughout the entire materials flow process—from purchased materials to shipment of

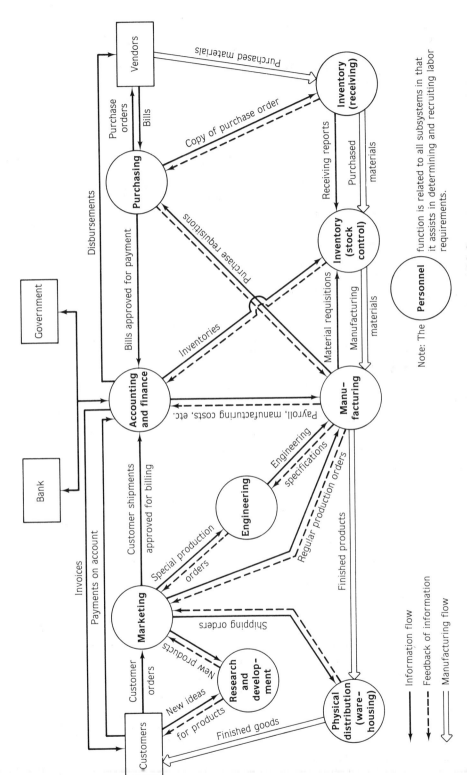

figure 3.3 An overview of the major subsystems of the ABC Company.

finished goods. With the integrated flow of essential information, management and operating personnel can make adjustments swiftly and effectively in response to the ever-changing business environment. The materials information flow approach, then, is an essential part of the interactive management information system planned for the ABC Company.

The information flow is not restricted to the materials area only. In fact, there may well be more information generated for activities that are not directly related to the materials flow process (figure 3.3). Many subparts of the marketing, research and development, engineering, accounting, and finance subsystems are not related directly to the manufacture of the final product. No matter what the source or need of information is, the new interactive MIS must be **open-ended.** This approach provides flexibility so that activities can be linked at minimum cost and effort. But more importantly, the open-ended approach allows for changing the direction and speed of information flow in response to management and operating personnel needs. More will be said about the information flow in subsequent chapters.

MAJOR SUBSYSTEMS IN INDUSTRIAL ORGANIZATIONS— ABC COMPANY

The preceding overview of the present and new systems serves as a background for presenting the various subsystems of the ABC Company. In a manufacturing environment, at least two basic subsystems (business functions) are designated as vital to the operation of the organization: **marketing** and **manufacturing.** Marketing is essential because it creates values of **time, place,** and **possession** for customers, and manufacturing creates values of **form.** However, a third subsystem must be identified—**finance**—since the foregoing utilities of time, place, possession, and form are not possible without the appropriate level of financing. In addition, the foregoing subsystems must be supplemented by supportive subsystems which include research and development, engineering, purchasing and inventory, physical distribution, and personnel. Thus, many basic and complementary subsystems are to be found in industrial organizations like the ABC Company. Each of these subsystems will be explored below.

Marketing Subsystem

The prime job of the marketing subsystem of the ABC Company is to contact potential customers and sell merchandise through salespeople, distributors, advertising, and special promotions. If products are sold by salespeople, a multicopy **sales order form** is originated by either these individuals or the order section of the marketing department. Copies of the sales order are distributed as follows: original copy to customer for acknowledging order, and duplicate copies to the salesperson and the manufacturing, stock control, shipping, and accounting departments. In addition, the marketing subsystem may prepare other forms, such as **contracts, bids, back orders,** and **change orders.**

Sales to employees other than regular charge transactions to customers are recorded normally on a **sales slip** and are prepared by a salesperson at the warehouse. The original copy is given to the employee and other copies are distributed to the shipping, stock control, and accounting departments. In essence, the marketing section usually starts the product information flow in the ABC Company, thereby being an initial source of data.

Research and Development, and Engineering Subsystems

Often, the marketing subsystem receives inquiries regarding a new product. It will initiate a **research and development order** (R & D order) for its development if marketing prospects look promising. In like manner, the marketing research section may have its own ideas on what products should be developed. This, too, will result in an R & D order. Other times, another organization will want to utilize the company's talents and expertise for pooling research talents. This is necessary in order to obtain a large sales order which neither organization is capable of fulfilling individually. In the final analysis, research and development can play a vital role in the ABC Company's growth rate.

Before manufacturing operations can commence, it is necessary to design the products as agreed upon by the marketing section in conjunction with research and development. This means designing the product from scratch. The **engineering blueprints** (manufacturing specifications) are forwarded to the manufacturing subsystem. The blueprints form the basis for producing manufactured parts, such as sub-assemblies, frames, and the like. The requirements for the product to be manufactured are summarized on a **bill of materials.** This forms the basis for "exploding" a bill of materials, i.e., determining the number of detailed items for the production order. With engineering specifications and appropriate bills of materials prepared, these data can be forwarded to the production control department of the manufacturing subsystem.

Manufacturing Subsystem

The manufacture of finished consumer products for the ABC Company involves many steps. Not only must plant, equipment, and tools be provided, but also appropriate personnel hired and trained to utilize the manufacturing facilities. Raw materials and goods-in-process must be available as needed. Production must be planned, scheduled, routed, and controlled for output that meets certain standards. The effective management of this area, then, is a study unto itself.

The company's products can be produced in anticipation of demand, on receipt of customer orders, or in some combination of these two. If goods are being produced to order, a sales order copy may be the **production order** (regular or special) in many cases. The usual arrangement is to have the production planning and control department initiate action on factory orders. The production order is also distributed to the stock control, shipping, and accounting departments. The original copy is kept in the production department files.

Referring to figure 3.3, manufacturing materials can be obtained from inside or outside the company. If a **purchase requisition** is prepared, it is forwarded to purchasing for vendor purchase. If, on the other hand, materials are available from the stock control department, a **materials requisition** is prepared. Other records and forms, found within the manufacturing function, are

    periodic production reports
    tool orders
    material usage reports
    material scrappage reports
    inspection reports
    labor analysis reports
    cost analysis reports
    production progress reports

Purchasing and Inventory Subsystems

The purchasing function is concerned with procuring raw materials, equipment, supplies, and utilities, as well as other products and services required to meet the ABC Company's needs. The procurement process begins with the completion of the purchase requisition, prepared in duplicate. One copy is forwarded to purchasing and the other retained by the originator. Based upon the purchase requisition form, a buyer locates the appropriate contract and orders from the supplier(s). If a contract has not been signed for the desired goods or services, the buyer sends a **request for quotation** to prospective vendors. Once the outside supplier has been determined, based on an analysis of total delivered cost, quality, and service, a **purchase order** is typed and mailed to the vendor. It contains the items to be shipped, prices, specifications, terms, and shipping conditions. The original is forwarded to the vendor while duplicate copies are distributed to the purchasing, receiving, stock control, accounts payable, and preparing departments.

As soon as goods are received from suppliers, they are checked and verified against the copy of the original purchase order by the inventory or receiving department. Once the receiving clerk is satisfied that the goods correspond to those on the purchase order, the individual prepares a **receiving report,** noting any discrepancies between the order and actual material received. Depending upon the type of goods, an **inspection record** is prepared along with the receiving report. Copies of both are sent to those departments—namely, purchasing, stock control, manufacturing, and accounts payable—that have need of such information. A carbon copy is retained by the receiving department.

Goods are delivered by the receiving department to the stock control department or any other department which has ordered them. The department that physically takes possession of the materials acknowledges receipt by signing a copy of the receiving report.

The function of stock control or stores department within the ABC Company is to store and protect all materials and supplies that are not required for current usage. The transfer-in of materials from an outside vendor to the stores department is documented by the receiving report mentioned earlier. The transfer-out of goods to the manuacturing departments is authorized by the materials requisition, also noted previously. Materials are issued in response to current manufacturing needs.

A most important source of data is the **inventory records** or **stock records** maintained by the department. Since it is responsible for materials in and out of stock, it has the necessary data for determining inventories. Stock records are in the form of visual records, backed up by magnetic tape files.

Stock control has the added function of replacing stock when it reaches a minimum level, often referred to as a reorder point. The stores clerk prepares a purchase requisition for the specific materials and forwards it to the purchasing department. In essence, the stock control department is an important source of data to keep the many subsystems operating in a manner compatible with the ABC Company's objectives.

Physical Distribution Subsystem

Once the customer order has been manufactured or finished goods are available in the field warehouse, they are ready for shipment. The finished products must be packed, labeled, and transported to the customer. The **order and shipment processing form** which authorizes shipment is delivered with or in advance of the goods. If pickup is made by the customer, receipt of goods will be acknowledged by signing a copy of the shipping order which is then filed in the ABC Company's shipping office.

Shipments that are made via public carriers must be accompanied by a **bill of lading** which is actually a contract between the consignor and the carrier. It serves as a description of the shipment and its contents. One copy each is for the customer and the public carrier, a third copy is filed in the shipping department as proof of shipment.

Accounting and Finance Subsystems

After the required business operations have been performed by the shipment of finished goods, the accounting department must prepare **customer invoices**. These serve not only as a record of charges, but also are the basis for which the seller can legally claim payment for goods and services. The first copy is sent to the customer while remaining copies are distributed to the salesperson, field warehouse, and the accounting department's accounts receivable file.

Depending upon the terms of the invoice, payments received from customers are deposited in the ABC Company's bank account. These payments are recorded in the **cash receipt journal** as documented evidence of their receipt. Periodically, a statement of account is mailed to each customer.

In addition to billing and collecting, the department is concerned with disbursing funds, mainly for payroll and goods and services. **Time cards** are the originating source for paying salaries and wages. They may also be used for making labor distribution charges to various departments and products. **Payroll checks, payroll registers,** and **W-2 forms** are the net result of payroll procedures.

The second type of disbursements is payment of vendor invoices. This initially involves checking the vendor's invoice against the purchase orders and receiving reports. Upon approval of payment by the purchasing department, **voucher-checks** are prepared. A voucher-check is a check which has an attached voucher containing sufficient space for date, purchase order number, vendor number, description, amount, discount, and net payment. The first copy is mailed to the vendor on the day of the month designated on the invoice, and duplicate copies are used for data processing. When processing is complete, the vouchers are filed.

The foregoing functions are not complete until all legitimate governmental forms have been prepared and the proper voucher checks drawn for the respective amounts due. Federal, state, and local governments require the preparation of specific tax forms, which include

- federal income tax returns
- reports on social security taxes withheld (employer and employee)
- federal and state unemployment compensation returns
- state income tax returns
- personal property tax returns
- city income tax returns

Other governmental information is required of the ABC Company which forms the basis for the statistical data on the United States. In the final analysis, government requirements can constitute a substantial load, over and beyond the normal data needed for the company's internal operations.

Personnel Subsystem

Although the personnel function is not shown as an integral part in figure 3.3, it is interconnected with all the major subsystems shown. Its basic task is to determine personnel needs in the present and future, thereby aiding in the recruiting of necessary employees. Placing the right person in the right job can be an arduous task; however, it can be facilitated by a computerized personnel and payroll magnetic tape file which can assist in recruiting internally for new positions before going to outside sources. Various personnel forms can be employed for this function, such as **personnel history and promotion records** and **personnel requisition forms**.

SUMMARY: Major Subsystems for the ABC Company

Marketing Subsystem   Initiates the product information flow. Its prime job is to contact potential customers and sell finished products through salespeople, advertising, and special promotions.

Research and Development, and Engineering Subsystems   Develop new products vital to the company's growth and design the new products from scratch.

Manufacturing Subsystem   Schedules and manufactures finished products whether in anticipation of demand, upon receipt of customer orders, or some combination of these two.

Purchasing and Inventory Subsystems Procure the raw materials, supplies, equipment, and other purchase requirements necessary to meet the company's operating needs. They also receive, store, and protect all materials and supplies not required for current use, and issues items for production to the authorized manufacturing departments.

Physical Distribution Subsystem Authorizes finished products to be shipped to customers from company-owned warehouses or via direct shipment orders.

Accounting and Finance Subsystems Receive information from other subsystems that form the basis for the company's financial statements and finance the company's activities.

Personnel Subsystem Determines and recruits the personnel needed in the present and the future to meet company objectives.

MAJOR SUBSYSTEMS IN SERVICE ORGANIZATIONS

In a preceding section, the basic subsystems found in industrial organizations included marketing, manufacturing, and finance. Although these are basic for a manufacturing firm like the ABC Company, they may not be basic for other types of organizations, in particular, service organizations. The basic subsystems (business functions) of a department store are considered to be marketing, purchasing, and finance. For a railroad, they are traffic, maintenance, and finance. Additional complementary subsystems are necessary, similar to those found under the ABC Company. In contrast, the basic functions found in a university are education and research, supported by additional complementary functions. Thus, an organization's objectives will determine the basic and complementary subsystems.

RELATIONSHIP OF SYSTEMS ANALYSIS AND DESIGN TO MAJOR SUBSYSTEMS

The foregoing exposition of major subsystems sets the stage for systems analysis of the present system, the subject matter for part two, Systems Analysis of Present Business Information System. In like manner, an understanding of these subsystems serves as a required background for designing the order-entry system—the subject matter for part three, Systems Design of New Management Information System. Hence, this presentation on the ABC Company gives the reader a useful background for the remainder of the text.

When contrasting the analysis of the present batch processing information system (systems analysis) to the design of the proposed interactive management information system (systems design), one central idea will emerge time and time again: interactive processing represents more than just the traditional flow of information through the organization. It is structured so that important information flows from one subsystem to another in order to provide the information needed, when it is needed, and where it is needed. In a real sense, an interactive management information system represents the internal communications network of the business, providing the necessary management information to organize, direct, and control according to corporate plans. Stated another way, it is a reorientation of traditional information flow which expands the basic purpose of recording "what happened" to include new purposes of telling "how and why it happened," the "amount of deviations from the plans," and the "effects on the plans." A history-keeping orientation is replaced by a management planning and control orientation. In essence, an interactive MIS processing environment utilizes integrated plans, policies, guidelines, methods, and procedures; mixes them with people; adds timely information; and comes out with the organized use of managerial and operational information for better company decisions.

CHAPTER SUMMARY:

After distinguishing between the basic types of business organizations, the chapter focused on an overview of the ABC Company, the text's master case study. Essentially, the various subsystems were explored within a traditional approach, that is, on an integrated basis within a batch processing mode. As was pointed out, the company has initiated a study to determine the feasibility of adopting an interactive processing MIS environment. Assuming the feasibility study is approved by top management, the first area to be converted will be the order-entry system—the subject matter for those future chapters that focus on the systems design of the ABC Company. Thus, this first chapter on the ABC Company sets the stage for presenting the step-by-step analysis that is required for developing an interactive order-entry processing system. Similarly, this presentation serves as a useful background for assisting the reader in fulfilling the requirements of the case studies ahead: designing the finished product inventory system and the accounts receivable system.

Questions

1. a. Distinguish between industrial and service organizations.
   b. Which type is the ABC Company? Why?
2. Suppose the ABC Company was a much larger firm, would the data processing organization chart be much different in an interactive processing mode?
3. a. Distinguish between the materials flow concept and the information flow concept.
   b. Which type is the ABC Company? Why?

Self-Study Exercise **77**

4. What problems can be expected by the ABC Company when changing from a batch processing information system to an interactive management information system?
5. What basic characteristics are to be found in an interactive processing environment for the ABC Company that are not presently found in the batch processing mode?
6. What changes will be taking place in the various subsystems of the ABC Company as it changes from a batch processing information system to an interactive management information system?

Self-Study Exercise

True-False:

1. _____ An industrial organization centers primarily on the sale and distribution of goods and services.
2. _____ Universities are considered to be service organizations.
3. _____ An important reason for implementing a new system is improved customer service along with improved selling efficiency.
4. _____ Generally, the data-entry terminal supervisors report directly to the vice-president of the management information system.
5. _____ The ABC Company is not an example of a "materials flow company."
6. _____ The basic subsystems of any service organization are marketing, manufacturing, and finance.
7. _____ The accounting subsystem starts the product information flow in the ABC Company.
8. _____ Bills of materials in the ABC Company are prepared by the engineering subsystem.
9. _____ The procurement process for raw materials begins with the purchase order in the ABC Company.
10. _____ The personnel subsystem is interconnected with all of the major subsystems in the ABC Company.

Fill-In:

1. Typically, _____ organizations are more complex in terms of organization structure than _____ organizations.
2. Since the ABC Company manufactures products for the consumer market, it can be called an _____ _____.
3. Within the data processing organization chart of the ABC Company, the _____ _____ reports to the system and programming manager.
4. Within a _____ _____ company, purchased raw materials and parts as well as manufactured materials flow into the various stages of the production process.
5. The three basic subsystems (business functions) of any industrial organization are marketing, _____, and _____.

78   Major Subsystems—ABC Company

6. The primary job of the _____ subsystem in the ABC Company is to contact potential customers and sell merchandise.
7. _____ _____ form the basis for producing manufactured subassemblies in the ABC Company.
8. The procurement process ends with the preparation of a _____ _____ to an outside vendor.
9. The function of _____ _____ in the ABC Company is to store and protect all materials and supplies that are not required for current usage.
10. An _____ processing system is capable of providing the managerial and operational information needed, when needed, and where needed.

Problems:

The ABC Company, like most other manufacturing firms, can operate in a batch or an interactive processing environment. The problems below treat an overview of these two types of information systems. Also, the first problem surveys the sources of data for the basic business operations of the ABC Company.

1. Using the overview of the major systems as a guide (refer to figure 3.3) for the ABC Company, insert the appropriate business functions (marketing, research and development and engineering, manufacturing, information systems, purchasing, accounting and finance, or personnel) in the overall system flowchart shown below.

Overall System Flowchart

2. a. Using a batch processing information system for the ABC Company, fill in the appropriate names for old and new files (magnetic tapes) that would be found in the accounting system.

Accounting System

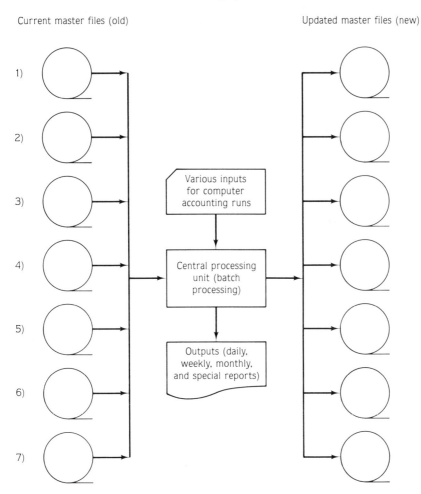

b. Typical weekly accounting reports for each of the files in (a) are:
1) _____
2) _____
3) _____
4) _____
5) _____
6) _____
7) _____

3. a. Using an interactive information system for the ABC Company, fill in the blank spaces with the appropriate data base elements that would be stored on line for the accounting system.

80 Major Subsystems—ABC Company

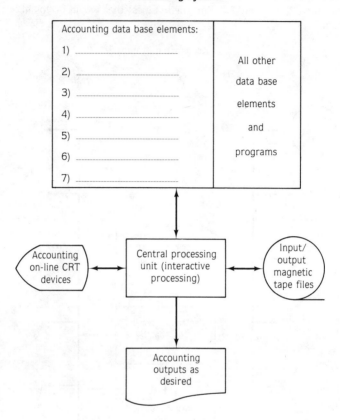

b. Typical information that can be extracted from the common data base elements in (a) are:
   1) _____
   2) _____
   3) _____
   4) _____
   5) _____
   6) _____
   7) _____

part two

systems analysis of present business information system

chapter 4

initiation of a system project

OBJECTIVES:

- To explore the important organizational factors that affect a typical system project.
- To set forth the major steps of a system project as undertaken by industrial and service organizations.
- To examine the need for an introductory investigation of a system project.
- To present the major components of the initiation of a system project.

IN THIS CHAPTER:

SYSTEM PROJECT AFFECTS THE ENTIRE ORGANIZATION
    Personnel
    Subsystems (Business Functions)
    Communication Process
    Organization Structure
    Financial Resources
MAJOR STEPS OF A SYSTEM PROJECT
INTRODUCTORY INVESTIGATION
ORGANIZATION OF SYSTEM PROJECT
    Establishment of Executive Steering Committee
    Selection of MIS Task Force
DEFINITION OF THE SCOPE OF SYSTEM PROJECT
    Selection of Desired Objectives
    Definition of the Problem
DETERMINATION OF THE SCHEDULE OF SYSTEM PROJECT
CHAPTER SUMMARY
QUESTIONS
SELF-STUDY EXERCISE

The implementation of a new system like a management information system is not a simple task. In order to install a new system effectively, a detailed and carefully laid out plan must be initiated by management and followed by all personnel involved. The large outlays for equipment, programming, conversion, and related activities require a systematic procedure for implementation. Otherwise, vast sums of money can be wasted. The purpose of this chapter and the next one is to explore the initial steps in implementing and controlling a system project.

In this chapter, the major areas affected by a system project are explored initially. Then, the real essence of the chapter is discussed, namely, the initiation of a system project. This will include establishing an executive steering committee to provide direction for the MIS task force, defining the scope of the project, and determining a realistic schedule. These three elements constitute the first step in the development of a large, fairly sophisticated system project, such as changing from a batch processing mode to an interactive MIS processing mode for the one or more subsystems found in a typical organization like the ABC Company.

SYSTEM PROJECT AFFECTS THE ENTIRE ORGANIZATION

The introduction of automated data processing or the extension of newer DP equipment and procedures has a far reaching impact on the entire organization. The decision to study the feasibility of a new system will affect more than the established methods and procedures. The most important areas of the organization that will be affected are personnel, subsystems (business functions), the communication process, the organization structure, and financial resources.

Personnel

The most profound changes, in many cases, occur in the area of personnel. The work habits of individuals accustomed to doing things a certain way over the years may be changed. In a similar manner, the methods and procedures to be followed by organizational employees may change substantially. Present employees may have to be shifted to different jobs; new employees may be required to fill jobs that require a certain technical ability. Also, the work areas of employees may be altered, which may not be to their liking. The decision to undertake a system project, then, can have a great impact on personnel.

Subsystems (Business Functions)

An advanced computer system may drastically affect an organization's subsystems or business functions. Some of the manual tasks of subsystems can be relegated to the computer; marketing analysis, inventory and production control, and flexible budgeting are logical candidates. In an expertly designed system, the output from one computer area may serve as input for another. This succession of data processing activities provides an excellent way to accomplish and coordinate the organization's business functions.

Communication Process

The question of centralization and decentralization is brought into focus by the possibility of immediate communication between central headquarters and outlying areas under a new system. Established lines of communications may be revamped, and new lines initiated. This is especially true with the current vogue to **distributed processing** which focuses on local and regional processing with only summary information forwarded to central headquarters. Thus, the new system project may well stress the need for more computer processing at multiple locations where the source data are first encountered.

Organization Structure

In light of computer replacement of employees and the need to add DP personnel, the organization structure may be modified. This is particularly true when an advanced system is being contemplated. A flattening of the structure may occur since jobs in several organizational levels can be relegated to the computer. In effect, newer types of organization structures are and will continue to be an integral part of newer DP systems.

Financial Resources

The development of a new data processing system has a substantial financial impact on an organization. The expenditure of large sums over a long period of time means that cash resources must be available when needed. During the programming and testing phases, rental is paid on DP equipment which is generating little or no computer output for operational use. Also, the conversion phase from the present system to the new system can make great demands on an organization's cash resources.

Monetary outlays are often overshadowed by personnel considerations. An organization can change and control the physical requirements for an advanced computer system, but the human factor cannot be as easily managed. The human factor can make or break any system project if not handled properly.

> **SUMMARY: System Project Affects the Entire Organization**
>
> **Personnel** The most profound changes are in this area where the human element can make or break any system project.
>
> **Subsystems (Business Functions)** Many of the manually oriented aspects of business functions (subsystems) are prime candidates for computer implementation, such as marketing analysis, inventory control, and flexible budgeting.
>
> **Communication Process** The current trend toward distributed processing affects the amount of data communications between the home office and the local and regional units.
>
> **Organization Structure** The implementation of newer DP systems may allow for the reduction in the number of organization levels.
>
> **Financial Resources** An important underlying factor is the cash requirement during various stages of a system project.

MAJOR STEPS OF A SYSTEM PROJECT

The analysis, design, and implementation of a new system may be difficult and costly. The complex relationships that exist among the factors above demand a carefully structured and analytic approach. Within a system project, these must be decided: **what** must be done, **how** to do it, **who** will do it, and **when** to do it. At any time, an organization is generally constrained in what it can do by limitations in capabilities and/or resources. Many times, it does those things that organizational personnel know how to do, ignoring other important areas, without weighing alternatives for best results. To avoid this narrow field of action, a systematic approach that encompasses the study and implementation of a system project is required.

An overview of the time, stated in percents, for the various steps in a system project—feasibility study and system implementation—is given in figure 4.1. These percents should not be taken as absolute, but as a general guide for a typical system project of approximately two years. Depending on its complexity, the Gantt chart or PERT/time can be used to control the activities. Note that the steps in figure 4.1 have been grouped and that many subactivities are related to each one.

An examination of the time allocations in figure 4.1 indicates that the feasibility study comprises about one third of the total time for a system change. This may seem excessive on the surface, but an understanding of what a feasibility study is indicates the need for such time. Reducing the time of the feasibility study generally leads to poor results.

The period required for system implementation comprises the remaining two thirds of the total time required to develop a new system. Management is often in a hurry to get the equipment installed and running during the system implementation phase. This is particularly true when the rental on the equipment starts and management see no productive output for the large cash outlays.

I. **Feasibility Study**	
A. Introductory investigation	4%
B. Systems analysis (including an exploratory survey report to top management)	8%
C. Systems design	15%
D. Equipment selection	10%
II. **System Implementation:**	
A. Preparatory work of new system:	
1. Training personnel (programmers and all others)	8%
2. Flowcharting and using decision tables	10%
3. Programming and desk checking	10%
4. Program compiling and testing	20%
B. Operation of new system:	
1. Parallel operations for checking new system	5%
2. Final conversion to new system	10%
	100%

figure 4.1 Major steps of a system project, stated as a percent of total time.

The initial step of system implementation should be the training of personnel—in particular, the programmers. Programmers should not be sent to the manufacturer's programming school until the equipment is selected, the final step of the feasibility study. Likewise, the equipment cannot be chosen until the organization has decided upon the systems design, which forms the basis for the equipment bids. The logical sequence is obvious. Failure to follow this sequence is likely to lead to incomplete, inefficient, and ill-conceived programs that are a disappointment to all involved. Benefits, both tangible and intangible, will never be realized under these unfortunate circumstances.

Although not noted in figure 4.1, system improvements are continuous. Review of the installed system checks its relevance to the current business environment. Thus, as time passes, a periodic review of the present system is necessary. If new system changes can be justified, the steps set forth in figure 4.1 should be followed.

INTRODUCTORY INVESTIGATION

The introductory investigation, the first step in figure 4.1, is the starting point on any system project. This is particularly true for a company which requires interface between management, especially at the higher levels, and computer technicians. As noted in the previous chapter, top management and the ABC Company's systems analysts must form a team to reap the full benefits of a new system: the computer technicians would not understand the place of the system within the overall organization, and, of course, management alone would not have the necessary expertise in computer systems. If both management and computer personnel work together effectively, the full potential of the new system can be realized.

The essential parts of an introductory investigation are:
    organization of a system project
    definition of the scope of the system project
    determination of the schedule of the system project

These basic items, the subject of the remainder of the chapter, constitute a logical framework for beginning a system project and serve as a starting point for preparing an exploratory survey report to top management.

A thorough introductory investigation, vital to the future well-being of an organization, must be done accurately and conscientiously. A fast, half-hearted attempt may lead the organization to install a new system incapable of meeting the needs of the organization and/or with higher data processing costs than expected. A careful and systematic approach leading to a definitive exploratory survey report to top management is a must for installing a successful system; there are no legitimate shortcuts.

ORGANIZATION OF SYSTEM PROJECT

The establishment of two groups—the executive steering committee and the management information system (MIS) task force—is a sound basis for organizing a system project. The executive steering committee is composed of executives whose departments and functions are affected by the study. Its basic functions are to give direction to the MIS task force and to report on the progress of the system project to all concerned personnel, including the board of directors. The task force is made up of personnel involved in the problem under study and reports to the executive steering committee.

This form of organization allows the two groups to jointly identify the relationships between processes and departments necessary to a successful problem solution. They also help to smooth the way for computer personnel by enlisting the cooperation of operating personnel, thus boosting the chances of acceptance of the task force's role.

As indicated previously, a very important reason for getting top-level executives involved in a system project is based on past experience. Top management has come to realize the full significance of the statement, "Management must learn to control the computer or the computer will control it." There are numerous case histories which disclose major failings in information systems simply because top management did not become involved. Quite often, managers fear that information systems are too complicated and too technical for them to understand. When not significantly involved, management cannot possibly be aware of the direction the computer technicians are taking in establishing priorities and emphasis on applications. Furthermore, it is almost a foregone conclusion that without direct management influence, the applications will be slanted toward programmers' interpretation of management needs and not the actual needs of management and operating personnel.

Establishment of Executive Steering Committee

The purpose of an executive steering committee is to give direction to the MIS task force. Members of the executive committee may include several of the following: president, executive vice-president, and vice-presidents of marketing, manufacturing, purchasing, finance, engineering, personnel, and research and development. This committee sets basic long range objectives to be met, in part, through the development of future systems. The committee also reviews what the MIS task force is doing. The executive steering committee, due to its high rank and status, can greatly enhance the chances of success by communicating their strong support of the system study.

Having established who will comprise the executive committee, its initial task is to issue a written statement that a MIS task force has been formed to study the feasibility of implementing a new system. It should be realistically stated that adjustments in personnel and jobs may be required to make the change, but that no one will be fired or asked to resign. Some employees directly affected by the system may be retrained and reassigned to new jobs. It is quite likely these employees will find their new responsibilities more rewarding since the system will eliminate the routine and repetitive aspects of their old jobs. This first memo to all employees should indicate that periodic written statements will be issued regarding the progress of the system project. The failure to issue a memorandum of this type will cause undue concern about the possibility of data processing equipment replacing people.

Determining overall time and budget guidelines One of the important duties of the executive steering committee is to develop overall time and budget guidelines. The work required for the implementation of a management information system is based upon the scope of the system project. Obviously, the more comprehensive the system is, the higher the cost will be. Also, the less time permitted to get the system operating, the more personnel and money will be required. The importance of time is often overlooked in planning a complex system.

The starting point is a schedule developed from the initial estimate of funds available and/or a date by which management wants the project completed. There is a strong possibility that the initial time and/or budget guidelines cannot be met; in such cases, the executive steering committee must modify the guidelines. The constraining factor(s) must be revised until there is agreement between the budget and a feasible time schedule. Thus, time and budget guidelines in a system project are interrelated and must be resolved by the executive committee working with the MIS task force. As mentioned previously, time and budget guidelines are developed and reviewed from an overall viewpoint. In later phases of the system project, they become more detailed.

Researching past management information systems reveals two important shortcomings to avoid when organizing MIS programs. Almost universally, organizations indicate that construction of an MIS system will take longer and cost more money than is originally budgeted. Development of an effective management information system takes several years and large expenditures.

Issuing written MIS progress reports One of the best ways for the executive steering committee to insure a successful system project is to issue written memorandums to all employees, as indicated previously. In the first memo, it is wise to include a statement on the scope and objectives of the study. Also, the names of the executive steering committee and MIS task force should be included. Subsequent memos issued by the executive committee should be written in a manner not only detailing the progress of the system project, but also enhancing the stature of the MIS task force. The content of these executive memos should clarify to all concerned that top management wants effective results from the system project.

After the first MIS memorandum, the MIS study manager should meet with departmental managers to outline the work required of them in analyzing, designing, and implementing the new system. Also, the names of the departmental personnel who will be responsible for making information available to the assigned analysts should be noted by the MIS manager. In essence, the MIS manager is the liaison person for the task force.

Selection of MIS Task Force

A task concurrent with the preparation of the first written memorandum from the executive steering committee is the selection of the MIS task force, sometimes called the MIS committee or study group. The number of participants will depend on several factors, most importantly the organization's size, the number of divisions and departments, the degree of centralization or decentralization, the number of business functions, budget constraints, and time considerations.

Since a management information system may impact the entire organization, a member from each of the organization's functional areas should be selected. If at all possible, each person selected should have several years of experience in his or her respective area, be objective, and be capable of creative thinking. Also, the individual should be familiar with the major problems of his or her functional area. The person who has this background and holds a responsible position generally has not had the opportunity to keep up with the latest developments in data processing technology, so it is necessary to have people with computer and programming experience within the MIS task force. This may mean going outside the company for qualified systems personnel and/or consultants. Whether the individuals selected are from inside or outside, the task force will function best when personnel with the required knowledge are present.

It is recommended that one person within the organization be named manager of the system project. This individual will direct the project and maintain its momentum. Other members may work full time or part time, depending upon the needs of the project and their current work assignments.

Organization of MIS task force The leader of the MIS task force should hold a high rank, comparable to that of a vice-president. A survey of many organizations indicates a positive correlation between the success of the study and the rank of the system project manager. A rank high on the executive ladder, where the individual has immediate access to top management, gives the system project a much better chance of success.

Depending upon the size of the undertaking, the MIS task force can range from five people to a dozen or more from inside or outside the organization. It is advisable to keep the number as small as possible to avoid the problems associated with large groups. Although all members will not be working on the study continuously, they will be engaged as needed. For the most part, the task force should be split into two or three smaller groups, sometimes called study or task groups. This will allow each group to investigate different aspects of the project simultaneously, thus saving much time.

Authority delegated to MIS task force Authority must be clearly delegated to the MIS task force so that all those with whom the group will work fully comprehend the need for their help and complete cooperation. Without the authority delegated officially to the group, many managers might not participate in a constructive way because they question the validity of newer data processing equipment, they do not understand the new system change, or they harbor some bias against the system study. The authority delegated from top management may have to be used at times to overcome the resistance which happens anytime there is a change in the system. The study group's tact and diplomacy can do much to avoid the resentment detrimental to the entire system project. Also, having the authority and the corresponding responsibility for a system project, MIS personnel will be held accountable for their actions.

Training of MIS task force   The management information system task force is a heterogeneous group in that some are knowledgeable about the operations of an organization's functional areas while others are data processing experts. Individually, each lacks the necessary knowledge to carry a system project to a successful conclusion; for this reason, both groups must receive additional training. Personnel from the functional areas should attend intensive data processing courses given by computer manufacturers, consulting firms, or software firms. The growth of computer courses as part of a university degree will tend to alleviate this general background problem in the future. Either approach should emphasize the background necessary to carry a project to a successful completion. At the same time, computer personnel and outside consultants will be reviewing and understanding the present operations. Generally, this is the minimum training the study group should have before it can get started on detailed investigation of the present system.

It is advisable to go beyond the minimum training requirements set forth above. In order for the task force to contribute more effectively to the needs and desires of management, it must have a broad knowledge of the organization's objectives. Plans must be such that they can be interpreted in terms of information system needs for the long term as well as for the short term.

DEFINITION OF THE SCOPE OF SYSTEM PROJECT

The question of what areas, that is, subsystems, will be included in the system project is up to the executive steering committee. This group might ask the MIS task force to direct its activities initially to the production function, the marketing function, or the finance function. When the scope of the study is broader and includes several or even all functional areas, it is better to implement it in manageable pieces, considering both the vertical and horizontal aspects of such an undertaking. From a horizontal approach, a marketing information subsystem may be implemented first, followed by a production control subsystem, and so forth. The vertical structure approach considers satisfying the informational needs of managers at different levels.

Definition of the study's scope can be approached from another viewpoint. Management might want the task force to investigate those opportunity and cost areas critical to the organization. These are ones where successful action is essential for business survival, where the greatest opportunities are for profits, or where the greatest costs are incurred. Thus, those areas that represent the best investment of time and money in the new system should be selected for implementation.

The scope of the system project, then, must be clearly defined from the onset. Otherwise, analysis may extend to areas that are not of prime interest in developing the new management information system. Limiting the field of analysis at this beginning stage reduces the time required to analyze existing procedures and indicates those areas requiring extended analysis. In addition, it will assist in revealing functional areas where the new system is either infeasible or too costly. In essence, the executive steering committee should look upon the scope of the study as a flexible set of instructions, subject to revision.

Redirection of the depth of analysis and areas of study can occur when the information being obtained by the MIS task force reveals previously undisclosed problems. Many times, the MIS group will ask to enlarge or contract the area under investigation; this can either take place when the problem is initially defined or when it is redefined at a later stage in the system project. Thus the executive steering committee sets forth the overall scope at this time and directs more attention to details of the problem definition at a later time.

Selection of Desired Objectives

The formulation of objectives is a joint effort of the executive steering committee and the MIS task force. Clearly stated objectives not only force top management to think seriously about the organization's future, but also bring to light problems that might otherwise have been overlooked. For both groups, they provide a framework in which to operate; the constraints and limitations within which the project must function are clearly stated. Experience has indicated that a system project has gone much more smoothly when a formal statement of objectives has been clearly delineated.

The objectives desired by management can take many directions, as depicted in figure 4.2. If they center around cost savings, consideration should be given to tangible and intangible benefits in order that the evaluation is complete and realistic. Other objectives can emphasize faster and more timely information for management decisions. In reality, this approach is aimed at cost reduction as well as faster service for customers. Ideally, a new management information system is one which is able to meet as many objectives as possible and, at the same time, reduce organization costs.

Better information to meet the organization's long-range planning needs.
Increased efficiency in organizational operations.
Reduction in data processing costs.
More timely information for management decisions.
Improved information to meet the planning and controlling needs for daily operations.
Improved customer service and relations.
Increased flow of data for meaningful information.
Uniform and accurate handling of data with a minimum of human intervention.
Elimination of conflicting and overlapping services within the organization.
Improved internal control (internal accounting control and internal check).
Improved operations through greater utilization of mathematical models.
Improved employee and public relations.
Increased managerial development and efficiency.
Reduction in collection of duplicate data.
Efficient utilization of personnel and equipment.
Increased overall net income from operations.
Reduction in data recording and manipulation errors.

---

figure 4.2 Selection of the desired objectives of an MIS project.

In addition to reduction in costs, more timely decisions, and improved service to customers, the following are also representative of objectives desired by both the MIS task force and the executive steering committee. The utilization of faster data processing equipment provides instantaneous information on the status of customer orders and allows the organization to schedule production more effectively, notify the customer of any shipping changes or delays, and expedite billing procedures. The uniform and accurate handling of data is available without the problems of human intervention. All these benefits improve customer relations and enhance the organization's competitive position.

Top management might consider other objectives, such as the elimination of conflicting and overlapping services. A very important objective is the employment of balances (internal accounting control) and checks (internal check) as the data are processed, eliminating the need for manual checking.

Another desirable objective might be improved operations through greater use of mathematical models, which can reduce the amount of inventory carried as well as the production costs. Mathematical models have been employed successfully to increase managerial efficiency, to effect better utilization of plant personnel and equipment, and to improve employee and public relations. Major exceptions to normal operations can be extracted and reported to a responsible person for immediate action.

While the foregoing objectives revolve around using the computer as a managerial tool, the executive steering committee must demand that output be oriented toward the user. Never should the strict technical requirements of the hardware be favored if service to company personnel is impaired. Research has shown that, during a system change, problems involving people are much more extensive and more deeply rooted than those involving technical systems. New methods and procedures can fail miserably if it becomes apparent that the computer is being favored to the detriment of user needs. Within the overall context, objectives that consider the **human element** must be incorporated into the user–machine operating mode of the computer system.

The above listing of objectives is by no means complete, but is a representative sampling of objectives that may be desired. In the final analysis, selecting objectives is the prime responsibility of top-level executives. Generally, objectives are unlikely to be sufficiently ambitious and demanding if not set forth by the executive steering committee. Thus, project objectives provide a frame of reference which can always be changed, if they are found to be unrealistic, at a later phase of the system project.

Definition of the Problem

Once the objectives have been agreed upon by both groups, the MIS task force defines the problem more precisely. As noted previously, the scope of the system project has been stated in general terms by the executive steering committee. It is the job of the task force to specify in greater detail the functional areas (subsystems) that will be explored. The group must make sure that their scope is compatible with the objectives and, if the desired objectives and the scope of the system project are in conflict, a conference between the two groups should resolve the problem.

Having defined the problem as accurately as possible within the scope of the project and the desired objectives, the study team should have little doubt about the areas to be covered by the investigation. It is advisable to state these findings in writing. A written memorandum by the MIS task force ensures the accomplishment of what was originally intended and reduces the chance of going off on a tangent. A carefully laid out plan, backed by the executive steering committee, indicates where the project will cut across organizational lines. Likewise, it indicates where authority is needed for changes in systems, methods, procedures, forms, reports, or organization for the functional areas to be explored.

DETERMINATION OF THE SCHEDULE OF SYSTEM PROJECT

The final part of the initial investigation is the preparation of a time schedule for the entire system project. Generally, a management information system study takes place over several years. Experience has shown a tendency to underestimate the time necessary for both the feasibility study and system implementation. The time factor is a function of the objectives desired and problem definition. If time becomes the important consideration, an optimum decision for an organization's DP needs will probably not be made.

When developing a time schedule, the MIS task force must determine the amount of work involved in each step of the system change and the personnel and skills resources needed. Consideration must also be given to the areas of training, programming, program testing, equipment delivery, physical requirements and installation of the equipment, file development, delivery of new forms and supplies, and conversion activities.

A member of the study group responsible for scheduling should not only prepare a realistic time table, but also be in a position at all times to report whether the study is ahead, behind, or on schedule. The manager of the MIS task force should issue reports periodically to the executive steering committee on the status of the project. Included in the reports should be problem areas, delays, and other information critical to the study. Utilization of the "exception principle" in progress reports is needed to control the project.*

---

*The "exception principle" refers to highlighting for investigation those items which have exceeded established limits by a specified amount. If necessary, action is undertaken to correct ongoing deficiencies.

SUMMARY: Introductory Investigation

Organization of System Project:
- **Establishment of executive steering committee** Its purpose is to give direction to the MIS task force. It consists of top management—president, executive vice-president, and vice-presidents from the functional areas.
- **Selection of MIS task force** Its purpose is to undertake the system project,

namely, the feasibility study and system implementation. It consists of systems analysts and personnel from the functional areas.

Definition of the Scope of System Project:
- **Selection of desired objectives** To provide a framework for both the executive steering committee and the MIS task force. The objectives establish the constraints under which the project must function.
- **Definition of the problem** To specify what functional areas (subsystems) will be explored in detail, based upon the specific objectives set forth.

Determination of the Schedule of System Project:
- When developing a realistic schedule, the MIS task force must determine the amount of work involved in each step of the system project and what resources in terms of personnel and skills will be needed. Likewise, utilization of the "exception principle" in progress reports is needed to control the project.

CHAPTER SUMMARY:

The introductory investigation is the first and most important part of any system project. Its importance cannot be over-emphasized since a thorough analysis of all items relevant to beginning a project is undertaken. It permits the MIS task force to comprehend what is expected of them by the executive steering committee. A quick and unsophisticated undertaking for this initial step will generally result in a mediocre management information system. Hence, there is a great need for a thorough introductory investigation before proceeding to the next step—a detailed investigation of the present system, the subject of the next chapter.

Questions

1. Of the many factors affecting a system project, which one can make or break the project? Why?
2. a. Why is it that the growth of a data processing system will lag behind the corresponding growth of an organization?
   b. What effect does this have on a system project?
3. a. What are the major steps involved in a system project?
   b. Can the feasibility study phase be eliminated? Why or why not?
4. Why is it necessary to establish an executive steering committee before initiating a system project?
5. Explain the relationships among the following:
   a. the scope of the system project
   b. the selection of desired objectives
   c. the definition of the problem
6. What is the importance of obtaining the backing of top management in a system project?

7. If you were assembling an ideal MIS task force for a system project in a typical manufacturing firm, who would be the members?
8. Of what importance to top management is the development of a realistic schedule for a system project?

Self-Study Exercise

True-False:

1. _____ The most important factor affecting a system project is the decision of whether or not to go to distributed processing.
2. _____ Monetary outlays of a system project are generally overshadowed by personnel considerations.
3. _____ System implementation takes considerably more time than the feasibility study.
4. _____ Within a system project, final conversion to the new system is the most time-consuming of all steps.
5. _____ The introductory investigation provides a starting point for a system project.
6. _____ The MIS task force oversees the efforts of the executive steering committee.
7. _____ Overall time and budget guidelines are the main concern of the MIS task force.
8. _____ It is advisable to keep the MIS task force small in order to avoid the problems associated with large groups.
9. _____ Top management involvement in a system project should be minimal.
10. _____ Defining the areas to be investigated is possible only after the desired objectives have been set forth clearly.

Fill-In:

1. An advanced computer system will generally have a dramatic impact on an organization's business functions or _____.
2. Although an organization can change and control the physical requirements of a new system, the same cannot be said for the _____ _____.
3. Within a _____ _____, systems design consumes more time than systems analysis or equipment selection.
4. Developing and testing _____ _____ are the most time consuming tasks during system implementation.
5. A logical beginning point for getting started on a system project is called the _____ _____.
6. The _____ _____ _____ gives direction and guidance to the MIS task force.
7. The MIS task force should include system analysts as well as participants from the organization's _____ areas.
8. The backing of _____ _____ is necessary for a system project since it cuts across the entire organization.
9. The _____ of the system project must be set forth initially to limit the possible areas for a system change.
10. Selection of desired _____ provides the framework within which the system project will operate.

chapter 5

systems analyis of present system

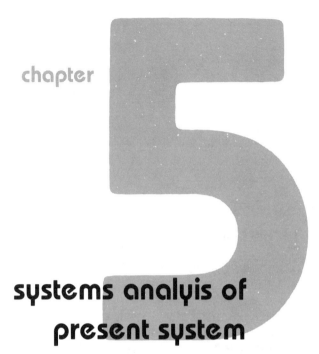

OBJECTIVES:

- To explore the more widely used methods of systems analysis for understanding the present system.
- To set forth the areas for which a detailed investigation is necessary to understand the present system.
- To examine the important aspects of each feasible system alternative, namely, tangible benefits, costs, and intangible benefits.
- To enumerate a suggested listing to be included in the exploratory survey report to top management.

IN THIS CHAPTER:

OBJECTIVES OF SYSTEMS ANALYSIS
METHODS OF SYSTEMS ANALYSIS
   Utilization of Interviews
   Review of Organization Systems and Procedures Manuals
   Obtaining Copies of Operating Documents and Reports
SYSTEMS ANALYSIS—DETAILED INVESTIGATION OF PRESENT SYSTEM
   Content of Detailed Investigation
   Historical Aspects
   Analysis of Inputs
   Review of Data Files Maintained
   Review of Methods and Procedures
   Analysis of Outputs
   Review of Internal Control
   Other Analyses and Considerations
   Documenting the Present System
   Analysis of Present Work Volume
   Analysis of Current Personnel Requirements
   Analysis of Present Benefits and Costs
SYSTEMS ANALYSIS—CONCLUDING INVESTIGATION
   Feasible MIS Alternatives
   Benefits and Costs for Each Feasible MIS Alternative
   Intangible Benefits for Each Feasible MIS Alternative
   Comparison of Present System to Each Feasible MIS Alternative
SYSTEMS ANALYSIS—EXPLORATORY SURVEY REPORT TO TOP MANAGEMENT
   Feasibility of New MIS Alternatives
   Recommending a New System from the MIS Alternatives
   Infeasibility of New MIS Alternatives
CHAPTER SUMMARY
QUESTIONS
SELF-STUDY EXERCISE

In the prior chapter, the mechanism for getting started on a system project was discussed. Specifically, this centered on the introductory investigation; its major parts are organizing, defining the scope, and scheduling the system project. Within this system structure, the next step of the feasibility study—namely, systems analysis—can be started and represents approximately eight percent of the total project time. The real work of the MIS task force, then, can begin now that an operational framework for a system project has been developed.

This chapter delineates the objectives and methods of systems analysis and the areas for which a **detailed investigation** is needed to understand the present system. The end result of this intensive analysis, commonly called the **concluding investigation,** is an examination of the benefits and costs of the present system in order to determine if one or more new system alternatives are more beneficial than the present one. A detailed examination of those important items is contained in the exploratory survey report to top management, the final subject matter of systems analysis. Thus, the feasibility or infeasibility of applying newer data processing equipment and procedures is established for an organization.

OBJECTIVES OF SYSTEMS ANALYSIS

During the systems analysis phase, the study team will gather information about the equipment, personnel, demands, operating conditions, and limitations of the present system. However, the primary objectives of systems analysis are

- to define the objectives that the new system must meet
- to identify any constraints on the development and operation of the system
- to specify the conditions under which the system will operate

Systems analysis serves as a basis for designing and installing a system, providing it is economically feasible to do so. **Systems design,** on the other hand, is concerned with system development. The new system design must be based on the facts obtained in the systems analysis phase and must lie within the framework of the system project. Further, the equipment selected must be within the constraints of the new system designed. The system implementation stage builds upon the systems analysis, systems design, and equipment selection phases by devising programs, methods, and procedures; recruiting and training qualified personnel; installing the equipment; and putting the new system into operation. Even though systems analysis, systems design, equipment selection, and system implementation are discussed separately, they are intimately related in that each has some effect on the other. To isolate one important phase from another will generally result in a mediocre system.

METHODS OF SYSTEMS ANALYSIS

Members of the MIS task force use several tools to obtain the necessary data during this study phase: interviews; company systems and procedures manuals; and company forms, documents, and reports.

Utilization of Interviews

The best way to understand what is transpiring is by talking directly with the individuals who are responsible for getting the job done. The analyst should study the activities of an organizational unit as set forth by the unit manager. Attention should also be focused on subordinates' work to confirm what has been said by their immediate supervisor.

For a successful interview, the analyst should be formally introduced to all supervisory personnel. From the outset, this approach establishes the level for the entire study as an official undertaking of the organization, backed by the executive steering committee as well as the board of directors. The interview should be friendly and informal. Notes about the interview should be taken as it progresses. The analyst should ask the supervisor for ideas on how the current system can be improved. In fact, systems personnel should actively solicit suggestions and try to incorporate many of these ideas in the new system. This is extremely important because the supervisors are in the best position to evaluate the current system and provide new ideas. Further, the supervisors will get a feeling of participation that will facilitate their acceptance of system changes at a later date. They should be reassured that any changes made will be discussed with them before implementation.

Review of Organization Systems and Procedures Manuals

Another excellent tool for the task force is reviewing the organization's systems and procedures manuals. Reference to inputs, outputs, and files employed in any given procedure will be helpful in documenting current operations. Analysts should be aware, however, that conditions may have changed since the material was written, and that existing procedures may differ from those actually used without management's knowledge. When the procedures manuals are nonexistent or out-of-date, the analyst will have to spend more time during the interviews to obtain all necessary information.

Obtaining Copies of Operating Documents and Reports

Most systems and procedures manuals contain actual copies of forms, documents, and reports used. If this is not the case, samples can be obtained during the interview period. Again, analysts should be aware that documents and forms may have gradually changed, and those in current use may be very different from samples in the manuals. A review of all samples for an organization may indicate a need to combine, eliminate, or design new ones.

SUMMARY: Methods of Systems Analysis

Utilization of Interviews This is the best method to understand what is transpiring in the present system. The activities of supervisory personnel and their subordinates should be studied firsthand to keep misunderstandings to a minimum.

Review of Organization Systems and Procedures Manuals This method gives the analyst an overview of what the present system is all about. Reference can be made to the specific inputs, methods, procedures, files, outputs, and internal control for a specific subsystem being studied. The information gathered can be used for analyzing the benefits and costs of the present system.

Obtaining Copies of Operating Documents and Reports This method centers on the output of the present system. A review of operating documents and reports by a skilled analyst reveals the need to combine, eliminate, or design new ones.

SYSTEMS ANALYSIS— DETAILED INVESTIGATION OF PRESENT SYSTEM

During the entire system project the executive steering committee and the MIS task force are in communication. The steering committee has selected areas likely to derive the greatest benefits from a new system; the task force studies these areas in great detail. If functional areas included in the original scope of the study are found to be poor candidates during the problem definition phase or later, or are found to be incapable of meeting the organization's desired objectives at some time during the study, there must be a meeting of minds to resolve these problems. The importance of communication between the two groups cannot be overemphasized.

Detailed investigation of the present system, a lengthy process, involves collecting, organizing, evaluating facts about the system and the environment in which it operates. There should be enough information assembled so that a qualified person could understand the present system without visiting any of the operating departments. Generally, the study group must devote its full time to this undertaking. Survey of existing methods, procedures, data flow, outputs, files, inputs, and internal control, to name the more important ones, should be intensive in order to fully reveal the present system and its related problems. No area should be excluded unless it has no relationship to the scope of the system project in terms of desired objectives and problem definition. The search of the task group should be comprehensive and far-reaching.

Each functional area reviewed, commonly referred to as a **subsystem project,** requires the assignment of a leader. The overall project should then be subdivided into specific tasks, each with a separate responsibility assignment and time schedule. The first phase of each task is to review the area in depth. As will be discussed later in the chapter, the second phase involves documentation.

In order to perform a thorough analysis of the present system, it is necessary to have accurate information from the operating areas. This means that operations personnel become involved almost from the beginning; if they are not, current system specifications will not be complete.

Content of Detailed Investigation

The essential elements of systems analysis that are preparatory for documentation are detailed in Figure 5.1. They are
1. historical aspects
2. analysis of inputs
3. review of data files maintained
4. review of methods and procedures
5. analysis of outputs
6. review of internal control
7. other analyses and considerations

Reviewing the present work volume, analyzing current personnel requirements, and preparing a schedule of present system costs and benefits are also considered part of systems analysis.

Historical Aspects

A brief history of the organization is a logical starting point for an analysis of the present system. The historical facts should identify the major turning points and milestones that have influenced its growth. A review of several annual reports can provide an excellent historical perspective. Emphasis should be placed on the industry in which the organization is operating, and its markets, distribution channels, competitors, organization structure, future trends, objectives, policies, unions, and government regulations. An historical review of the organization chart will identify the growth of management levels as well as the development of the various functional areas and departments. Not only must historical data be analyzed, but also current and future plans must be examined in order to understand the future thrusts of the organization and their corresponding

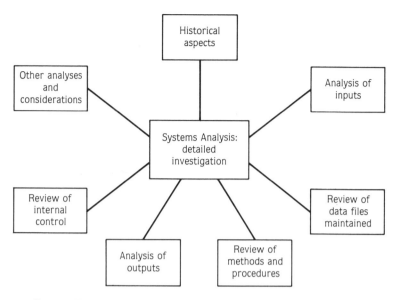

figure 5.1 The essential elements of systems analysis for a detailed investigation of the present system, in preparation for documentation.

implications. The MIS task force should investigate what system changes have occurred in the past. These would include operations that have been successful or unsuccessful with DP equipment and techniques.

Analysis of Inputs

A detailed analysis of present inputs is important since they are basic to the manipulation of data. Source documents are used to capture the originating data for any type of system. The task force should be aware of the various sources where data are initially captured, recognizing that outputs for one area may serve as inputs for other areas. The output of sales forecasting, for example, may be the input for determining the levels of finished goods inventory, which in turn, establishes the level of goods to be manufactured and materials to be purchased.

Data processing personnel must understand the nature of the form, what is contained in it, who prepared it, where the form is initiated, where it is completed, the distribution of the form, and other similar considerations. If these questions are answered adequately by the task force, they will be able to determine how these inputs fit into the framework of the present system. Thus, inputs along with data files are processed by methods, procedures, and equipment in order to produce meaningful output for current and future operations.

Review of Data Files Maintained

The task force should investigate the data files maintained by each department, noting especially their number and size, where they are located, who uses them, and the number of times per given time interval they are referenced. Information on common data files and their size will be an important consideration when designing a new system. This information may be contained in the systems and procedures manuals. Both static and constantly changing files are reviewed on the basis of storage media used.

A review of on-line and off-line files maintained will reveal information about data not contained in an organization's outputs. It is always better to have more than less data on file. This is particularly true when specialized studies are undertaken periodically. The gathering and retaining of all potentially useful data has some merit. However, the related cost of retrieving and processing the data is another important factor to consider. Files of information held for a long time may be difficult to store at a low cost and result in a high processing cost for preparing meaningful reports. The maintenance of large data files for extended periods of time is a function of their ultimate value in terms of future reports, storage costs, and requisite legal requirements.

Review of Methods and Procedures

Methods and procedures transform input data into usable output; this is reason enough for their review. A **method** is defined as a way of doing something; it proscribes how an action is undertaken. On the other hand, a **procedure** is a series of logical steps by which a job is accomplished. A procedure specifies what action is required, who is required to act, and under what condition the action is to be taken. A procedure is larger in scope than a method.

A procedures review is an intensive survey of the methods by which each job is accomplished, the DP equipment utilized, and the actual location of operations. Its basic objective is to eliminate unnecessary tasks or, stated another way, to perceive improvement opportunities in the present system. While an understanding of the present system is necessary for the system project, both tasks can be accomplished at the same time. Many times, the study group will make recommendations for immediate improvements in certain areas since the actual installation of the new system may be years away. This allows immediate cost reduction and increased efficiency within the respective departments.

Analysis of Outputs

The outputs produced by the various departments of an organization become the inputs for other areas. For this reason, the outputs or reports should be scrutinized carefully by the MIS task force in order to determine how well they are meeting the organization's needs. The questions of what information is needed and why, who needs it, and when and where it is needed are ones that must be answered. Additional questions concerning the sequence of the data, how often the form is used, how long it is kept on file, and the like must also be investigated. Often, too many reports are a carry-over from earlier days and have little relevance to current operations. During the interview, the many levels of management will request that certain reports be dropped and/or combined with others. The systems analyst will hear a loud call for **timely** management exception reports that detail only the significant deviations from the standard or budget.

All reports for internal use should be timely, accurate, complete, concise, and useful. Attention to standardization of forms should be considered. The cost of preparation should not exceed their value. Reports for all areas of an organization should be carefully scrutinized. In summary, the various study groups should not accept the current reports as ideal or even usable in their present format for the new system, but keep an open mind during this investigatory phase. When the time comes for designing new reports, they will be objective. In this manner, problems associated with the outputs currently being generated will be eliminated.

Review of Internal Control

A detailed investigation of the present system is not complete until internal control is reviewed. Many systems personnel fail to spend adequate time on internal control of the present system because they have not been trained to appreciate its importance. Locating the control points helps the analyst visualize the essential parts and framework of a system. At these critical points, some type of comparison, edit, or validation is performed. An examination of the present system for effective internal control may indicate weaknesses which should not be duplicated in the new system. The utilization of advanced methods, procedures, and equipment means that greater control over the data is available as opposed to the less effective, visual control many times found within the present system.

Other Analyses and Considerations

Other analyses and considerations of the present system should include the effects of seasonal or other cyclical characteristics, management questions that cannot be answered due to system inadequacies, and current financial resources and their future trends. Some of this information can be obtained from the financial statements. Data on land, buildings, equipment, and tools, representing an organization's production facilities, should be included in the analysis by the system group, as well as plans for new facilities and equipment. Physical inventories should be ascertained regarding total investment, number of items, turnover, and like considerations. Of special interest to the study team is the data processing equipment currently in use, including communication facilities. Equipment utilization charts and tables should be reviewed to measure its efficiency and effectiveness.

Documenting the Present System

As inputs, methods, procedures, data files, reports, internal control, and other important items are reviewed and analyzed, the process must be documented. Documentation can take several forms. A most common method is using system flowcharts, which should be drawn while the data are fresh in the analyst's mind: from the first overall flowchart, additional ones can be drawn so that each major operation is broken down into its subprocedures. The detailed procedures can then be related to each other and to the entire system by the information flow among them.

The flowcharting of present operations not only organizes the facts for the MIS task force, but also helps disclose gaps and duplication in the data gathered. It allows a thorough comprehension of the numerous details and related problems in the present operation. In essence, the knowledge gathered to date is brought together in a meaningful relationship for members of the various study teams. It should be noted that flowcharting alone, like any other documentation technique, need not be undertaken separately, but generally is combined with the preceding parts of systems analysis.

SUMMARY: Systems Analysis—Detailed Investigation of Present System

Historical Aspects   Centers on the major turning points and milestones in the growth of an organization.

Analysis of Inputs   Focuses on originating data in terms of the nature of input forms, what is contained in them, who prepares them, where they are initiated, where they are completed, their distribution, and the like.

Review of Data Files Maintained   Relates to the data files maintained by each department, their number and size, where they are located, who uses them, and the number of times per given interval they are referenced.

Review of Methods and Procedures   Refers to an intensive survey of methods and procedures by which data processing tasks are accomplished, the equipment used, and the actual location of data processing operations.

Analysis of Outputs   Refers to outputs and reports produced by the present system, the information contained, who needs them, when and where they are needed, how often they are used, how long they are kept, and the like.

Review of Internal Control   Focuses on locating the control points which assist in visualizing the essential parts and overall framework of a system.

Other Analyses and Considerations   Relate to an organization's financial resources, plant facilities, machinery and equipment, inventories, and the like. Also, they include reviewing present DP equipment and communication facilities and their utilization.

Documenting the Present System   Is the final step for bringing together the detailed investigation of the present system. System flowcharts are extremely helpful in breaking down each major data processing operation into its component parts.

Analysis of Present Work Volume

Many organizations require their departmental managers to keep an accurate count of the volume of inputs processed, files maintained, and outputs prepared within their respective departments. Others go a step further and compute an average cost for these items. Time can be saved if data on past and present work volumes are available, otherwise, it must be compiled. It is important for the study group to determine reliable figures for average and peak days as well as the work load at month's end. Accurate figures are necessary for the past five years to ascertain growth or reduction in volumes. These data will be used to determine the present cost of operations and the projected savings and costs for the new system.

Careful analysis of work volumes can be beneficial to the entire study. Where work volumes are small and the processing procedures involved, the feasibility of applying newer DP equipment is unlikely. On the other hand, when large volumes of work require routine and straightforward processing, the likelihood of employing advanced DP capabilities is very good. In addition, work volume analysis is helpful to the task force in determining whether a particular work station is a control point, a storage area, or a terminal point. These observations can be translated into needs for visual display, storage, or processing capabilities when the system is designed.

An examination of existing on-line and off-line files and their size may well serve as a guide for their eventual reorganization. Current files may contain a certain amount of common data. Each time the common data are changed, certain files must be changed. An example of this file problem is found in inventory files using a batch processing mode: separate magnetic tape inventory files that vary in content are maintained for the marketing, manufacturing, and accounting departments. Generally, this redundancy should be overcome with the new system, which has one inventory file available for on-line processing by all organization personnel.

Analysis of Current Personnel Requirements

Current personnel requirements should be analyzed by type, skill, and related cost. The efficiency of personnel working with or without equipment should be measured and analyzed. Most of these data can be obtained from personnel records, departmental statements, and departmental interviews. A summary of an organization's personnel resources can be utilized to appraise the costs and benefits of the present system. Information on the availability of data processing specialists in the local labor market may be needed if the skills of current personnel resources are limited. Employee turnover, fringe benefits, and management attitudes toward personnel and labor unions should be collected and considered for any new DP system. Every effort should be made to understand those elements which can make or break a system project.

Analysis of Present Benefits and Costs

One of the major reasons for reviewing the present operation is to determine its benefits and costs. Present system benefits include the level of service to customers, value of reports, return on investment and profit, ability of the present system to grow with the organization, and inventory turnover. Many of these benefits can be measured precisely, whereas others are, by nature, intangible and require subjective evaluation. For the final evaluation of the present system and the proposed alternatives, tangible benefits and costs are compared first. If the selected alternative system meets the established return on investment, there is no problem. However, if the proposed alternative falls below the acceptable level for capital investments, intangible factors are critical to the investment decision.

Costs should be analyzed by department since this is the most common basis for reporting and provides an excellent means of comparing costs to a new system. Typical components of departmental costs are found in figure 5.2.

Salaries and wages
Payroll taxes
Fringe benefits (life insurance, hospital care, pensions)
Equipment rental and/or depreciation of equipment
Repairs and maintenance of equipment
Facilities rental
Training costs

File maintenance costs
Personal property taxes on equipment
Insurance on equipment
Forms and supplies
Utilities
Outside processing costs by service bureaus and computer utilities
Other departmental costs

---

figure 5.2 Feasibility study—sample listing of an organization's departmental costs.

Usually, costs for the present system are projected for a five-year period, starting with implementation of the new system. The rationale is this: if a computer is selected, it will not be processing on a daily basis for about a year from the day of equipment selection (the final step of the feasibility study). Also, the equipment must be capable of handling an organization's work load from at least three years up to about five years. Thus, a five-year cost projection that starts with the completion of the feasibility study is a realistic approach for the present and new systems. Attempting to go beyond five years is undesirable in view of the rapid changes in data processing technology. Even though some studies have included a longer time frame, the results are much too unreliable.

SYSTEMS ANALYSIS—CONCLUDING INVESTIGATION

The final phase before preparing the exploratory survey report to the executive steering committee is the concluding investigation by the MIS task force. Since each functional area of the present system germane to the study has been carefully analyzed, a feasible set of processing alternatives from an overview standpoint must be developed, although not to the same depth as was necessary to understand the present system.

Feasible MIS Alternatives

Proposed system specifications must be clearly defined before feasible MIS alternatives can be developed. These specifications, which pertain to each functional area of the system report, are determined from the desired objectives set forth at the onset of the study. Likewise, consideration is given to the strengths and the shortcomings of the

existing system. Required system specifications, which must be clearly defined and in conformity with the project's objectives, are as follows:
1. outputs produced with great emphasis on timely managerial reports that utilize the "exception principle"
2. data base maintained with great accent on on-line processing capabilities via CRT terminals
3. input data prepared directly from original source documents for processing by the computer system
4. methods and procedures that show the relationship of inputs and outputs to the data base
5. work volumes and timing considerations for present and future periods, including peak periods.

The starting point for compiling the specifications above is output. After outputs have been determined, it is possible to infer what inputs, data base, methods, and procedures must be employed. The output-to-input process is recommended since the outputs are related directly to the objectives of the organization, the most important consideration of a project. The future work load of the system must be defined for the inputs, data base, and outputs in terms of average and peak loads, cycles, and trends.

Flexible system requirements The requirements of the new system may appear on the surface to be fixed. A closer examination often reveals that the task force should think of these specifications as flexible. For example, the objectives set forth in the study state that certain data must be updated once a day. Perhaps, the best solution is to incorporate the data in a data base which is updated as actual transactions occur; this approach is within the constraints as initially set forth and introduces a new way of maintaining files. The important point is that alternative methods are available in DP areas, even though they appear inflexible. With this approach in mind, it is possible to design a number of different systems with varying features, costs, and benefits. In many cases, more data processing systems will be investigated and analyzed when flexible system requirements are considered.

Consultant's role in feasible MIS alternatives A clear understanding of the new system requirements is the starting point for developing feasible MIS alternatives. This phase is by far the most important and difficult undertaking of the project to date. The experience of the outside consultant is of great value to the study group, for this individual's knowledge of many installations can immeasurably reduce the effort required to identify the most promising solutions. Too often, a study group goes off on a tangent about a specific system approach which should have been initially discarded as infeasible. The outside consultant can exercise the necessary influence to make certain that the study group does not get bogged down in time-consuming trivia and can point out the shortcomings of a certain approach. The consultant can act with the head of the study group to resolve conflicts resulting from members who like their own approaches over the others presented. The consultant's objectivity can enhance an organization's chances of selecting an optimum system when judging the merits and weaknesses of the alternatives. Thus, the key to developing promising MIS alternatives and selecting the optimum one is to employ the talents and experience of the MIS task force to their fullest.

Benefits and Costs for Each Feasible MIS Alternative

After developing feasible MIS alternatives, the estimated savings and incremental costs for each alternative are determined. Major areas for **estimated savings,** sometimes referred to as **cost displacement,** are enumerated in figure 5.3. **Incremental costs** are segregated into two categories, one-time costs and additional operating costs. These are listed in figure 5.4. The difference between the estimated savings, on one hand, and estimated one-time costs and additional operating costs, on the other, represents the estimated net savings (or losses) to an organization before federal income taxes.

Accurate figures for a five-year period are of great importance, indicating the need for accounting department assistance. Many times, the best way to increase the accuracy of the figures compiled by the study group is to have the outside consultant assist in reviewing the data. This individual's knowledge of current DP equipment and ready access to equipment rental and purchase costs will save time in this phase of the study. Likewise, exposure to similar cost studies will add creditability to the final figures in the exploratory survey report to top management.

Reduction in the number of personnel: lower salaries and wages.
Lower payroll taxes and fringe benefits with fewer people.
Sale or elimination of some equipment: depreciation and/or rental are no longer applicable.
Reduction in repairs, maintenance, insurance, and personal property taxes.
Lower space rental and utilities.
Elimination or reduction in outside processing costs.

---

figure 5.3 Estimated savings.

Estimated One-Time Costs:
Feasibility study (includes systems analysis, systems design, and equipment selection).
Train programming and operating personnel.
Document all feasibility study applications.
Program these applications.
Program assembly and testing of programs for new system.
Data base conversion.
Site preparation (includes construction costs, remodeling, air conditioning, and power requirements).
Additional computer time (in excess of hours alloted free of charge).
Parallel operations (the old and the new system operate concurrently).
Conversion activities (from existing system to new system).
Other equipment and supplies (includes form-handling equipment, files, magnetic disks, magnetic tapes).
Estimated Additional Operating Costs:
DP (computer and related) equipment: monthly rental and/or depreciation.
Maintenance of equipment (if not included above).
Program maintenance (programmers).
Wages and salaries of data processing personnel (direct supervision, equipment operation, and other data processing jobs), payroll taxes, and fringe benefits.
Forms and supplies (for new data processing equipment).
Miscellaneous additional costs (such as power costs and insurance, repairs, maintenance, and personal property taxes on equipment purchased).

---

figure 5.4 Estimated one-time costs and additional operating costs.

Five-Year Study Period When computing the estimated savings and incremental costs, it is not sufficient to base these estimates on the present DP workload. Rather, the MIS task force should review the operating work volume compiled during the detailed investigation. The trend of growth or cutback in an organization's workload should be analyzed and projected for the next five years. These data can then be utilized to project savings and costs, similar to the analysis in table 5.1 (p. 114). For alternative 3, consideration has been given to higher future costs. Realistic increases in salaries, wages, and employee benefits must be included in these estimates.

The starting point of this projection is the day of system implementaion. The net savings (the difference between the two sums after taking into account federal income taxes) should be discounted back to the present time. The purpose of the discounted cash flow is to bring the time value of money into the presentation. This is shown in table 5.2 for system alternative 3. Notice that the net savings after federal income taxes of $175,396 over the five-year period (the anticipated life of the system), when discounted, shows a negative present value for this alternative of $27,229. On the basis of a discounted twenty percent return on investment for this alternative, it should not be undertaken because the company's cut-off point for capital investments is twenty percent. Because the revised discounted rate of return is approximately sixteen percent (based upon present value factors), consideration should be given to additional benefits.

In addition to the approximate sixteen percent return on investment, important tangible benefits may be available to justify the MIS project. These include:
1. reduced investment in the amount of inventory
2. less spoilage and obsolescence of inventory
3. lower purchasing costs through automatic reordering
4. lower insurance costs and taxes on inventory
5. smaller warehouse area
6. lower transportation costs
7. less interest charges on money needed to finance inventory

More effective inventory control can have a pronounced effect on inventory—a large balance sheet item. In a similar manner, a more accurate projection of an organization's cash position will reduce its needs for short-term financing. Other large asset items should be evaluated in order to determine if tangible savings are available through more effective control. All these values discounted to the present should be added to table 5.2. If the present value of net savings for a system alternative is still negative after adding these tangible benefits, the intangible benefits must be explored.

Intangible Benefits for Each Feasible MIS Alternative

A number of intangible benefits, that is, qualitative factors, will be uncovered by studying the potential contributions of each system alternative. Generally, the use of interactive equipment indicates that many qualitative benefits will accrue to a typical organization. A list of these may be found in figure 5.5 (p. 116). Even though qualitative factors are nonquantifiable initially, their ultimate impact is in quantitative terms, reflected in an organization's financial statements.

table 5.1 Net savings (or losses) before federal income taxes for a five-year period (on a rental basis). (Robert J. Thierauf, *Systems Analysis and Design of Real-Time Management Information Systems*, © 1975. Reprinted by permission of Prentice-Hall, Inc., Englewood Cliffs, New Jersey.)

The American Company: SYSTEM ALTERNATIVE #3
Estimated net savings before federal income taxes (on a rental basis)—five-year period

	Years from start of system implementation					Five year total
	1	2	3	4	5	
Estimated savings:						
Reduction in personnel	$120,200	$400,500	$440,300	$490,500	$540,500	$1,992,000
Sale of equipment	120,000					120,000
Rental savings	25,000	51,000	54,500	58,000	61,800	250,300
Elimination of rental equipment	2,050	4,380	4,690	5,000	5,300	21,420
Other savings	3,000	3,060	3,210	3,370	3,540	16,180
TOTAL estimated savings	$270,250	$458,940	$502,700	$556,870	$611,140	$2,399,900
Estimated one-time costs:						
Feasibility study	$95,000					$95,000
Training	50,000					50,000
Systems and programming	255,500					255,500
Data base conversion	272,500					272,500
Other conversion costs	75,500					75,500
Site preparation	55,400					55,400
Other one-time costs	22,300					22,300
TOTAL estimated one-time costs	$826,200					$826,200
Estimated additional operating costs:						
DP equipment rental	$110,000	$120,800	$127,400	$134,100	$141,000	$633,300
Additional personnel	34,000	60,700	62,300	63,400	64,600	285,000
Program maintenance	20,000	30,700	32,200	33,800	36,000	152,700
Forms and supplies	10,000	21,500	23,000	24,500	26,000	105,000
Other additional operating costs	4,400	12,400	12,800	13,200	17,600	60,400
TOTAL additional operating costs	$178,400	$246,100	$257,700	$269,000	$285,200	$1,236,400
NET SAVINGS (losses) before federal income taxes	($734,350)	$212,840	$245,000	$287,870	$325,940	$337,300

table 5.2 Discounted cash flow on a 20-percent return after federal income taxes (on a rental basis). (Robert J. Thierauf, *Systems Analysis and Design of Real-Time Management Information Systems,* © 1975. Reprinted by permission of Prentice-Hall, Inc., Englewood Cliffs, New Jersey.)

The American Company: SYSTEM ALTERNATIVE #3
Discounted cash flow—20 percent return after federal income taxes (on a rental basis)—five-year period

Year	Net savings (losses) before federal income taxes table 5.1	Federal income tax at 48 percent rate	Net savings (losses) after federal income taxes	Present value of $1.00	At 20 percent Present value of net savings (losses)
1	($734,350)	($352,488)	($381,862)	.833	($318,091)
2	212,840	102,163	110,677	.694	76,810
3	245,000	117,600	127,400	.579	73,765
4	287,870	138,178	149,692	.482	72,152
5	325,940	156,451	169,489	.402	68,135
TOTALS	$337,300	$161,904	$175,396		($27,229)

116 Systems Analysis of Present System

An analysis of figure 5.5 indicates that the intangible benefits of an interactive system offer two major results: increased revenues and decreased operating costs. Better customer service and relations should enhance an organization's chances of increasing sales to its present customers, as well as many potential ones who are looking for these characteristics in their vendors. An interactive MIS environment not only affects an organization externally, but also internally in terms of faster and more frequent reporting. In addition to accuracy, speed, and flexibility, this processing mode allows management more time to plan and organize activities and, in turn, direct and control according to the original plan. This is in contrast to other systems which do not always facilitate the functions of management. In summary, the qualitative factors, upon close examination, can have a pronounced effect upon the evaluation of each system alternative.

Comparison of Present System to Each Feasible MIS Alternative

Once a thorough analysis of the important factors has been completed, the MIS task force is in a position to compare the alternative systems to the existing system. Although several approaches are possible, only one—the decision table—will be presented for identifying and listing the relevant benefits (tangible and intangible) and costs.

Improved customer service by using better techniques to anticipate customer requirements, which results in fewer lost sales and less overtime in the plant for rush orders, among other advantages.
Better decision-making capability in the areas of corporate planning, marketing, research and development, engineering, manufacturing, inventory, purchasing, physical distribution, finance, accounting, and personnel, through more timely and informative reports via I/O terminals.
More effective utilization of management's time for planning, organizing, directing, and controlling because of the availability of timely data and information.
Ability to handle more customers faster with more automatic data processing equipment.
Closer control over capital investments and expenses through comparisons with budgets or forecasts.
Improved scheduling and production control, resulting in more efficient employment of personnel and machinery.
Greater accuracy, speed, and reliability of information handling and data processing operations.
Better control of credit through more frequent aging of accounts receivable and analysis of credit data.
Reversal of trend to higher hiring and training costs arising from the difficulties in filling clerical jobs.
Ability to utilize the many standard mathematical models and techniques.
Improved promotional efforts to attract new customers and retain present ones.
Greater ability to handle increased workloads at small additional costs.
Enhanced stature in the business community as a progressive and forward-looking organization.

---

figure 5.5 Intangible benefits with advanced MIS.

Decision table to evaluate MIS alternatives The decision table to evaluate MIS alternatives is shown in table 5.3. The conditions in the upper part of the table represent the important facts assembled in the study, and the lower part contains the possible courses of action. Each rule or system alternative represents a set of actions corresponding to a certain set of conditions. Rules 3 and 4 (system alternatives 3 and 4) indicate the highest returns when compared to the other alternatives. The important action required is a reevaluation of their intangible benefits.

table 5.3 Decision table for appraising feasible MIS alternatives. (Robert J. Thierauf, **Systems Analysis and Design of Real-Time Management Information Systems,** © 1975. Reprinted by permission of Prentice-Hall, Inc., Englewood Cliffs, New Jersey.)

DECISION TABLE

Table name: FEASIBLE MIS ALTERNATIVES—EXPLORATORY SURVEY
Date: February 25, 198_   Preparer: Robert J. Thierauf

	\multicolumn{12}{c}{Rule number}
	1	2	3	4	5	6	7	8	9	10	11	12
Condition												
Tangible Benefits:												
Meets return on investment criteria—20% after taxes*	N	N	N	N	N							
Lower order processing costs	Y	Y	Y	Y	Y							
Lower investment in inventory	Y	Y	Y	Y	Y							
Less future cash requirements	N	Y	Y	Y	N							
Intangible Benefits:												
Improved customer service	N	Y	Y	Y	Y							
Improved promotional efforts	Y	Y	Y	Y	Y							
Ability to handle more customers faster	N	N	Y	Y	Y							
Better decision-making ability	Y	Y	Y	Y	Y							
More effective utilization of management's time	Y	Y	Y	Y	Y							
Improved scheduling and production control	Y	Y	Y	Y	Y							
Closer control over capital investments and expenses	Y	Y	Y	Y	Y							
Better control of credit	N	N	Y	Y	Y							
Ability to handle more volume at lower costs	Y	Y	Y	Y	Y							
More accuracy and reliability of data	Y	Y	Y	Y	Y							
Greater utilization of mathematical models and techniques	Y	Y	Y	Y	Y							

*1–14%, 2–14%, 3–16%, 4–15%, 5–13%

table 5.3 Decision table for appraising feasible MIS alternatives (continued)

	Rule number											
	1	2	3	4	5	6	7	8	9	10	11	12
Action												
Utilizes an interactive system	X	X	X	X	X							
Utilizes remote job entry processing	—	—	X	X	X							
Minor changes of inputs and outputs	X	—	—	—	—							
Substantial changes of inputs and outputs	—	X	X	X	X							
Need for a data base	X	X	X	X	X							
Moderate revision of methods and procedures	X	X	—	—	—							
Complete revision of methods and procedures	—	—	X	X	X							
Employment of an additional consultant for study	—	—	X	X	X							
Recruitment of new data processing personnel	—	X	X	X	X							
Reevaluation of intangible benefits	—	—	X	X	—							
Other Information:												

SUMMARY: Systems Analysis—Concluding Investigation

Feasible MIS Alternatives Centers initially on developing proposed system specifications—inputs, the data base, methods and procedures, outputs, work volumes, and timing considerations. Based upon these specifications, alternative MIS systems are developed within the constraints of the system project.

Benefits and Costs for Each Feasible MIS Alternative Focuses on developing detailed savings and costs for each MIS alternative in order to derive net savings (losses) before federal income taxes. In turn, these data are discounted back to the present time at a certain rate and compared to an organization's cut-off point for capital investments. If the return is unsatisfactory, the intangible benefits are investigated.

Intangible Benefits for Each Feasible MIS Alternative Relates to studying the qualitative factors for each MIS alternative in order to justify undertaking one of the alternatives. Thus, all benefits (tangible and intangible) and costs are thoroughly investigated for each MIS alternative.

Comparison of Present System to Each Feasible MIS Alternative Refers to a comparative evaluation of proposed MIS alternatives to one another, such as that found in a decision table. In turn, the best MIS alternative can be compared to the present system.

SYSTEMS ANALYSIS—
EXPLORATORY SURVEY REPORT TO TOP MANAGEMENT

At the conclusion of the foregoing studies, ample information should have been accumulated to make a final recommendation to top management. The exploratory survey report, authored and signed by the MIS task force, is presented to the executive steering committee. It should be financially oriented since large sums of money are involved; any information with direct or indirect bearing on finances must be included. Generally, the approval of the recommendations presented in the report must come from the board of directors or top management.

The contents of this report must be as objective as possible to ensure that the best system is selected. The equipment should meet the needs of the system that has been developed (rather than the system's having been altered to meet the capabilities of certain equipment). Consideration must be given to the fundamental fact that a computer-oriented system can more readily absorb growth in volume with a slight increase in operating costs than other systems.

Feasibility of New MIS Alternatives

The feasibility of changing to a new management information system is a difficult undertaking when numerous alternatives are available. An analysis of the facts for one system alternative is a job in itself; the comparison of many proposed systems is formidable. Using the data in table 5.3, the feasibility of a new system is promising for proposed system alternatives 3 and 4. When consideration is given to important intangible benefits, the feasibility of applying newer (interactive) equipment and procedures has been established. The question, then, becomes one of determining which proposal is best when all critical factors are appraised.

Recommending a New System from the MIS Alternatives

As has been demonstrated, the recommendation of new DP equipment and procedures is determined by weighing quantitative and qualitative factors, with emphasis on an organization's future growth pattern and related problems. Now a comprehensive report must be prepared that states this recommendation. A suggested listing of items to be covered in the final exploratory survey report is depicted in figure 5.6. The report gives management an opportunity to examine the data and appraise their validity and merit. It also provides management with a sound basis for constructive criticism of the system project.

An examination of table 5.3 indicates that system alternatives 3 and 4 are best; now the question is to choose one of these alternatives. On the surface, both have about the same benefits, except that alternative 3 gives a higher return on investment. A closer inspection, however, reveals that only alternative 4 utilizes "intelligent" terminals (those capable of being programmed). Conversion today will mean no or minimal conversion costs in the future for intelligent terminals. With this added advantage, the MIS task force feels the future cost savings justify accepting a lower return. Therefore, its recommendation to top management has been finally determined.

Scope of the study, in which the objectives are stated and the problem is clearly defined.
Overview of the existing system, pointing out its weaknesses and problems.
Adequate description of the recommended system alternative, indicating its tangible and intangible benefits, its superiority in eliminating or reducing the deficiencies of the present system, and its general impact on the organization.
Financial data on the recommended system alternative, similar to that found in tables 5.1 and 5.2.
Reference to other feasible system alternatives which were investigated, giving reasons for their final rejection. A decision table, similar to table 5.3, should be included.
Financial data on system alternatives that were not selected, similar to that found in tables 5.1 and 5.2.
Schedule of funds required for specific periods of time during system implementation.
List of additional personnel needed to implement the new system and personnel requirements during conversion.
Accurate time schedule for the remainder of the system project.
Other special analyses and considerations.

---

figure 5.6 Suggested listing for an exploratory survey report to top management.

Infeasibility of New MIS Alternatives

A considerable expenditure of time, effort, and cost on the MIS exploratory survey may result in determination of the infeasibility of a new system. This conclusion may be the result of limiting the scope of the study to areas where progress in terms of new technical improvements has been slow or nonexistent, an area which does not lend itself to newer DP equipment, or some other reason. When an opportunity exists for technical improvement, a broad approach is desirable, for the most part. Setting too narrow a scope on the study can be avoided with the help of outside consultants or DP personnel who are knowledgeable about newer systems and capable of suggesting fertile areas for a feasibility study. To start blindly on a MIS exploratory survey without a high expectation of success is wasteful and indicates lack of clear identification of important corporate objectives by top management.

Waiting for future hardware and software developments may be a major reason why the study group does not recommend a change in the present system. Overcaution should be avoided because an organization's competitors are not standing still and will try to capitalize on each opportunity for system improvements. If immediate benefits are improved customer service, better managerial reports, and other operational improvements, waiting for newer developments may be very costly to an organization in the long run.

CHAPTER SUMMARY:

Whereas the previous chapter focused on the **introductory investigation** of a system project, this chapter concentrated on the **detailed investigation** of the present system, a lengthy process known as **systems analysis**. This phase was immediately followed by the **concluding investigation** which centers on the analysis of costs and benefits, both tangible and intangible. In turn, this information forms the basis of the exploratory survey report to the executive steering committee. In this comprehensive report, the feasibility or infeasibility of applying newer DP equipment and procedures is explored. Thus, systematic application of the investigation phases (introductory, detailed, and concluding) ensures top management that the best management information system has been selected for implementation.

Questions

1. Of the various methods given for systems analysis, which one is the best for obtaining accurate and complete information?
2. What part of the detailed investigation of the present system is most important from a management point of view?
3. a. What is the purpose of reviewing data files maintained?
   b. What is the purpose of reviewing internal control?
4. What is the rationale for analyzing present work volume?
5. What are the problems associated with calculating net savings after federal income taxes for feasible MIS alternatives?
6. What questions must the systems analyst answer if he or she is going to improve the present system?
7. Why have an exploratory survey report? Why not save this expense and procure the necessary equipment for the new system?
8. What are the essential contents of an exploratory survey report to top management?
9. What is the important relationship between the introductory investigation and the exploratory survey report?
10. Many organizations have found that initial estimates of new system costs are too low and benefits too high. What are the major factors contributing to this condition and how may they be overcome?

Self-Study Exercise

True-False:

1. _____ The best method for determining what is transpiring in the present system is by interviewing.
2. _____ The collection, organization, and evaluation of facts about the present system is called systems development.

3. _____ The least important phase of systems analysis is review of internal control.
4. _____ Investigation of data files maintained by an organization refers to those stored on line only.
5. _____ A procedure is a series of logical steps by which a job is accomplished.
6. _____ Where work volumes are small and the processing steps are complex, the feasibility of applying advanced data processing equipment is somewhat limited.
7. _____ Tangible benefits, by their very nature, require subjective evaluation.
8. _____ Generally, one-time costs include DP salaries, program maintenance, maintenance of equipment, etc.
9. _____ The time period for savings and costs of a system project is typically ten years.
10. _____ Even though intangible benefits are initially nonquantifiable, their ultimate impact is in quantitative terms.

Fill-In:

1. A review of an organization's _____ and _____ manuals is one way to understand the present system.
2. Examining the essential elements in a _____ _____ of the present system is commonly referred to as systems analysis.
3. _____ _____ are used as input to capture the originating data for any type of system.
4. The MIS task force is required to investigate _____ _____ in terms of their number and size, where they are located, who uses them, and the number of times they are referenced.
5. The purpose of reviewing _____ _____ is to help the MIS task force visualize the essential control points and framework of a system.
6. When large volumes of work require _____ _____ processing, the likelihood of employing newer DP equipment and procedures is quite high.
7. If the return on a proposed system alternative falls below the acceptable level for capital projects, the _____ factors are critical to the final decision.
8. The starting point for compiling new system specifications is to work with the _____ first.
9. A _____ _____ is an effective method for evaluating system alternatives.
10. The final system recommendation to top management is the essential focus of the _____ _____ report to top management.

Problem:

1. The American Company, which is currently switching from an integrated information system (batch processing) to an interactive information system, conducted a feasibility study some time ago. The three major steps included systems analysis, systems design, and equipment selection. The problem below centers on the first step—in particular, the financial aspects of the exploratory survey.

   In the exploratory survey report to top management, financial data on six feasible system alternatives were developed for the company. The data below were used in calculating the net savings (losses) before federal income taxes for system alternative 6. The company expects a twenty percent return after federal income taxes, discounted back to the present time.

Estimated savings—1st year, $100,000
Estimated savings—2nd year, $310,000
Estimated savings—3rd, 4th, and 5th years, add 5% to previous year
Estimated one-time costs—1st year, $505,000
Estimated additional operating costs—1st year, $105,000
Estimated additional operating costs—2nd year, $145,000
Estimated additional operating costs—3rd, 4th, and 5th years, add 5% to previous year
Federal income tax rate—48%
Present value of $1 at 20%—1st year, .833
Present value of $1 at 20%—2nd year, .694
Present value of $1 at 20%—3rd year, .579
Present value of $1 at 20%—4th year, .482
Present value of $1 at 20%—5th year, .402

Using these facts, determine whether or not the company should explore intangible benefits; that is, the present value of net savings at twenty percent for system alternative 6 is negative after considering all tangible benefits. Assume a five-year savings and costs projections that start after the completion of the feasibility study.

chapter 6

systems analysis of the batch order-entry system— abc company

OBJECTIVES:

- To initiate a feasibility study that will determine the practicality of an improved order-entry system.
- To undertake a detailed investigation of the present, batch-oriented order-entry system.
- To explore the feasible order-entry system alternatives in terms of their benefits and costs.
- To present a comprehensive report to top management regarding the feasibility of a new order-entry system.

IN THIS CHAPTER:

INTRODUCTORY INVESTIGATION OF NEW ORDER-ENTRY SYSTEM
    Organization of System Project
    Definition of the Scope of System Project
    Determination of the Schedule of System Project
    Presentation of Results of Introductory Investigation
SYSTEMS ANALYSIS—DETAILED INVESTIGATION OF PRESENT BATCH ORDER-ENTRY SYSTEM
    Overview of Present Batch Order-Entry System
    Historical Aspects
    Analysis of Inputs
    Review of Data Files Maintained
    Review of Methods and Procedures
    Analysis of Outputs
    Review of Internal Control
    Other Analyses and Considerations
    Analysis of Present Benefits
SYSTEMS ANALYSIS—CONCLUDING INVESTIGATION OF NEW ORDER-ENTRY SYSTEM
    Feasible Order-Entry System Alternatives
    Tangible and Intangible Benefits Versus Costs for Each Alternative
    Comparison of Present Order-Entry System to Each Alternative
SYSTEMS ANALYSIS—EXPLORATORY SURVEY REPORT TO TOP MANAGEMENT
    Recommendations for New Order-Entry System
    Presentation of Recommendations for New Order-Entry System
CHAPTER SUMMARY
QUESTIONS
SELF-STUDY EXERCISE
CASE STUDY 1—FINISHED PRODUCT INVENTORY
CASE STUDY 2—ACCOUNTS RECEIVABLE

In the prior two chapters, the proper approach for initiating a system project, including the preparation of an exploratory survey report to top management, was treated. Within this chapter, an identical approach is taken to investigating the old and new order-entry systems of the ABC Company. The feasibility of a new order system in terms of systems analysis is explored for the text's master case study, with consideration given to those factors relating to the two case studies located at the end of the chapter, namely, finished product inventory and accounts receivable. Thus, a comprehensive presentation of systems analysis for an order-entry system gives the reader an opportunity to visualize the first part of a system project. This **experiential** approach prepares the individual for undertaking the systems analysis steps required for the end-of-chapter case studies.

After a discussion of the launching of the ABC Company's order-entry system, a detailed investigation of the present system operating in a batch processing mode is undertaken. This intensive system review is followed by a concluding investigation whereby a comparison is made between feasible order-entry system alternatives and the present system. This information provides a basis for preparing an exploratory survey report to top management regarding the feasibility or infeasibility of the new order-entry system. In effect, the ABC Company has evaluated all the pertinent factors deemed necessary in reaching a "go" or "no-go" decision on a new order-entry system.

INTRODUCTORY INVESTIGATION OF NEW ORDER-ENTRY SYSTEM

Because of the complexity and cost of most system projects, the MIS vice-president of the ABC Company has decided to employ a systematic approach to the study of a new order-entry system. As a means of getting the project off to a good start, a variation of the introductory investigation steps set forth in chapter 4 is used below.

- organization of system project
- definition of the scope of system project
- determination of the schedule of system project
- presentation of results of introductory investigation

Each of these important phases of the introductory investigation is discussed below.

Organization of System Project

A study conducted by the MIS vice-president of the ABC Company had identified the following key objectives for future systems work:
- provide better customer service
- improve selling efficiency
- help production more closely match output to customer demand
- provide more timely information for analysis of the company's present and future operations, both at plant and corporate levels
- improve coordinated control of the overall company

Upon further discussion with the company's vice-presidents and several departmental managers, the first step toward accomplishing these objectives was identified—upgrade the current order processing system and related procedures. In essence, the company's order-entry system is to be studied in great detail to accomplish the foregoing objectives.

The board of directors has established an executive steering committee, consisting of the four vice-presidents whose sections are affected by the order-entry system: marketing, manufacturing, physical distribution, and accounting (and finance). In turn, these executives will be giving direction to the MIS task force, comprised of the sales manager, the field warehouse managers, and the systems and programming manager. This group is to study the feasibility of upgrading the current order processing system. Further recognized was the need to involve others in the study as it progressed; the manufacturing plant managers, plant and field warehouse managers, controller, and system personnel were all asked to provide their help as needed.

Definition of the Scope of System Project

The MIS task force visited each of the five field warehouses to interview key personnel on the possible objectives of a new order processing system. In particular, the task force was especially interested in identifying the good and bad aspects of the existing system.

The MIS task force also discussed customer service needs with several former and current major customers. All indicated that they were satisfied with the product quality of the ABC Company and that their prices were highly competitive; however, they were dissatisfied with the level of customer service. When asked for details, former customers indicated their primary reasons for no longer buying from the ABC Company were that they expected and were currently receiving from ABC's competition
- an average order cycle time of seven days
- a level of 92 percent of all ordered items filled
- an error rate on billing of approximately 3 percent

Last, the MIS task force added their in-depth knowledge of the current order processing system to the information gathered from these interviews. Key information about the current system is summarized in figure 6.1.

The vice-president of marketing was very excited about the possibility of the new order-entry system to help regain the 3 percent of the market that he felt had been lost in

Average daily order volume at each field warehouse:	
Orders per day	30 orders
Item lines per order	6 items
Average total order-entry processing cost per order:	$11.50
Orders received:	
By mail from salespeople or customers	95%
By telephone from salespeople or customers	5%
	100%
Average order size:	
Dollar value	$1923
Pounds	4380 lbs.
Average order cycle time:	
Time required to receive order at field warehouse	2 days
Internal order processing time	8 days
Outbound shipment transit time	2 days
	12
Customer service level (in 12 days):	
Percent of items filled	84%
Percent of orders incorrectly billed	7%

figure 6.1   Current order processing information—average data of five field warehouses of the ABC Company.

the past three years. The ABC Company market share was currently 18 percent of a $420 million per year market. Thus, regaining 3 percent of the market share would contribute an additional $12.6 million per annum in sales, or approximately $600,000 per year in increased earnings based on the current profit margin of 4.5 percent.

In reference to the earnings increase, the vice-president of accounting argued that full achievement of all customer service goals would require an improved inventory control system and production planning system in addition to the proposed new order-entry system. He also informed the project team that, in order to receive the approval of the executive steering committee, the cost to design and implement any system as well as purchase or rent necessary equipment must be recovered in less than three years. This set an upper limit of $1.8 million for the cost of a complete MIS for order entry, inventory control, and production planning. Additionally, the vice-president insisted that the order-entry system provide timely data for better control of finished product inventory. He felt that an on-line inventory control system coupled with centralized management of all finished product inventory would enable the ABC Company to meet its customer service goals subject to these constraints.

The vice-president of manufacturing agreed with the need for improved customer service. He felt the key to improve the order cycle time was better coordination between production scheduling and field warehouse inventory control. He insisted that field warehouse inventory control clerks needed to know the lead time required to fill a replenishment order. Lead time varies based upon where the plant is in its monthly production cycle when it receives the replenishment order. Since major products were produced on a two-week cycle and minor products on a four-week cycle, the lead time could vary significantly. The failure to account for the variation in lead time was felt to be the major reason for out-of-stock conditions in the field warehouses.

Needless to say, the MIS task force held several meetings with the executive steering committee to present the facts they had gathered. Together, they developed a set of objectives for the new order-entry system. These objectives are thought to be very challenging, but obtainable. Furthermore, if the new order processing system could meet these objectives, the ABC Company would be the industry leader in customer service—lowest average order cycle time, highest percent of items filled, and lowest error rate on billing. This, coupled with their competititive pricing and high quality, would help them to regain that share of the market they had lost. These objectives are presented in figure 6.2.

Average total order-entry processing cost per order: (present cost of $11.50 plus increased cost for new system of $5.13)	$16.63
Orders received:	
By mail from salespeople or customers	20%
By telephone from salespeople or customers	80%
	100%
Average order cycle time:	
Time required to receive order at field warehouse	1 day
Internal order processing time	2 days
Outbound shipment transit time	2 days
	5 days
Customer service level (in 5 days):	
Percent of items filled	96%
Percent of orders incorrectly billed	1%

figure 6.2 Objectives of the new order-entry processing system for the ABC Company.

The executive steering committee agreed to pay up to $200,000 per year in additional order processing costs (represents an increase of $5.13 per order) to meet these objectives. However, the vice-president of physical distribution agreed to these objectives subject to the following conditions: total dollar investment in inventory would not be increased, total cost to deliver finished products to the customer would not increase, and no additional field warehouses would be opened.

Determination of the Schedule of System Project

The executive steering committee strongly emphasized the importance of this project to improve the company's competitive position and urged the MIS task force to set a completion date as early as possible. The MIS task force, in turn, prepared a list of the major activities to be performed during the systems analysis and systems design phases and estimated the weeks required to complete each task. From this list, the Gantt chart shown in table 6.1 was prepared. The study group concluded that the systems analysis study could be completed by two people in about eight weeks. They estimated the

table 6.1 Gantt chart for the order-entry processing system of the ABC Company.

The ABC Company: Feasibility Study—ORDER-ENTRY PROCESSING SYSTEM

Activity	Time period (week) 1-27
Introductory Investigation	weeks 1–2
Systems Analysis:	
Detailed investigation	weeks 4–5
Concluding investigation	weeks 5–7
Exploratory survey report to top management	weeks 7–8
Systems Design:	
System outputs	weeks 10–16
Data files	weeks 12–16
System inputs	weeks 10–16
Methods and procedures	weeks 14–17
System controls	weeks 16–20
Equipment Selection	weeks 19–22

Date: January 18, 198_

Preparer: MIS Task Force
G. W. Reynolds

systems design phase and equipment selection could be completed by two people in about eighteen weeks.

The MIS task force felt it was too early to identify the time required for each specific implementation task. They knew from past experience that the total effort required for system implementation is approximately three times that required to complete the systems design and equipment selection phases. They, therefore, estimated that implementation would require approximately 108 weeks (3 × 2 people × 18 weeks) of effort.

Presentation of Results of Introductory Investigation

The MIS task force and the executive steering committee met with the executive vice-president to present the results of the introductory investigation. The main focus of the meeting was that poor customer service had led to the loss of 3 percent of the market share or $12.6 million in annual sales volume. The data in figure 6.1 was compared to customer expectations of service to highlight the magnitude of the problem. The task force then explained the concept of an integrated order-entry, inventory control, and production planning system. They presented the information contained in figure 6.2 as the potential that could be achieved and, with the achievement of this potential, the recapture of 3 percent of the market share.

In conjunction with the factors above, the MIS task force outlined the approach they thought would prove most successful. It began with a complete analysis of the present order-entry system, summarized in a study report presented to top management for review. With management's approval, the design phase would begin. The system would be designed to interface efficiently and effectively with the future inventory control and production planning systems. Implementation of the order-entry system would begin upon management approval. Although the executive vice-president was disappointed to hear an estimated completion time of one full year for the order-entry processing system, he accepted the task force's explanation that a substantial change such as this takes time to be implemented properly.

The task force went on to explain that the inventory control and production planning systems must take second priority since they require data from the order-entry system. These systems could be partially designed concurrent with the design of the order-entry system, but the current number of MIS department personnel is insufficient to allow the simultaneous development of all three systems.

Even though the executive vice-president gave his approval, he asked that the MIS task force be alert to opportunities for integrating the new order-entry system with other areas, such as accounts receivable. He further recommended that a project description be written which communicates to company personnel that a systems analysis study of the order-entry system is being conducted. The project description is shown in figure 6.3. The executive vice-president also expressed appreciation to all parties involved in the presentation and took the opportunity to inform the MIS task force that the project was vital to the future success of the ABC Company.

Project Description

A. **Project Title:** Systems Analysis Study of a New Order-Entry System
B. **Project Objectives:**
   1. To identify requirements for any future order-entry system.
   2. To evaluate the economic and technical feasibility of a new order-entry system.
C. **Approach To Be Taken:**
   1. Interview personnel and customers to gather information on the existing order-entry system.
   2. Analyze inputs, methods, procedures, data files, outputs, and internal controls of the current system.
   3. Identify and evaluate alternative systems.
   4. Prepare recommendations and present exploratory survey report to top management.
D. **Potential System Benefits:**
   1. Reduce order cycle time and percentage of errors in billing; this improvement in customer service may enable the company to improve its market share.
   2. Provide an efficient order-entry system as the basic module to an integrated network of systems: order entry, finished product inventory, production planning, and possibly accounts receivable.
E. **Schedule and Effort Required:**
   Project start date: 3/15/8_
   Expected completion date: 5/15/8_
   This project will require two full-time people for eight weeks.
F. **Approved by:** J. W. Bluster, Executive Vice-President

figure 6.3 Project description of the order-entry system for the ABC Company.

SUMMARY: Introductory Investigation of New Order-Entry System

Organization of System Project The executive steering committee consists of four vice-presidents—marketing, manufacturing, physical distribution, and accounting—and the MIS task force includes the sales manager, field warehouse managers, and the systems and programming manager, as well as other system and organization personnel as needed.

Definition of the Scope of System Project The project is to center on a new order-entry system that will reduce average order cycle time from twelve days to five days, increase customer service level from 84 to 96 percent of orders filled, and reduce the percent of orders incorrectly billed from 7 to 1 percent. Additionally, the project is to be the initial major module (or building block) for adding inventory control, production planning, and accounts receivable modules at a later date.

Determination of the Schedule of System Project The scheduling of the order-entry system indicates that systems analysis will take approximately eight weeks and systems design will extend to about eighteen weeks. Overall implementation of the new system would take approximately one year.

Presentation of Results of Introductory Investigation The main thrust of the presentation to the executive vice-president by the executive steering committee and the MIS task force is that the new system should allow the company to improve its market share by three percent, thereby increasing sales by about $12.6 million and profits by about $600,000 at the current annual rate. Based upon the data presented, the executive vice-president gave approval to the order-entry system project.

SYSTEMS ANALYSIS—DETAILED INVESTIGATION OF PRESENT BATCH ORDER-ENTRY SYSTEM

In order for the MIS task force to comprehend what transpires in the present batch order-entry system, there is need for a detailed investigation. This includes the many steps of systems analysis set forth in the prior chapter, which are
- historical aspects
- analysis of inputs
- review of data files maintained
- review of methods and procedures
- analysis of outputs
- review of internal control
- other analyses and considerations

In addition to examining each area in detail, an overview section is presented and should be sufficiently comprehensive that any knowledgeable reader can envision the scope of the order-entry system and the relationship of its parts to the preceding areas. An analysis of the present benefits of the batch order-entry system is presented at the end of this section.

Overview of Present Batch Order-Entry System

Once the sales order is received at the field warehouse—by mail, telephone, or hand carried—the first major activity is a credit check. If the accounting clerk questions the ability of the customer to pay the ordered amount, the order is sent to marketing management for further checking and a decision is made. In most cases, credit clearance is obtained and the necessary order entry cards are keypunched and verified. Warehouse shipment cards, reporting shipments to customers or other field warehouses, are keypunched, verified, and merged with the order entry cards. On the other hand, if credit is not approved, the order is returned to the customer.

New orders are placed in the open order file. Shipment cards trigger the purge of an order from the open order file and creation of a shipment record for the shipment file. The order-entry system creates a multicopy Order and Shipment Processing Form for each new order. A Shipment by Carrier Report and an Orders Ready for Billing Report (for corporate headquarters at St. Louis) are also produced daily. An overview of this process is shown in figure 6.4.

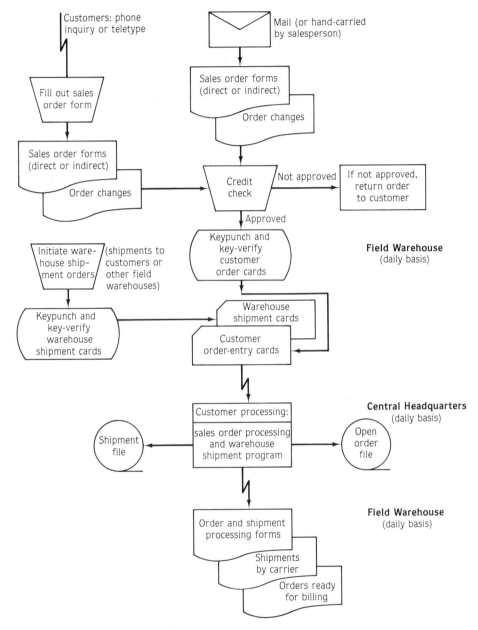

figure 6.4 The daily flow of sales order processing and warehouse shipment data for the ABC Company.

Historical Aspects

During the recent chaotic economic period, the company's inventories had skyrocketed. Although market conditions have since stabilized, finished product inventory remains too high in relation to sales. The current order processing system and its procedures are based on providing input to and receiving output from a batch processing order-entry system that is run nightly at the central headquarters. The result is that company inventory and order status information can be two days behind actual orders and shipments.

In spite of high inventories, the ABC Company's customer service is below customer expectations and industry standards. In fact, the vice-president of marketing has gone on record that the company has lost market share because of poor customer service—slow processing of customer orders and a low percentage of complete orders shipped. Furthermore, the current order processing system involves extensive paperwork, requires considerable time to match manually input and output documents, has a large number of orders rejected due to keypunch errors, and is incapable of providing timely information for analyzing company sales.

Analysis of Inputs

The key input to the order-entry system is the order form shown in figure 6.5. The information from this form is keypunched onto cards for input to the system (figure 6.6, page 138).

For each order, the following order-entry cards are punched:
1. The 201 card identifies the customer name and address to which the order will be shipped.
2. The 202 card specifies the date the order is taken by the salesperson and the desired shipping date to meet the customer's requirements. The salesperson generally assumes two working days from shipment date to delivery date.
3. The 203 card specifies the item number and quantity on the order. A separate card must be prepared for each item. The TC (Type Change) is always A for entry of a new order. The 203 card is also used for making corrections to information on the open order file. A card whose TC is C (Change) corrects the quantity for any item on the open order file; a card punched with a D (Delete) will delete an item from the order.
4. The 204 card is used to cancel an entire order. In this case, all information about this order is removed from the open order file.

At the time of shipment, the order is loaded onto a truck at the field warehouse for delivery to the customer. A shipment processing form is prepared by the shipping clerk. On this form, the shipping clerk identifies the customer order number; assigns a six-digit shipment number; enters the shipment date, carrier name, trailer number of the carrier's trailer, and the terms of the shipment—P for freight paid by the company or C for freight collect. The information from this form is punched into a 205 card, illustrated in figure 6.7, and entered into the order-entry system. Receipt of this card triggers the order-entry system to remove the associated open order record from the open order file and pass that record onto the shipment data file.

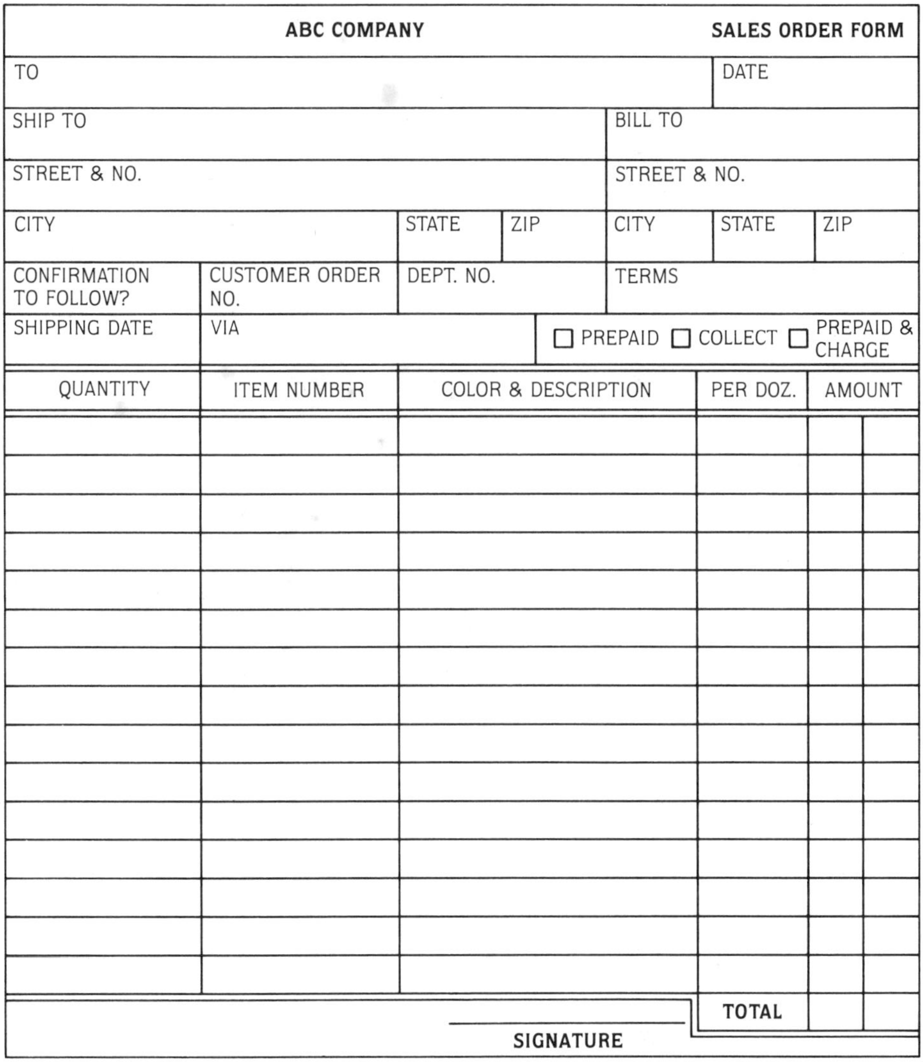

Systems Analysis—Detailed Investigation of Present Batch Order-Entry System

ABC COMPANY					SALES ORDER FORM		
TO					DATE		
SHIP TO					BILL TO		
STREET & NO.					STREET & NO.		
CITY		STATE	ZIP	CITY		STATE	ZIP
CONFIRMATION TO FOLLOW?	CUSTOMER ORDER NO.	DEPT. NO.		TERMS			
SHIPPING DATE	VIA			☐ PREPAID ☐ COLLECT ☐			PREPAID & CHARGE
QUANTITY	ITEM NUMBER	COLOR & DESCRIPTION			PER DOZ.		AMOUNT
						TOTAL	
		SIGNATURE					

figure 6.5 A sales order form for the ABC Company.

Review of Data Files Maintained

Two basic data files are used in the current order-entry system, the **open order file** and the **shipment file.** The data elements stored on each file are shown in figure 6.8. The open order file is created from the order form; the key to the file is the six-digit customer order number. This sequential file is stored on magnetic tape.

The shipment file, on the other hand, is a magnetic tape file created from the open order file. As shipment data are input to the order-entry system, the order is removed from the

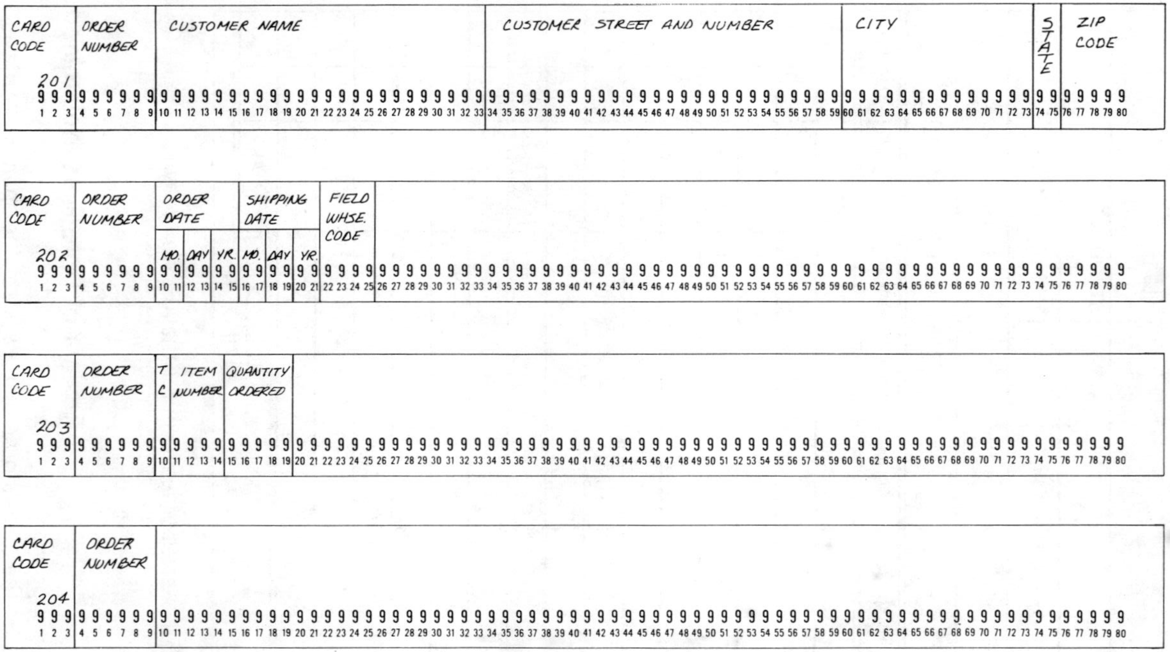

figure 6.6 The card input required to enter each new order of the ABC Company.

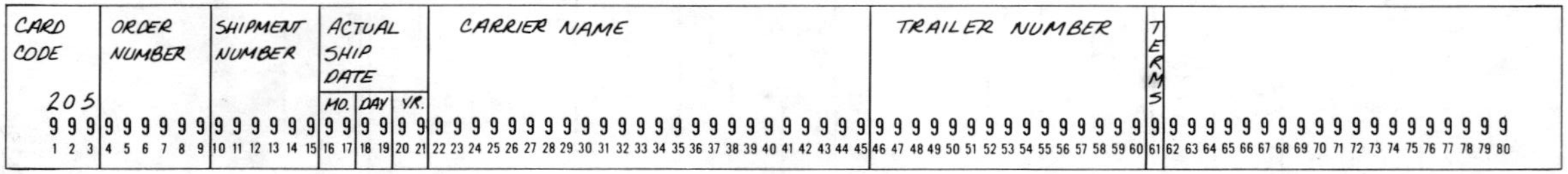

figure 6.7 The card input required to move an open order record to the shipment file of the ABC Company.

Systems Analysis—Detailed Investigation of Present Batch Order-Entry System

	Input Document		Data File (number of characters)	
Data Element	Order-entry cards	Warehouse shipment cards	Open order file	Shipment file
Record Identification			3	3
Order Number	X	X	6	6
Shipment Number		X		6
Field Warehouse Code	X	X	4	4
Customer: Name	X	X	24	24
Address	X	X	26	26
City	X	X	14	14
State	X	X	2	2
Zip Code	X	X	5	5
Order Date	X		6	6
Ship Date: Desired	X		6	6
Ship Date: Actual		X		6
Shipping Terms		X		1
Carrier Name		X		24
Trailer Number		X		15
Order Line Item (up to 10):				
Item Number	X	X	40	40
Quantity	X	X	50	50
Total Number of Characters			186	238

figure 6.8 Order-entry system—data files of the ABC Company, open order and shipment.

open order file and placed on the shipment file. The key to the shipment file is a six-digit shipment number assigned by the clerk at the time of shipment. Several orders may be included in a single shipment and all such orders will be assigned the same shipment number.

Review of Methods and Procedures

After the MIS task force reviewed the present methods and procedures of the order-entry system, they were flowcharted (figure 6.9). A detailed description of these procedures is found below.

On the **first** day, an order is taken by a salesperson, or a customer mails an order to the nearest field warehouse. By the **third** day, the order has been received by the field warehouse. Provided it is accurate and complete and credit is approved, the order is transcribed onto the sales order form (figure 6.5) for keypunching; this becomes input to the order-entry processing system. Order input is grouped together and read into a card reader. The data are transmitted to the computer at company headquarters (St. Louis) for processing by the order-entry system. Any data transmitted after 5:30 P.M. (St. Louis time) will not be in time for that night's processing.

On the **fourth** day, multicopy order and shipment processing forms are printed on a small 300 line per minute printer at the field warehouse from data forwarded from the central headquarters. The red copy of this form is kept by the field warehouse account-

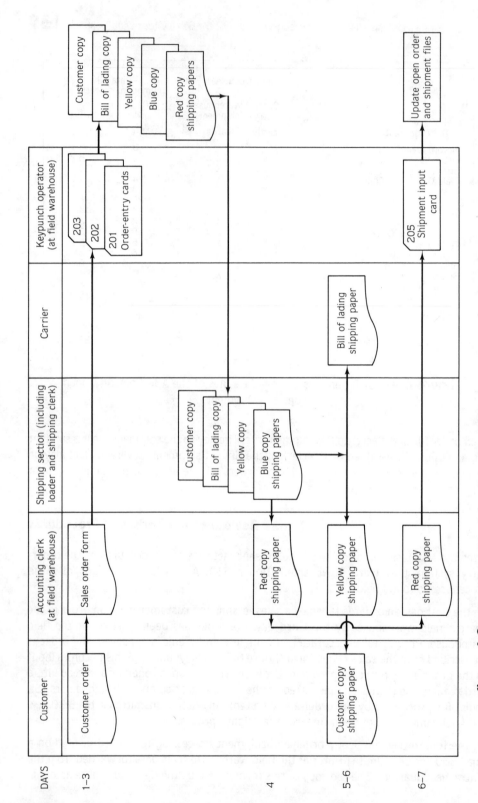

figure 6.9 Document flowchart of the current order-entry processing system of the ABC Company.

ing clerk to answer questions from the salesperson or customer regarding the order. One copy of this form is mailed to the customer as confirmation that the order has been received and is being processed. Three copies of the form are sent to the shipping section. The shipping clerk determines the number of trucks needed for loading the next day's shipments, contacts the carriers to arrange a pickup time, and notifies the field warehouse manager of the next day's loading schedule.

On the **fifth** day, the blue copy of the shipment processing form is filed by the shipping clerk and two copies are given to the warehouse loader. As items are picked to fill an order, the loader notes any items which must be short-shipped due to insufficient inventory or which are out-of-stock entirely. One copy of the duely noted shipment processing form is given to the carrier as the shipment bill of lading. The last annotated or yellow copy is then returned to the shipping clerk.

On the **sixth** day, the shipping clerk pulls the blue copy of the shipment processing form, matches it to the yellow copy, makes any corrections noted by the loader, refiles the blue copy, and sends the yellow copy to the field warehouse accounting clerk. Here the red copy is pulled and matched to the yellow copy; any corrections noted by the loader are made to the red copy. The yellow copy is then filed; the red copy is sent to keypunch for preparation of input cards to the order-entry processing and accounts receivable systems. Separate cards are prepared for input to these two systems.

All shipment input cards are grouped together and read into a card reader. Data are transmitted to the computer at central headquarters and processed by the order-entry system. Any data received after 5:30 P.M. St. Louis time cannot be processed until the next evening.

On the **seventh** day, the accounts receivable section at St. Louis prepares several copies of the customer invoice. The original copy is mailed to the customer for remittance of funds and the duplicate is filed at company headquarters. This information is transmitted from St. Louis to the field warehouse that originally took the order. Upon receipt of the customer invoice, the field warehouse accounting clerk pulls the yellow copy of the shipment processing form to check for any discrepancies. If there are billing errors, the field warehouse notifies the accounts receivable department in St. Louis, and the appropriate credit memorandums are issued.

Ideally, customer orders should be delivered by the **seventh** day. In practice, however, an average of twelve days is required from order placement to order delivery. An order is not delivered by the seventh day for one or more of the following reasons:

1. One or more items were out of stock.
2. The field warehouse erred when transcribing the original order onto a keypunch form.
3. A keypunching error was made.
4. The order reached St. Louis too late for the 5:30 cut-off for input into order processing.
5. The shipping clerk initially held the order until enough merchandise to be shipped to that area of the country had accumulated to make a full truckload (which allows a lower delivery rate).

6. The shipping date specified by the salesperson was later than the fifth day of the order cycle.
7. The order received from the salesperson was inaccurate and/or incorrect and had to be held until he or she could be contacted.

Analysis of Outputs

There are three key output documents from the current order-entry system. The first is a multipart order and shipment processing form produced on the printer located at the field warehouse (from which the order is filled). This form is printed in five parts on different colored paper to aid in the proper paperflow (figure 6.10).

Order Number:		Carrier:		Ship Date:	
Order Date:	Warehouse:	Truck or Car Number:		Ship Terms:	Shipment Number:
Consigned To:					
Destination:	City:		State:	Zip Code:	
Quantity		Brand Description		Unit Weight	Short-Ship Code/Qty.
Bill of Lading					
No. of Pkgs.		Description for Bill of Lading			Weight

figure 6.10 Multipart order and shipment processing form for the ABC Company.

In order to assist in controlling shipments by carriers, a second report of shipments by carrier is produced daily on the printer at the field warehouse (from which the shipments are made). This report is filed and used to check freight bills received from carriers prior to payment. A sample report is shown in figure 6.11.

Systems Analysis—Detailed Investigation of Present Batch Order-Entry System

	SHIPMENTS BY CARRIER					DATE: 06/15/8-
	LOS ANGELES WAREHOUSE					
CARRIER NAME	SHIPMENT NUMBER	TRAILER NUMBER	ORDER NUMBER	WEIGHT	FREIGHT	
ALLEN BROS. TRUCKING	645378	TNX1034	432819	4010	C	
			432831	1250	C	
			432850	2630		
			432897	5280		
			432903	3770		
CALAHAN FREIGHT	645376	RON4513X5	432840	2600		
			432862	4350		
			432865	6030		
MCGINNIS CARRIERS	645373	FNR143TR7	432923	4520	C	
RYBOLT DELIVERY	645375	SR54TX51	432928	1050		
			432929	2500		
			432930	4950	C	
			432931	1030		
			432937	1070		
			432939	4750		
WALIN'S WHSE & TRUCKING			433003	6030		
			433005	1050		
			433009	2500	C	
			433011	1030	C	

figure 6.11 A shipments-by-carrier report for the ABC Company.

The last output document is a listing of orders ready for billing. It is printed daily at central headquarters in St. Louis. This listing is given to the accounts receivable department, where it is used to prepare customer invoices and make the appropriate accounting entries. A sample report is illustrated in figure 6.12.

Review of Internal Control

Order input received from the salespeople is often inaccurate. Although the customer's name is always correct, the address information is often incorrect. The color and description information is generally correct, but often the four-digit item number is either omitted or incorrect. The six-digit customer order number consists of a two-digit salesperson identification number followed by a sequentially assigned order number; the salesperson always records the first two-digit identifier correctly, but frequently miscodes the four-digit order number. In fact, the salesperson with the highest sales volume in the company always assigns the same six-digit customer order number to all orders. The foregoing problems with accuracy and completeness of the original order form result in a two- to three-day delay in the processing of approximately twenty percent of all orders.

The order-entry system performs various edits on all input. For example, only numeric and alphabetic data may be present in numeric and alphabetic fields, respectively, and the item number must be present in a table of valid item numbers representing all the various products sold by the ABC Company. If any input associated with entry of a new order, change of an order, or entry of shipment data is incorrect, then that order or

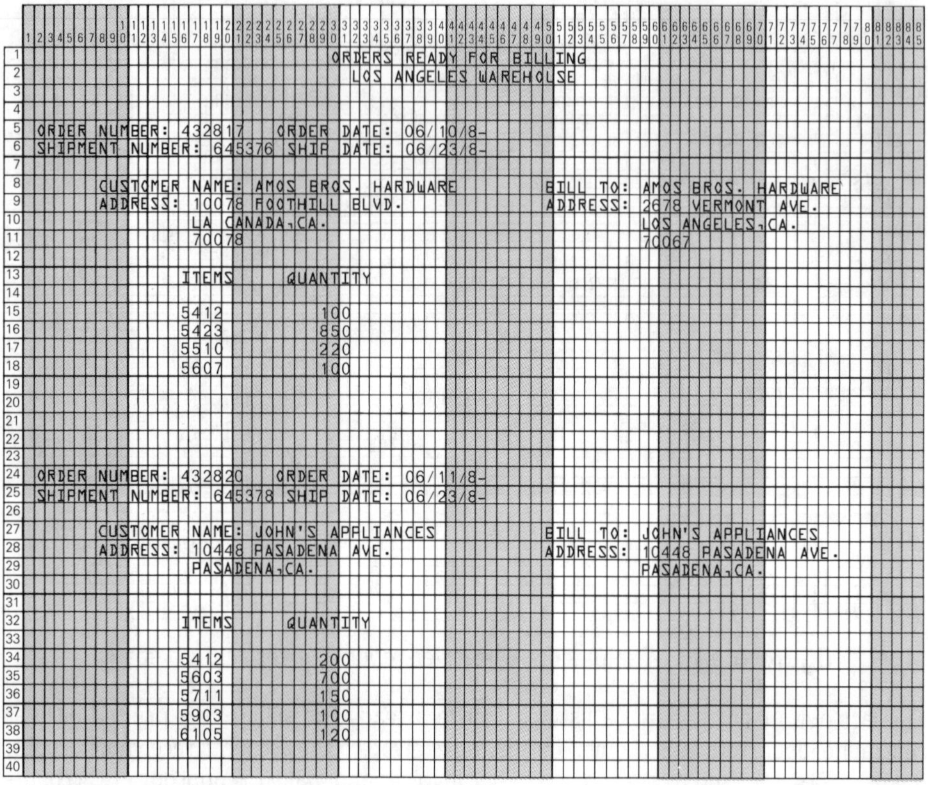

```
                                    ORDERS READY FOR BILLING
                                    LOS ANGELES WAREHOUSE

ORDER NUMBER: 432817      ORDER DATE: 06/10/8-
SHIPMENT NUMBER: 645376   SHIP DATE:  06/23/8-

         CUSTOMER NAME: AMOS BROS. HARDWARE         BILL TO: AMOS BROS. HARDWARE
         ADDRESS: 10078 FOOTHILL BLVD.              ADDRESS: 2678 VERMONT AVE.
                  LA CANADA, CA.                             LOS ANGELES, CA.
                  70078                                      70067

         ITEMS     QUANTITY
         5412      100
         5423      850
         5510      220
         5607      100


ORDER NUMBER: 432820      ORDER DATE: 06/11/8-
SHIPMENT NUMBER: 645378   SHIP DATE:  06/23/8-

         CUSTOMER NAME: JOHN'S APPLIANCES           BILL TO: JOHN'S APPLIANCES
         ADDRESS: 10448 PASADENA AVE.               ADDRESS: 10448 PASADENA AVE.
                  PASADENA, CA.                              PASADENA, CA.

         ITEMS     QUANTITY
         5412      200
         5603      700
         5711      150
         5903      100
         6105      120
```

figure 6.12 Orders of the ABC Company which are ready for billing.

shipment data are rejected. The open order file and shipment file are not updated if input data are in error.

Other Analyses and Considerations

The ABC Company, with 18 percent of the total industry sales, ranks second. Over the next five years, annual sales are expected to grow from $75 to $100 million. This would represent a growth from 39,000 to 48,750 orders per year. To recapture 3 percent of the market in five years would represent $16.7 million in additional annual sales from 8,125 additional orders per year, thereby generating $750,000 in additional profits.

Fundamentally, the company is in a sound financial condition. It is noted for its strong management and its high profit. It is highly regarded by the investment community and has no problems obtaining credit or raising funds through stock or bond offerings. As shown in table 6.2, the ABC Company's current financial status is considered good to excellent by the investment community.

Overall, the executive steering committee is strongly committed to helping the company move ahead. Because the company has sufficient resources to be an aggressive competi-

table 6.2 Summary of current financial statements (income statement and balance sheet) of the ABC Company.

The ABC Company
Financial Statements for the Current Period

Income Statement

	Thousands
Sales	$75,000
Cost of goods sold	40,345
Gross profit on sales	$34,655
Distribution costs	$10,835
Selling and administrative expenses	6,750
Depreciation	8,500
Interest	2,520
Income taxes	2,675
NET INCOME after federal income taxes	$3,375

Balance Sheet
Assets:

Cash	$1,100
Receivables	9,100
Inventories	19,000
Plant and equipment	50,300
Other assets	500
TOTAL assets	$80,000

Liabilities and Capital:

Current liabilities	$5,000
Long-term debt	25,100
Shareholder equity	49,900
TOTAL liabilities and capital	$80,000

tor, the executive vice-president and the executive steering committee have agreed that improvements to the order-entry processing system be of a sufficient magnitude that the company can regain its competitive edge. The executive steering committee, then, is eager to review feasible order-entry system alternatives that the MIS task force will identify and evaluate in their exploratory survey report to top management.

Analysis of Present Benefits

Based upon the foregoing analysis of the present system, the current order-entry system is barely meeting its minimum requirements. Although the system handles the current volume of orders at a cost of $11.50 per order, its main limitations are the long average order processing time (twelve days), low percent of order items filled (84 percent), and high percent of orders incorrectly billed (7 percent). For costs incurred, benefits are not generally commensurate with those found in an efficient and effective system.

SUMMARY: Systems Analysis—Detailed Investigation of Present Batch Order-Entry System

Historical Aspects   Places great emphasis on declining market share because of poor customer service—slow processing of customer orders and a low percentage of complete orders shipped.

Analysis of Inputs   Specifies types of punched card input for order-entry system: the customer name and address (201) card, the order and shipping date (202) card, the item number and order quantity (203) card, the cancellation (204) card, and the shipment (205) card.

Review of Data Files Maintained   Refers to the major magnetic tape files maintained for the order-entry system: the open order file, containing 186 characters of information, and the shipment file, containing 238 characters of information.

Review of Methods and Procedures   Focuses on current order-entry methods and procedures that result in an average order cycle time of twelve days: two days to receive an order at a field warehouse, an average of eight days for order-entry processing procedures, and two days for the average outbound shipment transit.

Analysis of Outputs   Relates to the output documents for the order-entry system: the multipart order and shipment form, a report of shipments by carrier, and a listing of orders ready for billing.

Review of Internal Control   Refers to inaccurate information that originates outside and inside the company. Problems abound with incorrect customer order numbers and incorrect data found in the open order file and shipment file.

Other Analyses and Considerations   Relates to the financial condition of the company, considered good to excellent for this type of industry. In addition, the executive vice-president wants the company to regain its competitive edge by improving customer service.

SYSTEMS ANALYSIS—CONCLUDING INVESTIGATION OF NEW ORDER-ENTRY SYSTEM

Now that each area of the present batch order-entry system has been carefully analyzed, a feasible set of order-entry system alternatives must be developed in order to select the best one. Before the best system alternative can be selected, the following areas are examined:

feasible order-entry system alternatives
tangible and intangible benefits versus costs for each alternative
comparison of present order-entry system to each alternative

Feasible Order-Entry System Alternatives

The first requirement of the new order-entry system is to reduce the order cycle time. The MIS task force examined the feasibility of establishing WATS telephone lines that allow salespeople to call order information directly to the field warehouse. Not only would the time to receive the order be reduced greatly, but also the salesperson would provide order information directly to a trained order-entry clerk. Problems of inaccurate or incomplete orders would be greatly reduced. An additional advantage would be that the WATS lines would provide an inexpensive way to improve the communications between sales management and salespeople in the field.

The second major requirement of the new order-entry system is to improve internal order processing. It is clear that significant changes must be implemented to reduce the internal order processing time from an average of eight days (refer to figure 6.1) to just two days (refer to figure 6.2). The current system has too many built-in delays associated with the central processing of all order-entry data. There are delays because of transmitting order-entry data to St. Louis, delays in processing the data, and delays in transmitting the results to the field warehouses. It is obvious that there is a better way to manage the information processing requirements of the company. In effect, there is need for distributing the company's information processing to more than one computer system.

The MIS task force has identified the following basic alternatives for distributed processing:

1. **Distribute Remote, Non-Intelligent Interactive Terminals**   The installation of non-intelligent interactive terminals at each of the field warehouses would allow user access to the central computer system in St. Louis via direct lines or public telephone lines. Data could be entered and processed by the central computer as orders are received, and the open order and shipment files could be updated continually throughout the day.
2. **Distribute Remote, Intelligent Interactive Terminals**   Intelligent terminals would be connected to a minicomputer or microprocessor that controls the terminals. This distributed arrangement would allow the utilization of terminals, even though not connected to the central computer at a given moment. The intelligent terminal could display input formats, perform error checking, and store data for later transmission to the central computer at St. Louis.
3. **Distribute Computer Processors and Storage**   A general-purpose minicomputer (or larger computer, if justified) would be installed at each of the five field warehouses. The central computer would be kept to perform the large batch processing jobs. The minicomputers would be used to meet the local information processing needs. Summary and even some detailed information would be passed from the field warehouse to the central computer for processing on an as-needed basis.

Tangible and Intangible Benefits Versus Costs for Each Alternative

Either full-time or measured-time WATS lines could be established for salespeople to call order information directly to the field warehouses. The costs associated with these two alternatives vary depending on the location of the field warehouses. For full-time WATS lines, the ABC Company would pay approximately $1500 per month for the first 240 hours of usage each month. For measured-time WATS lines, the cost is approximately $220 per month for the first ten hours of usage each month and $17 for each additional hour. Using these average costs, the break-even point is 75 hours per month. If the WATS lines to a given field warehouse are used less than 75 hours per month, then measured-time WATS lines are the lower cost alternative. If the usage exceeds 75 hours per month, then full-time WATS lines are the better alternative.

An estimated average of ten minutes per order would be required to communicate accurate and complete order information from the salesperson to the order-entry clerk. The volume of orders at a field warehouse ranges from 500 to 750 per month, with an average of 600 per month. Clearly the volume of orders already exceeds the breakeven point. The total annual cost for outfitting all five field warehouses with full-time WATS lines would be approximately $90,000.

The MIS task force was pleased to learn that only 83 to 125 hours per month of the 240 hours available would be required for order-entry use. The remaining hours could be used to improve communications between sales management and field salespeople. Assuming the salespeople could be trained to use the WATS lines for 80 percent of their orders (within one-half day of taking the order) and process the other 20 percent of orders routinely through the mail, the new order-entry system would require an average time of 0.8 day to receive an order. Even more exciting was the potential to begin processing an order on the same day the order was taken from the customer.

Distribute remote, nonintelligent interactive terminals The MIS task force uncovered several problems associated with this first system alternative. For one, the use of remote, non-intelligent interactive terminals would be a drain on the resources of the central computer, to the point where it would be necessary to acquire a more powerful central computer, capable of handling both interactive and batch processing. The cost of such a move is approximately $180,000 per year. Second, the cost to communicate between the central computer and the non-intelligent terminals would be very expensive. For the current batch processing system, the communication lines between the five field warehouses and St. Louis have to be open about half an hour per day. To allow continuous communication between the non-intelligent terminals at the field warehouses and the central computer, the communication lines must be open eight to ten hours per day. This would represent an increase of $1000 in communications cost per month for each field warehouse, or a total of $60,000 per annum. Third, any time the central computer is down or the communication lines are not operating, order processing could not continue at the field warehouse.

Although many vendors are capable of providing non-intelligent interactive terminals, the MIS task force did not feel that this type of terminal was appropriate for the new system. Typically, terminals cost between $100 to $300 per month to lease, depending upon the vendor, the model, and options selected.

Distribute remote, intelligent interactive terminals The MIS task force is impressed with the capability of intelligent interactive terminals to display input formats and to perform error checking. Use of these terminals by a trained order-entry clerk would greatly increase the accuracy of order processing. Also, the terminals have some storage capability so that order information can be saved and transmitted later. Thus, order processing could continue even though the central computer is down or the communication lines are not working.

For this order-entry system alternative, there are two basic types of intelligent terminals available. The first type of intelligent terminal includes its own minicomputer or microprocessor, plus its own storage capability. It is basically a "stand-alone" piece of equipment, that is, one requiring no additional equipment to perform its function. These terminals lease from $350 to $750 per month, depending upon the vendor, the model, and options selected. The second type of intelligent terminal requires three separate pieces of equipment: the terminal itself, the minicomputer or microprocessor, and additional storage capacity. Generally, four to thirty-two terminals, which lease from $125 to $400 per month, can be attached to a single minicomputer or microprocessor, an additional $300 to $800 per month. Additional storage capacity leases at approximately $50 to $100 per month for each 10K (1K = 1024 bits) increment of storage capacity.

Distribute computer processors and storage Adoption of this system alternative would mean each field warehouse would have its own computer and data processing resources, making it completely independent of the central headquarters, except for transmitting data periodically. Not only would the central computer's workload be reduced greatly, but also each field warehouse would have sufficient data processing resources to meet local information processing needs.

The first applications installed on each computer would be order entry, inventory, and perhaps, accounts receivable. These applications would be developed in a standardized way to meet the needs of the field warehouses. Thus, there would be only a single development cost associated with these key applications.

The MIS task force felt it necessary to hire a computer operator and a data-entry operator/assistant computer operator at each field warehouse. However, the additional data-entry operators at the field warehouses would offset the need for additional operators at the central computer site. They estimate a net increase of five personnel at a total cost (salary plus fringe benefits) of $100,000 per year.

If each field warehouse has its own computer, it is possible to batch process order-entry data several times each day. Any errors or problems could be detected quickly and appropriate corrective action could be undertaken on a timely basis. With the computer located at each field warehouse, there would be no delays due to transmission of data to the central computer, processing or data re-transmission. It would also be possible to connect either non-intelligent or intelligent terminals to the field warehouse terminal.

Comparison of Present Order-Entry System to Each Alternative

Although only three order-entry system alternatives are identified above, the second and third alternatives have two and three variations, respectively, that should be considered by the MIS task force for evaluation. Important information about the present system versus the six basic alternatives is summarized in table 6.3. Each of the six system alternatives assumes the installation of two terminals at each field warehouse—one to handle order-entry input preparation and the other to handle shipment processing input. It should be noted that the cost of full-time WATS lines is not included in table 6.3. The rationale is that present and future systems would benefit from the WATS lines; thus, the costs of $90,000 are dropped from inclusion in this illustration.

The average total order-entry processing cost per order was calculated by dividing the total additional cost for each alternative by 56,875, the estimated annual number of orders five years hence, assuming that 3 percent of the market is recaptured. This quotient was then added to the cost per order for the current system. The MIS task force had set a goal of a maximum of $16.63 per order (refer to figure 6.2) for the new system. If 3 percent of the market is to be recaptured, it would appear that the alternatives associated with distributing computer processors and data storage capability must be discarded. However, these alternatives have the important advantage of providing the field warehouses with the means to meet their managerial and operational information requirements. Because of this added advantage, the MIS task force elected to continue to evaluate these alternatives.

The average internal order processing time was estimated at two days for all alternatives (refer to figure 6.2). It was felt the limiting factors would be the availability of product and the probability of out-of-stock conditions, with the resultant delay in filling the order. Without the probabilistic delay, internal order processing could be completed in less than a day.

Because each of the six alternatives is capable of reducing order-entry processing errors, they have the capacity to lower the percent of orders incorrectly billed. The use of intelligent terminals with their input format display capability seem to have the highest potential for reducing errors. Although the additional costs associated with each alternative are approximate, the MIS task force recognizes that there is need to refine these costs before a final decision can be made.

The MIS task force feels that the staff personnel required to run the five field warehouses would be neither increased nor decreased as a result of any alternative, except if computer processors were installed. It is possible that the central staff would be reduced by changes to the current system, but it is very difficult to justify where and by how many. Thus, no personnel reductions are included as a benefit for any of the system alternatives.

Development costs for all alternatives are judged as approximately the same. The cost estimates per table 6.4 are based upon an amount of $400 per week for system personnel. These figures serve as additional input for preparing the exploratory survey report to top management.

table 6.3 Current order-entry system compared to six feasible alternatives for the ABC Company.

		System alternatives					
		Distribute remote, non-intelligent interactive terminals	Distribute remote, intelligent interactive terminals		Distribute computer processors and storage		
	Current system		Stand-alone devices	Clustered devices	Frequent batch processing at accounting	Use of non-intelligent terminals office	Use of intelligent terminals
		(1)	(2)	(3)	(4)	(5)	(6)
Average order-entry processing cost per order	$11.50	$16.14	$12.66	$12.63	$18.01	$18.43	$19.17
Average internal order processing time (days)	8	2	2	2	2	2	2
Percent of orders incorrectly billed	7%	Lower than 7%	Much lower than 7%	Much lower than 7%	Lower than 7%	Lower than 7%	Much lower than 7%
Additional cost to upgrade central computer per year	$0	$180,000	Not required	Not required	Not required	Not required	Not required
Additional cost to maintain continuous communication link with central computer per year	$0	$60,000	Not required	Not required	Not required	Not required	Not required
Additional personnel cost per year	$0	$0	$0	$0	$100,000	$100,000	$100,000
Additional cost for terminals per year	$0	$24,000	$66,000	$64,500	$0	$24,000	$66,000
Additional cost for distributed computers	$0	$0	$0	$0	$270,000	$270,000	$270,000
Total additional cost per year	$0	$264,000	$66,000	$64,500	$370,000	$394,000	$436,000
Total development cost	$0	$80,000	$80,000	$80,000	$80,000	$80,000	$80,000
Provide the means to meet a wide range of information processing requirements	NO	NO	NO	NO	YES	YES	YES
Capability to perform complete order processing functions if central computer is down	NO	NO	YES	YES	YES	YES	YES

table 6.4 Total development costs for the new order-entry processing system of the ABC Company.

Phase	Elapsed time in weeks	Full-time people required	Total weeks of effort	Cost of system work	Other costs	Total costs	Percent of total costs
Systems analysis	8	2	16	$ 6,400	$ 1,600	$ 8,000	10%
Systems design and equipment selection	18	2	36	14,400	3,600	18,000	23%
System implementation	27	4	108	43,200	10,800	54,000	67%
TOTALS	53		160	$64,000	$16,000	$80,000	100%

152

> **SUMMARY: Systems Analysis—Concluding Investigation of New Order-Entry System**
>
> **Feasible Order-Entry System Alternatives** Refers to three basic order-entry system alternatives that utilize distributed processing: distribute remote, non-intelligent interactive terminals; distribute remote, intelligent interactive terminals; and distribute computer processors and storage.
>
> **Tangible and Intangible Benefits Versus Costs for Each Alternative** Looks at all the important factors for the proposed order-entry system alternatives. Accent is placed on using full-time WATS lines for salespeople and various forms of distributed processing to improve customer service.
>
> **Comparison of Present Order-Entry System to Each Alternative** Centers on comparing the total average processing costs per order for each of the proposed alternatives (from $12.63 to $19.17) to the current system ($11.50).

SYSTEMS ANALYSIS—EXPLORATORY SURVEY REPORT TO TOP MANAGEMENT

This is the final step in systems analysis. This final section of the chapter explores the recommendations for a new order-entry system and their presentation. The completion of these activities concludes the systems analysis phase of the feasibility study.

Recommendations for New Order-Entry System

The MIS task force prepared an exploratory survey report summarizing their findings to the executive steering committee. The report contained these five key recommendations for management approval:

1. Begin immediately to establish a full-time WATS system to improve communications between sales management and salespeople. The annual expenditure of $90,000 for the WATS system is justified by the reduction from two days to less than one day in average time required to receive an order at the warehouse. This, in turn, will reduce the total average order cycle time from twelve days. This faster response time is required if the ABC Company is to move aggressively toward recapturing that 3 percent of the market lost principally by poor customer service.
2. Authorize the MIS task force to proceed with the design phase of a new order-entry system, which is technically and economically feasible, as determined during the exploratory survey. The design phase will cost an estimated $18,000 and require the full-time effort of two people for an elapsed time of eighteen weeks.
3. Develop the new order-entry system utilizing **distributed processing.** The built-in delays associated with the current centralized processing of order-entry data make it impossible to reduce the average order cycle time and improve the accuracy of input without moving to some type of distributed processing.

4. Use remote intelligent terminals, that is, system alternative 3 (clustered devices) per table 6.3, as the basis of the new order-entry system for the following reasons:
   a. Remote intelligent terminals provide very high potential for reduction of the order cycle time and reduction of input errors.
   b. Use of remote intelligent terminals is by far the lowest cost alternative for the ABC Company, as shown in table 6.3.
   c. No additional staff must be hired to install and use the remote intelligent terminals.
   d. Remote intelligent terminals provide the capability to continue order-entry processing even if the central processor is down.
   e. The information processing requirements of the field warehouses are insufficient to warrant the installation of computer processors and storage capability at each warehouse. However, remote intelligent terminals can be used to improve data input and information display capability. The bulk of data processing will be performed by the central computer in St. Louis.
5. Continue to examine the feasibility of new finished product inventory and accounts receivable systems concurrently with the design of the order-entry system. Ideally, these three systems would be designed in a modular fashion so that basic information processed by one system will be available to the other systems. This will require a closely coordinated effort with careful attention to the interaction among the three systems.

Presentation of Recommendations for New Order-Entry System

Based upon an oral presentation of the foregoing recommendations by the MIS task force, the executive steering committee concurred with all of the recommended items. Their acceptance and strong support of the recommendations was due in large measure to their high degree of participation in the exploratory study.

The MIS task force next met with the executive vice-president to present their recommendations for system alternative 3 (as set forth in table 6.3). Having been kept informed of the project's status by the executive steering committee, he was well prepared to discuss and review their recommendations. He was particularly pleased to see that the MIS task force had followed through on his suggestion to examine the accounts receivable function and how it might interface with the new order-entry system. Within two weeks, the MIS task force had his approval on all recommendations to proceed with the systems design phase.

SUMMARY: Systems Analysis—Exploratory Survey Report to Top Management

Recommendations for New Order-Entry System Focuses on the development of one selected system alternative, which includes the use of a WATS system, the distributed processing concept, and remote, intelligent interactive terminals (clustered devices). It is system alternative 3 in table 6.3.

Presentation of Recommendations for New Order-Entry System Relates to the presentation of the foregoing recommended system alternative to the executive steering committee and, in turn, to the executive vice-president. Approval by both concludes the systems analysis phase of the feasibility study.

CHAPTER SUMMARY:

Within this chapter, an introductory investigation of the order-entry system of the ABC Company was undertaken, followed by its detailed investigation. The next phase in systems analysis was presented, namely, a concluding investigation which focused on the analysis of costs and benefits for six proposed system alternatives. Based upon an extensive analysis of these system alternatives, alternative 3 was selected; thereby forming the basis for the exploratory survey report to the executive steering committee and the company's executive vice-president.

Fundamentally, the recommended order-entry system utilizes a WATS system, is a distributed processing system, and makes use of remote, intelligent interactive (clustered) terminals. It is capable of an average order processing cycle time of five days versus twelve days presently. Its average percent of items filled is 96, and its percent of orders incorrectly billed is one—versus 84 and 7, respectively, for the present order-entry system. To accomplish the foregoing improvements, the average order-entry processing costs will rise from $11.50 to $12.63. However, as indicated in the chapter, these higher processing costs will be offset by an increased market share. Finally, this system allows the possibility of adding new system modules—namely, finished product inventory and accounts receivable, the subject matter for the case studies to follow.

Questions

1. a. What are the important objectives of future system development for the ABC Company?
   b. Why were these objectives selected?
2. a. Who are the members of the executive steering committee and the MIS task force for the order-entry system project?
   b. Why were these particular company members selected?
3. a. What are the present causes of poor customer service?
   b. What can be done in the new system to improve customer service?
4. What important facts were uncovered during the introductory investigation phase?
5. State briefly the essential aspects of the present order-entry system.
6. a. What is the purpose of the TC field on the 203 card?
   b. Suggest another way of handling type change.

156 Systems Analysis of the Batch Order-Entry System—ABC Company

7. a. Describe the type of data files found in the present order-entry system.
   b. Describe the type of output from the present order-entry system.
8. What bearing does the present financial condition of the ABC Company have on the new system project, i.e., the order-entry system?
9. Discuss the advantages and disadvantages of the three basic alternatives for distributed processing in an order-entry system.
10. Why was the system alternative 3—distribute remote, intelligent interactive terminals—selected for inclusion in the exploratory survey report to top management?

Self-Study Exercise

True-False:

1. _____ The executive steering committee reports to the MIS task force on the new order-entry system.
2. _____ There is expected to be very little change in the customer service level with the new order-entry system.
3. _____ The number of orders incorrectly billed is expected to be reduced substantially with the new order-entry system.
4. _____ The time allotted to systems design is two system analysts for eight weeks.
5. _____ The present open order file contains 238 characters of information.
6. _____ The present internal order-entry processing procedures average eight days.
7. _____ The average cost of processing an order with the present system is $11.50.
8. _____ The only feasible order-entry system alternatives include distributing remote, non-intelligent terminals or intelligent interactive terminals.
9. _____ An integral part of the final order-entry system alternative is the use of a full-time WATS system.
10. _____ There is no need for the executive vice-president to approve the new order-entry system.

Fill-In:

1. The current customer service level is _____ _____ percent of items filled.
2. The average order processing cycle time is _____ days with the present system.
3. The _____ card is used to cancel a customer's order.
4. The _____ card is used to move the open order record to the shipment file.
5. The customer order becomes the basis for preparing the _____ _____ form.
6. The open order and the _____ files constitute data storage for the present order-entry system.
7. Problems abound with incorrect _____ _____ numbers in the present order-entry system.

8. A possible order-entry system alternative is to distribute _____ _____ and storage.
9. The recommended order-entry system alternative has average processing costs of _____ per order.
10. Part of the selected system alternative is a full-time _____ system.

case study

chapter 6

finished product inventory

CURRENT SYSTEM OVERVIEW

An integral part of the present order-entry system is the finished product inventory system. As noted in the chapter (figure 6.9), the loader in the field warehouse notes on the yellow copy of the order and shipment processing form exactly which products and quantities have been shipped. The annotated yellow copy is sent first to the shipping clerk and then to the accounting office so that their records may be corrected if necessary. A corrected red copy of the shipment processing form is sent to keypunching to prepare input for the order-entry system, while the yellow copy remains in the accounting office.

In addition to the order-entry open order and shipment files being updated (as noted in the chapter), the order-entry system creates a record for each shipped product which, in turn, is forwarded to the finished product inventory system for updating the computerized warehouse inventory records. The format of these updated records is shown in figure a; the contents of the finished product inventory (magnetic tape) file are set forth in figure b.

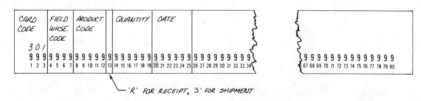

case study 1 Finished product inventory system receipt and shipment input
figure a for the ABC Company.

Receipts of finished goods from manufacturing plants to restock the warehouse inventory are handled in the following manner. A two-part shipping paper is sent with the finished product. The product is received by the loader/unloader at the warehouse, and the items plus quantities are received and checked closely against the shipping papers. Any differences are carefully noted on both copies of the shipping papers. Both copies of the annotated shipping papers are given to the accounting clerk at the field warehouse.

	Beginning position	Length (number of characters)
Record identification	1	3
Product code	4	5
Field warehouse code	9	4
Resupply plant code	13	4
Current inventory	17	6
Reorder point	23	6
Reorder quantity	29	6
Average lead time in days	35	4
Average volume of monthly shipments	39	6
Shipping weight (pounds per case)	45	5

case study 1 Contents of finished product inventory file for the
figure b ABC Company

This clerk files one copy, and sends the second copy to keypunch to prepare a 301 card, as shown in figure a. The card is keypunched and all 301 cards for each day are read into the card reader at the field warehouse for transmission each night to the central computer in St. Louis. These data are then used to update the finished product inventory system.

The finished product inventory system at St. Louis prints a daily report showing the status of each product stored in the warehouse, which is transmitted to the field warehouses each morning. Any product with an inventory level below the reorder point is highlighted on the report. The field warehouse accounting office prepares a stock replenishment order for each highlighted product and mails it to the plant responsible for resupplying the warehouse. Accounting office personnel, exercising their own judgment, usually order more than the recommended reorder quantity to reduce the possibility of an out-of-stock condition on a high-profit item.

CURRENT SYSTEM LIMITATIONS

Although finished product inventory is too high in relationship to sales, customer service is much slower than customer expectations and industry standards. It seems as if there is a continual mismatch between what is in inventory and what is ordered by the customer. The vice-president of physical distribution feels one reason for poor inventory management is that the field warehouse managers are too narrow-minded. They are primarily concerned with having sufficient inventory to meet customer demands for high-profit items. Customers, however, need a combination of products to meet their needs. Thus, customer orders are generally for several products at a time, including low-profit items. Frequently, these low-profit products are not carried in sufficient quantities. An out-of-stock condition occurs, resulting in a several day delay in filling orders.

A second reason for high inventory levels and poor customer service is the simplistic inventory control policy. The policy is based on reordering only when the inventory level falls below the reorder point. The lead time required to resupply each field warehouse is estimated and the average usage during lead time is set as the reorder point. The average volume of products shipped during the lead time plus the difference between the current inventory position and the reorder point is set as the reorder quantity. The past six months of shipment data are used to calculate the reorder point. The reorder point for each product is revised upward by 5 percent every six months in an attempt to keep pace with the general annual increase of 6 to 10 percent in actual shipments.

Since the time that this inventory policy had been implemented, manufacturing made a determined effort to reduce costs by making longer production runs. Major products are manufactured on a two-week cycle, and minor products on a four-week cycle. The reorder points are recalculated using the same historical shipment data (with volume adjusted by 5 percent for each six months) and new average lead times. Although it is recognized that longer production cycles should be a part of the inventory control policy, this approach would introduce an additional level of complexity into the inventory control policy.

OBJECTIVES TO BE MET BY NEW FINISHED PRODUCT INVENTORY SYSTEM

The primary objective of the new finished product inventory system is to provide the capability of raising the percent of orders filled without an out-of-stock condition from 84 to 96 percent. A secondary objective is to create the proper inventory balance between low-profit items and high-profit ones, such that an inventory reduction of $1 to $2 million can be realized. This reduction is felt to be achievable by establishing a proper inventory balance and by lessening the time required to receive and process stock replenishment orders at the field warehouses.

The vice-presidents of marketing and physical distribution had many discussions on how these objectives could be met. They finally agreed that the following steps were appropriate:

1. Develop and implement a more complete inventory control policy accounting for the longer production runs used in manufacturing. This policy would also recognize the need to keep a proper balance of low-profit and high-profit items at each field warehouse.
2. Consolidate total responsibility for all finished goods inventory under a single central group reporting to the vice-president of physical distribution. This group could oversee the ABC Company's inventory and physical distribution systems much more effectively than individual field warehouse managers operating on their own.
3. Review periodically current shipment data to revise the reorder point and reorder quantity for each product, to provide a systematic formula for adjusting the reorder point.
4. Establish an effective means of communicating the current inventory status of each item at each field warehouse to the new central group responsible for inventory control.
5. Develop computer system and/or communication system capabilities consistent with and supporting these objectives.

Assignment: Finished Product Inventory System

1. Using the outline suggested in figure 6.3, prepare a project description for the feasibility study of a new finished product inventory system.
2. Referring to the narrative description of the current finished product inventory system, prepare a document flowchart, similar to figure 6.9, for the processing of receipts of finished products from manufacturing plants to restock warehouse inventory.

case study chapter 6

accounts receivable

CURRENT SYSTEM OVERVIEW

An extension of the present order-entry system is the accounts receivable system. As stated in the chapter (figure 6.9), the loader in the field warehouse notes on the yellow copy of the shipment processing form exactly which products and quantities have been shipped. The annotated yellow copy is sent first to the shipping clerk and then to the accounting office so that their records may be corrected if necessary. A corrected red copy of the shipment processing form is sent to keypunch to prepare input for the order-entry and accounts receivable system. Separate cards are prepared for input to these two systems. The yellow copy of the shipment processing form remains in the accounting office. The input is batched, read into a card reader, and transmitted once each afternoon to the central computer in St. Louis for processing.

The accounts receivable system prepares a multicopy customer invoice. The original is mailed to the customer; one copy is kept in the accounts receivable department at St. Louis; and a second copy is transmitted from St. Louis back to the printer at the field warehouse that originated the order. The invoice normally arrives at the customer's place of business two or more days after the receipt of goods.

The accounting office at the originating field warehouse receives the copy of the customer invoice and pulls the yellow copy of the shipment processing form to check for any discrepancies. On about 7 percent of the orders, there are billing errors that must be corrected. In this case, the accounting office at the field warehouse notifies the accounts receivable department at St. Louis by mail of the necessary correction. The accounts receivable department then issues a debit or credit memo to the customer.

The ABC Company uses an open-item receivables accounting method. When remittance is received for specific billed items, it must be applied to specific open invoices so that the discount earned (or forfeited) can be determined. However, when remittance is received on account, the foregoing procedure is not followed.

Two reports are produced by the system to help monitor the level of accounts receivable and measure the efficiency of credit and collection. The first report—the number of days' sales outstanding in receivables—is calculated by dividing the most recent three months' average dollar sales volume per day into the dollar amount of receivables outstanding at

ABC COMPANY
Accounts Receivable Aging Schedule

Month Ending 2/28/-

Customer name	Customer number	Total balance	0–30 days	31–60 days	61–90 days	Greater than 90 days
A & B Stores	2055	$2,827.25	$2,827.25			
Ayres Stores	2150	5,109.66	2,742.00	$296.91		$2,070.75
All Stores	2257	5,261.70			$5,261.70	
Arnold's Stores	2350	1,129.00	1,129.00			
Big Stores	2435	1,550.90	1,550.90			
Campus Stores	2654	1,781.70	1,781.70			
Consolidated Stores	2750	1,915.15	812.65		1,102.50	
Downtown Stores	2810	1,399.00	1,399.00			
General Stores	3258	2,049.50	2,049.50			

case study 2—figure a Accounts receivable aging schedule for the ABC Company.

the time. The second report—the accounts receivable aging schedule—shows by customer the dollar amount and percent of total receivables in four categories: receivables 0–30 days old, receivables 31–60 days old, receivables 61–90 days old, and receivables greater than 90 days old. A sample aging schedule is shown in figure a.

CURRENT SYSTEM LIMITATIONS

The primary concern with the current accounts receivable system is the 7 percent error rate in billing. One of the principal reasons for the high error rate is that separate input is prepared from shipment processing information for the order-entry system and the accounts receivable system. Not only is this a duplication of effort, but also it creates situations where the order-entry system and accounts receivable system are in disagreement as to the status of an order (shipped or unshipped) and the actual products and quantities on a shipped order.

The days' sales outstanding index is not a reliable indicator of the efficiency of credit and collection for the ABC Company. Even though the company has a fairly stable collection experience with a fixed rate of customer payments, sales vary according to the following: 15 percent of sales during January, February, and March; 25 percent during April, May, and June; 20 percent during July, August, and September; and 40 percent during October, November, and December. During a period when sales are changing rapidly, the average sales per day is changing, although less rapidly because it is calculated as a three-month average. Furthermore, the sales during the most recent month generally contributes the largest percentage of receivables. Thus, in a poor sales month following two months of high sales, the receivables will be lower and the average sales per day higher than the current sales rate. The result is that the days' sales outstanding index will be low—completely independent of small variations in customer payment patterns. Similarly, during a good sales month following two months of low sales, the days' sales outstanding index will be high. The days' sales outstanding for the ABC Company varies between forty and sixty days, with the major portion of the variation due to changes in sales volume.

The credit manager is well aware of the impact of the variation of sales on the calculation of days' sales outstanding. He also reasoned that the variation in sales has a dramatic impact on the aged receivables report. The credit manager feels that the current accounts receivable system produces unreliable information to monitor credit and collection.

OBJECTIVES TO BE MET BY NEW ACCOUNTS RECEIVABLE SYSTEM

The primary objective of the new accounts receivable system is to reduce the billing error rate from 7 percent to 1 percent. An associated objective is to eliminate unnecessary duplication of effort associated with order entry and accounts receivable processing. A further objective is to provide better information for monitoring the credit and collection operations.

Assignment: Accounts Receivable System

1. Using the outline suggested in figure 6.3, prepare a project description for the feasibility study of a new accounts receivable system.
2. Referring to the narrative description of the current accounts receivable system, prepare a document flowchart, similar to figure 6.9, for the preparation of customer invoices.

part three

systems design of new management information system

chapter 7

overview of systems design

OBJECTIVES:

- To show why creativity and brainstorming are necessary for imaginative systems design.
- To set forth the essential characteristics of any well-designed information system.
- To demonstrate why the human factor must be considered as an essential element in systems design.
- To present the steps in systems design from an overview standpoint.

IN THIS CHAPTER:

APPROACHES TO SYSTEMS DESIGN
   Preferred Systems Design Approach
IMAGINATIVE SYSTEMS DESIGN
   Creativity
   Brainstorming
CHARACTERISTICS OF A WELL-DESIGNED SYSTEM
   Acceptability
   Decision-Making Ability
   Economy
   Flexibility
   Reliability
   Simplicity
   Modular Design Method
STEPS IN SYSTEMS DESIGN
   Consider the Human Factor in New System
   Review Appropriate Data Compiled to Date
   Determine Requirements for New System
   Design the New System
   Document the New System
CHAPTER SUMMARY
QUESTIONS
SELF-STUDY EXERCISE

Once a decision has been made to implement a new system, the system specifications must be developed. Imagination and creativity are a must for this phase; otherwise, some basic weaknesses of the existing system will be duplicated unconsciously by the MIS task force. The fundamentals of systems design, the second step of a MIS feasibility study, are covered in this chapter. Great emphasis is placed on the modular design approach in this chapter and succeeding ones.

Systems design is the creative act of inventing and developing new inputs, a data base, off-line files, methods, procedures, and outputs for processing business data to meet an organization's objectives. Systems design builds on the information gathered during systems analysis. Whereas systems analysis is an intensive study of the present facts, systems design is the creation of the specifications for a new system. The systems analyst is very much concerned about informational output required. In essence, the job of the systems analyst is an arduous task that requires special design skills and talents.

APPROACHES TO SYSTEMS DESIGN

There are three basic approaches to systems design, and some controversy over which is the proper approach. One extreme is that the work involved in understanding the present system is a waste of time and tends to make the systems analyst think too much in terms of the old system; this school of thought desires to develop an ideal system without regard to the present system. The other extreme is to study the details of the present system and, then, design an improved system. This approach relies too heavily on the present system, resulting in only marginal improvements; actually, very little creative work is performed since the system is practically a duplication of the existing one, except for more advanced data processing equipment.

Preferred Systems Design Approach

Each of the viewpoints above has inherent weaknesses and should be avoided. The first fails to take into account the exceptions and peculiar problems of specific areas, which may not be apparent unless some time is spent examining the present system. These realities can make the best designed system an undesirable one for any organization. The other approach is too conservative, lacking the creativity needed for an effective new systems design. Analysts capable of systems analysis and design are short-sighted if they allow the present system to cloud their thinking to the extent of inhibiting their creative talents.

In view of these difficulties, the best approach takes the desirable aspects of both. Fundamentally, it involves a thorough comprehension of the existing system with all its exceptions and problems. However, by refusing to allow these to be the overriding factors, system analysts retain sufficient latitude to design a new system imaginatively and creatively.

This combined approach is also preferred because details are of greater importance in designing an interactive system than a batch-type system. First, an interactive system combines various types of equipment which must be closely coordinated and timed with other equipment. Secondly, minor errors and omissions can make the system unworkable in a closely knit operational environment. Lastly, programs which are run separately must work together to handle an organization's data processing needs. Thus, there is a great need for meshing many functional areas with complex, detailed operations. Only by a close analysis of the exceptions and problems of the present system can this meshing be properly accomplished in an interactive system.

IMAGINATIVE SYSTEMS DESIGN

Even though system objectives and requirements have been established in the exploratory survey report, there are innumerable approaches available to the systems analyst within the selected framework, which makes the task a challenging one. The systems design should be developed from a broad knowledge of approaches that have been successful in other installations. For example, the decision to install an interactive management information system that cuts across the entire organization offers many possible design alternatives to the system analysts of the MIS study group.

The participants of the exploratory survey report to management are generally the ones to undertake the design of the recommended MIS alternative. If systems analysts are not part of the group, they should be recruited. Likewise, if additional personnel are needed to represent the various departments affected by the system project, they should be enlisted and made an integral part of the MIS task group because participation and cooperation of all functional areas, represented by departmental personnel, is the key to implementing a successful system. It is much easier to accommodate their suggestions in the initial design than to redesign at a later date. Too many installations have faced embarrassing situations only because appropriate departmental personnel were not given an opportunity to evaluate the systems design as it progressed.

Creativity

Creative design thinking definitely leads to new and untried approaches. The unusual approach and the exploration of nonstandard methods and procedures often bring systems analysts to the point where innovations can be devised for a proposed system. The incubation period of new design concepts will finally result in the illumination of creative systems design. Too often, time for creative thinking is unduly restricted because of the ever pressing demand for immediate results. This is typically the fault of top managers who do not understand the importance of this creative phase in relation to the overall quality of the resultant systems design.

Good systems design requires an analytical mind that can reduce a complex situation to its essential elements. The systems analysts must think logically and be highly imaginative in their design approaches, for they must be able not only to visualize the possibilities of system alternatives, but also to communicate these ideas clearly. Their abilities must go beyond the systems analysis phase in order to conceptualize new systems design. They must take the initiative when required and answer valid objections to their work. In essence, they must operate effectively in two worlds—the **conceptual** one of imaginative systems design and the **real** one of effective systems.

Those in the study group should possess the attributes above while selected personnel from each department should be a sounding board for systems designers. Departmental representatives, who have the answers to the proverbial questions of who, what, where, when, how, and why, should be able to recommend improvements over the present system. Likewise, they should be objective in their thinking and should accept worthwhile changes set forth by the MIS task force. Facts that may be needed subsequently can be gathered under their direction. The departmental representatives and systems analysts can complement one another to produce a more effective system than analysts working alone.

Brainstorming

The purpose of brainstorming is to produce as many solutions as possible to the problem at hand. Any idea is recorded, no matter how ridiculous it may appear, and no criticism allowed. One idea will often lead to others. The session continues until the group has exhausted its ideas.

Once the session has adjourned, those whose problem areas were discussed will evaluate the ideas presented. Most of the ideas will be rejected on the bases of common sense and logical judgment. In fact, it may be that none can be considered as potential solutions. However, the creative faculty of systems analysts may modify one or more ideas into workable solutions. A fresh look by an uninhibited and unrestricted free association can help solve problems where the proposed solutions are different from the existing ones.

CHARACTERISTICS OF A WELL-DESIGNED SYSTEM

Before examining the modular design method plus the essential elements and steps involved in systems design, it is helpful to review the characteristics of any well-designed system, such as an interactive management information system. Basically, characteristics associated with effective system operations are:

- acceptability
- decision-making ability

economy
flexibility
reliability
simplicity

Without these essentials, it is questionable that the right **information** will go to the right **person** at the right **time** in the right **format** and at the right (lowest) **cost.** In like manner, it is doubtful that the system will be accepted by those organization personnel who are to operate it. In designing an efficient and effective system, it is of utmost importance to consider the human factor and the equipment that these people will be required to use. Systems analysts must evaluate the capabilities and limitations of the personnel and corresponding factors of the equipment itself.

Acceptability

Success of a new system pivots on its acceptance or nonacceptance by organization personnel. If operating personnel are convinced that the new system will not benefit them, it is a poor one, it does not follow established company policies, or some other legitimate reason, the system is in serious trouble. To overcome this resistance, participation by operating personnel during all phases of the change-over is necessary because they constitute the organization which must use and live with the newly designed system.

Decision-Making Ability

An effective system produces not only information, but information at the lowest possible cost, pertinent and timely for making decisions. The system should be designed so that decision making is as automatic as possible. Those decisions which can be handled best by the system should be relegated to it; all other decision areas should be relegated to the appropriate level of management or nonmanagement personnel. Thus, in a decision-making environment, an organization's objectives will be easier to attain, especially when the "management by exception" principle is employed.

Economy

For economic operations, data should be captured as near to its source as possible and allowed to flow through the system automatically from that point. Activities which must be performed in sequence should be located as closely as possible, both organizationally and physically. Eliminating duplicate information files, reducing the provision for every possible contingency, and eliminating small empires in an organization's functional areas are other examples of improving operations in an economic manner.

The question of potential savings versus costs must be considered. Generally, no information or service should be produced that is not justified by its cost; all other system characteristics are subservient to this DP principle. Often, it is expensive to develop the functional area of the overall system with a greater capacity than its integrated and related parts, or, stated another way, there is need for a proper balancing of the subsystems and their related parts to effect economy of operation. For example, it makes no sense to develop elaborate order processing methods and procedures for fast customer service in the office if the plant operations are not comparably geared to providing prompt service.

In deciding whether to centralize or decentralize an organization's operations, the economy of centralized operations must be compared with the reduction of communication and paperwork costs under decentralized conditions. Also, the improvement in response time, increased flexibility, and greater unity of control and responsibility under decentralized operation must be considered.

Flexibility

To be effective, the system must be flexible—capable of adapting to changing environmental conditions. There will always be variations in products, manufacturing processes, or accounting procedures, necessitating the system's expanding or providing additional output. Under changed conditions, managers must be prepared to adjust their operations accordingly. Without the ability to absorb changes, an organization may lose customer goodwill as well as encounter problems with its own personnel. Thus, a well-designed system must be flexible if it is to continue meeting its desired objectives and organization goals. The modular approach to systems design can be employed to bring about a flexible system.

Reliability

Reliability of the system refers to consistency of operations. Are data input, processing methods and procedures, and informational output consistent over an extended period of operation? The degree of reliability can range from a constant and predictable mode of operation to a complete breakdown of the system, although most systems do not continuously operate at these extremes.

A high degree of reliability can be designed into the system through sound **internal control** by including numerous control points where variances from established norms and practices can be detected and corrected before processing continues. Control functions should be allocated to organizational units independent of the functions to be controlled. In all cases, controls should be incorporated into the system during the design phase.

Simplicity

The trademark of any effective system is simplicity. Simplicity can be affected by providing a straight-line flow from one step to the next, avoiding needless backtracking. Input data should be recorded at their source, or as close to the source as possible, to reduce or eliminate the need for recopying. Functions should be assigned to organizational units in such a way that coordination, communication, and paper work are reduced. Each organization group should have the authority and responsibility for its area and be held accountable for its performance to one superior only.

In general, a simple system can be better understood and more easily used than a complex one, and there will be fewer system problems since there are fewer things to go wrong. Personnel will have no need to question complicated procedures that are difficult to follow, thereby bolstering morale.

> **SUMMARY: Characteristics of a Well-Designed System**
>
> **Acceptability** Relates to acceptance or approval of the system by those personnel who will use it. To improve acceptability, organization personnel should participate in the development of the new system.
>
> **Decision-Making Ability** Refers to the capability of the system to produce timely and pertinent information, facilitating the decision-making process of organization personnel, at the lowest possible cost.
>
> **Economy** Centers on capturing control information in the system as near to the source as possible.
>
> **Flexibility** Focuses on the capability of the system to adapt to changing environmental conditions. Likewise, organization personnel must be prepared to adjust their operations to changing conditions.
>
> **Reliability** Refers to the consistency of the system to control operations. A high degree of reliability can be designed into the system by incorporating good internal control.
>
> **Simplicity** Stresses the capability of the system to provide a straight-line flow from one step to the next, thereby avoiding needless backtracking and duplication.

Modular Design Method

The design of a system and its related subsystems should be approached from the outset with **modularity**, or the **building block** concept in mind. This involves identifying all of the major subsystems which become the major modules of functional areas to be designed. At this point, all of the major modules are identified in functional block diagrams. Since they are the highest level functions, they represent a beginning point for a functional breakdown of a system into related parts, applying the process iteratively from the top down. The resulting analysis is represented by a tree diagram, wherein major functions at the top are successively broken into separate data processing functions in the lower branches of the tree.

As illustrated in figure 7.1, the major module—accounting—is broken down into seven intermediate modules—customer billing, accounts receivable, accounts payable, payroll, cost control, inventory control, and budgets. The payroll module is, in turn, divided into six minor modules—gross pay, mandatory deductions, voluntary deductions, net pay, labor cost distribution, and fringe benefits. These minor modules are further subdivided into basic modules. (Although not shown in figure 7.1, it is also necessary to break down the other intermediate modules.)

Through the subdividing process for a major module, two important phenomena are observed:
>   the branches begin to terminate—that is, they do not lend themselves to further breakdown.
>   some of the functions turn up in more than one place—these are duplicate modules.

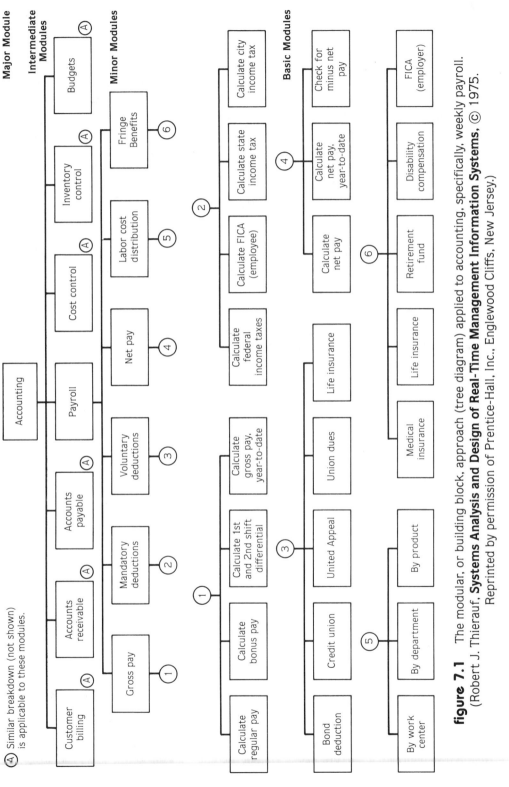

figure 7.1 The modular, or building block, approach (tree diagram) applied to accounting, specifically, weekly payroll. (Robert J. Thierauf, *Systems Analysis and Design of Real-Time Management Information Systems*, © 1975. Reprinted by permission of Prentice-Hall, Inc., Englewood Cliffs, New Jersey.)

When the process is complete, a thorough functional analysis is available even though the system has not been fully designed. It should be noted that an analyst often finds alternative ways to break down a function into its component parts, and may spend some time changing and filling in details considered unimportant on an earlier pass. This time is well spent since this step forms the heart of a system.

When the functional analysis is complete, the analyst creates a system structure for the functional modules which will operate within whatever hardware constraints are imposed. This approach allows the combining of individual modules, resulting in a reduction of the complexity of an overall system. The modular approach also facilitates the system modification or updating that may be necessary during succeeding stages of systems design and in the future.

An alternative to the tree diagram (figure 7.1) is the tabular approach found in table 7.1. The advantage of the tabular form over the tree diagram is that duplicate modules can be readily identified. Referring to table 7.1, the intermediate modules of both payroll and cost control have the same minor module, namely, labor cost distribution; as a result, their basic modules are the same. Therefore, the module needs to be programmed only once. If the remaining modules were completed in table 7.1, as well as in other modules for an organization's subsystems, the labor cost distribution module might appear several times, thereby indicating other duplicate modules. By going down to the lowest level of the modular pyramid, the analyst is forced to organize logically the data base elements and the source of raw data which must be analyzed and grouped for useful information.

No matter what modular approach is employed in the detailed design of the subsystems, it is necessary to prepare final system flowcharts for the recommended system without specifying the equipment to be used. Accuracy, simplicity, and ease of understanding are the essential components since nontechnical personnel will be reviewing and evaluating them. Hence, either the tree or tabular modular approach is a logical method for designing any type of new system that is ultimately captured in system flowcharts.

In summary, a modular design approach is a process of breaking down the system into its lowest component parts and examining them for duplication. This functional approach serves as a basis for designing subsystems from the highest to the lowest levels. Also, the modular subsystem approach is extremely helpful in integrating modules on the same level and on a related basis.

STEPS IN SYSTEMS DESIGN

Systems design is the most creative step in the entire system project. Consideration is initially given to the human factor that affects some aspect of the design process. Data on the present system and in the exploratory survey report—in particular—the new system recommendation—are reviewed. The requirements of a new system include: new policies consistent with organization objectives, output needs, methods and procedures, a data base structure, off-line files, planned inputs, a common data representation, internal control, system controls, and system performance. Based upon these requirements, important design principles are incorporated. Thus, a thorough analysis of the foregoing areas

table 7.1 Modular or building block concept (in tabular form) applied to accounting, specifically, payroll and cost control. (Robert J. Thierauf, **Systems Analysis and Design of Real-Time Management Information Systems,** © 1975. Reprinted by permission of Prentice-Hall, Inc., Englewood Cliffs, New Jersey.)

Intermediate modules	ACCOUNTING—MAJOR MODULE	
	Minor modules	Basic modules
Customer billing	(A)	
Accounts receivable	(A)	
Accounts payable	(A)	
Payroll	Gross pay Mandatory deductions Voluntary deductions Net pay Labor cost distribution Fringe benefits	Refer to figure 7.1 By work center ◁1▷ By department ◁2▷ By product ◁3▷
Cost control	Materials cost distribution Labor cost distribution Overhead cost distribution Cost exceptions	By work center By department By product By work center ◁1▷ By department ◁2▷ By product ◁3▷ By work center By department By product By work center By department By product
Inventory control	(A)	
Budgets	(A)	

(A) A similar breakdown is applicable to these modules.

gives the MIS task force a basis for designing and documenting a new system, such as an interactive management information system.

From an overview standpoint, the foregoing steps in the design process can be segregated as follows:
1. Consider the human factor in new system.
2. Review appropriate data compiled to date.
3. Determinine requirements for new system.
4. Design the new system.
5. Document the new system.

Starting point for systems design   A logical starting point for systems design may be one of the following: an activity that dominates the new system, an area that is the most costly for the new system, the most inefficient area of the present system, or an area that will reap many intangible benefits, such as better customer service and improved managerial reporting. Based upon the area or activity selected, each important element must be analyzed and alternative ways of data handling must be developed. As indicated previously, this involves creativity, and, if necessary, brainstorming. After a considerable amount of mental activity and sketching feasible systems within the framework of the selected system alternative, the newly designed system will be documented with flowcharts and/or decision tables.

Design system in general terms   System personnel should design a new system in general terms so that it is capable of operating on equipment from most manufacturers. To design a system with only one manufacturer in mind reduces the potential of the system and, many times, limits the ultimate success of the system project. Equipment is important at this stage, but only in general terms.

Consider the Human Factor in New System

The human factor is an extremely important consideration underlying the essential elements of systems design in coding and data representation. For example, research indicates that whereas a machine can read documents numbered "A4B" as easily as those coded "AB4," a human being cannot. When trade-offs are necessary, the human element should be given preference over the equipment.

Error rates with various coding schemes increase as the number of characters in the data code increases. It is suggested that longer codes be divided into smaller units of three (or four) characters, such as 123–456 instead of 123456, to increase their reliability. Characters used in the coding scheme should be in common use, and special symbols avoided. Also, functions wherein a number of data-entry errors has occurred in the past should be studied to see if there is a systematic pattern. Factors that contribute to the occurrence of systematic errors should be considered in designing data codes.

In view of the problems associated with coding and data schemes, the analyst should always examine their impact on organization personnel. System experts should have some knowledge about themselves and others since people view the same thing in different ways: what may be perfectly logical and rational for systems analysts may be just the reverse for operations personnel, who must utilize new methods and procedures. When designers specify the human resources to be employed, they should place themselves in the user's shoes when constructing work norms, procedures, and measurements for individual or group effort. The more the designers apply this "other person" approach, the better the final systems design will be.

A system, then, should be designed to interface with anyone who uses its results. This requirement applies equally to the needs of employees and the organization's customers. If employees can understand what the system produces, what it requires of them, and, at the same time, that it is helpful to them, the end result can only be improved operating efficiency for the organization. If customers can easily understand their bills, then improved customer relations will result.

Review Appropriate Data Compiled to Date

A necessary requirement for effective systems design is a review of the present system and the proposed system. The best way to acquire knowledge about the present system is to review the information accumulated during the systems analysis phase and the data contained in the exploratory survey report to top management. Both areas are investigated below.

Review data on present system Generally, many weeks have elapsed since the data were compiled on the present system. The purpose of the review is to recall pertinent facts about the system—in particular, its problems, shortcomings, and exceptions. A detailed listing of what should be evaluated in the review and avoided in the new system is contained in figure 7.2. Notice the preponderance of the word **eliminate.** Too often, the weaknesses of the present system are present in the new one since this phase of systems design was performed superficially or discarded entirely.

An evaluation of tangible and intangible benefits indicates too few for the present system, but will be increased within the new system.
The present uneven handling of exceptions will be eliminated.
Unnecessary inputs, reports, records, files, and forms of the present system will be eliminated with the new system.
Present duplication of operations, functions, and efforts will be avoided in order to standardize comparable data in the new system.
Excessive internal control or the lack of control procedures in the existing system will be remedied.
Unnecessary refinement in the quality of data and superfluous data on reports will be eliminated.
The flow of work will be reviewed by the systems analyst. Present peaks and valleys of data flow can be overcome by eliminating system bottlenecks, establishing new cut-offs, and rescheduling certain operations with the new system.
Excessive steps in the present methods and procedures will be reduced in the new one.
Waiting time between data processing steps will be reduced or eliminated.
Present system's inability to handle an organization's growth on a one-shift basis will be eliminated.
Overall, the systems analyst will evaluate how well or how badly the organization has met its objectives by considering the deficiencies and weaknesses of the present system.

---

figure 7.2 Systems design—review present system weaknesses in order to eliminate them in the new system.

An examination of figure 7.2 indicates certain questions in the analyst's mind for determining the deficiencies and shortcomings of the present system. Basically, they are:
1. Can inputs, files, outputs, methods, and procedures be improved in order to accomplish an organization's objectives to the highest degree possible?
2. Are all operations necessary? Does this result in duplicate or overlapping operations, files, and the like?
3. Is there a faster, simpler, and/or more economical way of processing the data?
4. Are data recorded in a manner compatible with their final use?
5. Is it possible to reduce the work volume by modifying or changing policies, the organization structure, files, departmental functions, or other established organization practices?
6. Can the system be improved through work simplification?

These may not be new or revealing to the reader, but they are necessary for an in-depth review. The systems analyst will ask these same basic questions repeatedly as the many parts of the system are devised, and many alternative subsystems will be discarded in order to devise more effective alternatives. The creative talents of the systems analyst can now be appreciated: the challenge to create a management information system that is within the scope of the study and that will provide a logical answer to the questions above is clearly a great one.

Review data on new system A review of data on the recommended MIS alternative will assist the systems analyst in obtaining an overview of the new system. The tangible and intangible benefits are explicitly enumerated in this exploratory survey report to the executive steering committee. All pertinent factors, such as the kind of changes in the organization structure and special factors to be considered, must be kept in mind.

The exploratory survey report contains a basic framework of the new system; the overall system and related subsystems are defined in general terms. Subsystems are generally flowcharted for each area of the study, and overall system flowcharts depict the relationships among the many areas.

Even though system flowcharts are prepared for each major functional area, the most meaningful ones to the systems analyst are those flowcharts depicting how the various subsystems are interconnected, since outputs from one subsystem become the inputs for another. For example, the output from sales forecasting and finished goods inventory becomes the input for materials planning whose output, in turn, becomes input for inventory control. This succession of input–output activities continues through production scheduling, production dispatching, data collection, and finally to operations evaluation which employs the "management by exception" concept. Only exception items, when compared to the current production schedule, budgets, standards, and comparable items, need be brought to the attention of management—no matter what level is associated with the activity. Thus, a review of all information, such as flowcharts, decision tables, and quantitative and qualitative factors, provides a sound basis for determining the requirements of the new system.

Determine Requirements for New System

No matter what area or activity is chosen as a starting point, prime consideration is given to policy and output needs, which are an integral part of the new system's tangible and intangible benefits. The basic questions used in determining the weaknesses of the present system, noted in an earlier section, help the systems analyst create detailed subsystems that start with outputs. These outputs are related to the alternative methods, procedures, and data files from which they are derived. Finally, all activities are directed back to inputs that are compatible with the new systems design.

Since the analyst specifies the new system requirements in more detail than contained in the exploratory survey report, it is possible that modifications to the initial requirements must be undertaken. If the changes are minor, the chances of their affecting the system's benefits are slight. On the other hand, if the revised design calls for major modifications, it is quite possible that the original savings and costs cannot be achieved and that certain conclusions are no longer valid. This situation calls for an immediate meeting between the executive steering committee and the MIS study group to determine the best course of action. It may be necessary to reevaluate the MIS exploratory survey report for a final answer. If the analysis was thorough, reexamination of all data compiled will be unlikely. However, if the report was deficient in critical areas of investigation, the GIGO (garbage-in, garbage-out) principle rises to the forefront. The accidental omission of essential activities or areas can be just as embarrassing as the willful omission.

Content of systems design Figure 7.3 illustrates the elements essential to systems design. Underlying them is a consideration of the **human element** in the new system. Together, the elements provide a basis for documenting the new system. Inasmuch as future chapters will concentrate on these design elements, only an overview is given below.

New policies consistent with organization objectives Systems analysts are often backed into a corner when constrained by organizational policies currently in effect. A close scrutiny of information on the new system will highlight those policies which are candidates for change and any need for further policy changes will become apparent as the design work progresses. Any policy changes approved by management should be consistent with an organization's objectives as delineated in the final system report.

Possible policy changes can be related to a number of areas. Administrative policies can sometimes be made more uniform in order to reduce the number of exceptions in the

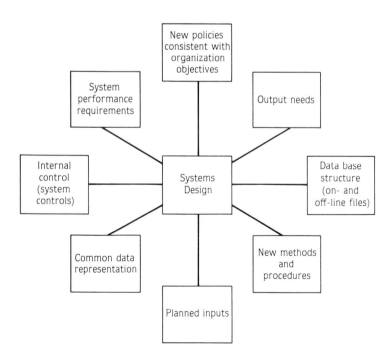

figure 7.3 The essential elements of systems design that are preparatory for documentation.

system. Examples are sales commission rates (which can be too complex for any system) and many pricing and discount policies. These complicated rate structures come into existence over the years to stimulate trade in special market segments; but with the passage of time, the reasoning originally given no longer applies. Unfortunately, the higher processing costs associated with the handling of these exceptions still remains. The problem can be resolved by determining if customers will be dissatisfied by simplification of the exceptions. If they are, the question of whether or not dissatisfaction outweighs data processing savings must be resolved.

A lucrative area for policy adjustment is a reduction in the work volume. A good example concerns the initial expensing of low-value purchased materials, rather than charging them to an inventory account and expensing them as they are withdrawn for production. Many organizations have found that the cost of maintaining inventory records on these items—pricing, making the entry for withdrawal, and charging it to a job—is greater than the value of the item. Another example is an accounts payable policy of investigating overshipments and overcharges. The labor time needed to straighten out these discrepancies can be higher than the benefits since errors on both sides tend to balance out over a period of time.

Output Needs Output needs cover more than reports, which are generated on demand, on schedule, and by exception. They include listings, summaries, documents, magnetic tapes, magnetic disks, magnetic drums, CRT displays, and responses from I/O terminals. Many of these outputs provide the link between the data processing section and its ultimate users, customers and vendors, as well as all levels of the organization's personnel. The systems analyst must develop output that meets user requirements; otherwise, output will probably not be available when needed or in the desired format. Information users or departmental managers and representatives, working with systems analysts, will specify the format, level of detail, degree of accuracy wanted, and frequency of the report. The systems analyst may be able to improve on the output requirements of the user through new equipment capabilities.

Once the systems analyst has defined clearly the legitimate output needs for a specific area, he or she is in a position to devise the methods, procedures, on-line files, off-line files, and inputs that will produce the outputs. Basic design alternatives should be considered and evaluated for the best output, and this is the point at which the true creative talents of the systems analyst are needed. The individual must keep in mind that output information must be planned and that the data required to produce the output must be captured and inputted to the system. Actually, inputs, data files, methods, and procedures limit the type of output. A management information system cannot supply the output needs unless it has read and stored the necessary input data, either on line or off line. Input data should be captured and stored in a format to minimize processing and data storage requirements.

Building upon this overview of output needs, chapter 8 will concentrate on the detailed design of system outputs and reports. Outputs, as indicated previously, provide a logical basis for getting started on the design of a new system.

Data base structure (on-line and off-line files) For the most part, prior system design approaches utilized files which were structured for individual applications. This individualistic approach resulted in a great deal of redundancy, that is, the same data were carried in a number of files. Common keys which provided a means of referencing file data, such as employee number, were mandatory since they were the only method of cross-referencing two or more files.

The duplication of information in a variety of files is costly in that excessive amounts of secondary storage capacity must be provided and a greater amount of magnetic tape, disk, or drum manipulation is required. A much more critical aspect, however, lies in the updating of the multitude of files. It is not difficult to conceive of one file being updated while the status of the same record on a second file, a third file, or other files remains unchanged because of an oversight or error in systems design, computer operation, or documentation. The **single record** concept overcomes these data file problems. Under this concept, all relevant information which formerly appeared in multiple data processing files now appears in one file of an organization's data base.

Conceptually, the data base of a management information system is a group of coordinated files of basic data elements in which duplication of data, if any, is kept to a minimum. Each file of individual data within the data base is a building block or module. In essence, the data base is a comprehensive store of up-to-date, accessible unit data that can serve as a basis for preparation of any required information or report.

In an MIS environment, all information need not be resident in one or more on-line files. In addition to data stored on line in magnetic disk, drum, mass storage, or comparable file devices, there is need for off-line files, such as magnetic tapes. These data files could be interrelated with on-lines files to permit the automatic updating and retrieval of information. Thus, data that are referenced periodically less frequently than weekly—are poor candidates for on-line storage.

The systems analyst should remember that more data should be provided in off-line files than required for present output needs. Efficient systems design dictates a minimum of storage data. However, this should be consistent with the ability to meet and satisfy anticipated output needs now and in the future, as well as those that arise unexpectedly. If output requirements are specified in detail as they should be, the question of what files to maintain and the appropriate information to be stored is no problem for the systems analyst. A problem occurs when output requirements are so inadequately specified that it is difficult to design organized off-line files to meet system needs.

An organization's data base structure is the subject of chapter 9, which delineates methods of file storage and important data base design criteria, emphasizing the integration of both types of files (on-line and off-line) such that an organization's information needs are met.

New methods and procedures Now that output requirements for the new system have been defined, the systems analyst's attention is focused on new methods and procedures to produce these outputs. Consideration at this time must also be given to on-line and off-line files as well as inputs since it is difficult to isolate methods and procedures from these basic system requirements. This phase requires intensive creativity from the system analyst.

The design of new methods and procedures is essentially a process of thinking logically and involves developing many system design alternatives for a specific area. The new methods and procedures are tested for practicability, efficiency, and low cost. Attention must be given to those designs that most closely meet the project's objectives. Once the many possibilities have been reduced to reasonable alternatives for one of the functional subsystems, these alternatives are examined thoroughly to choose the best one under the existing conditions. Included in these inventive steps is recognition that each functional area or activity is not isolated, but rather a part of the entire system. Thus, each related part must be considered in the final evaluation of the area under study. The addition of this dimension to the systems analyst's job increases the magnitude of the task.

This introduction to new methods and procedures is elaborated further in chapter 12, with emphasis on the steps in typical system procedures.

Planned inputs Inputs must be planned, examined, and evaluated carefully from many viewpoints by the systems analyst because much time and cost are involved in converting data to information. Inputs may be handled in a more efficient manner on an individual or group basis. It might be possible to capture the data initially in a machine processable form. The accuracy requirements of input data generally call for editing and verifying methods. In some cases, inputs can be processed on a random sequence basis; in others, there is need for a particular sequence before processing. The element of time constraints on the inputs and variations in input volume are valid considerations when designing a new system.

An important design consideration of data input is the provision for retention of raw data to provide a basis for preparing future reports and answering questions not presently formulated. A word of caution is necessary since the cost of storing, retrieving, and processing the data into meaningful output can be very high. The criterion used for storage of file information is applicable to inputs; that is, the potential value from the stored data should be greater than the related cost of storing, retrieving, and processing the data.

Chapter 11 specifies important material on input design, including new methods and procedures, on-line and off-line files, and outputs, as well as the interaction of the human element with source documents using data-entry equipment.

Common data representation One of the fundamentals of efficient processing is the automatic capturing of data in a common data processing language as a by-product of a previous phase. This means that captured data are in an acceptable format for subsequent handling on data processing equipment without human intervention. When one thinks in terms of handling large amounts of data, only a common language will permit processing in a fast and accurate manner. Common data representation includes print code (optical and magnetic ink), card code (eighty or ninety-six columns), and channel code (punched paper tape and magnetic tape).

Most large data processing installations use more than one code and this approach is encouraged where the various codes are handled by data processing equipment with no need for manual intervention. However, this should not be carried to the extreme: for example, the use of different channel codes is not recommended since it may be necessary to devise costly procedures for further processing. The best method is to design a system where a language will be common to all parts.

Internal control (system controls) No systems design job is complete without adequate provision for internal control. The systems analyst should make certain that the final design allows no one person full responsibility over an entire operation which provides an internal check. This should be apparent in such areas as cash and payroll where one person with complete responsibility can defraud an organization without much difficulty. Control points must be built into the system. Checks at control points ensure that what has been processed agrees with predetermined totals—internal accounting control. Controls of this type ensure accuracy at the input stage and during processing, resulting in reliable output.

Design of a new data processing system encompasses one or more of the following controls: MIS department, input, programming, output, batch/interactive, hardware, and security. The successful incorporation of these controls will keep auditing procedures to a minimum. Sufficient safeguards, built into the system, are necessary to avoid fraud and inaccuracies. In chapter 14, detailed information on internal control—more specifically, system controls—will be found.

System performance requirements Determining system performance requirements involves ascertaining the number of transactions that the computer must accept, process, and return within a certain time. It is not sufficient to design a system that can process forty transactions per minute if these forty transactions represent the volume expected shortly after the system is operative and will be increasing subsequently. Transaction volume must be projected not only for normal future processing loads, but also for peak periods—the time interval during which the largest number of transactions will be processed. An analysis of project volumes plus a statistical sample of transactions can be used to determine the number of peak periods and the extent of overlapping of peak periods. If the new system developed cannot handle present, and more importantly, normal and peak loads in the future, it is not a feasible system. The most economical system that meets this benchmark volume of peak loads is the one that an organization should probably implement.

An important performance capability of a management information system is its ability to reduce time delays. The prompt acceptance of input data whenever they are entered into the system reduces the time span from input through processing to output. Receiving inputs and returning processed outputs on this basis necessitates the acceptance of transactions whose time and rate of entry may be controlled.

The best method for developing system performance requirements is to follow a set of logical procedures:
1. Define and list all transaction types.
2. Determine normal and peak periods for all transactions.
3. Project all normal and peak periods.
4. Determine the overall level of throughput for normal and peak periods.

Regarding the **first procedure,** new transactions that come into being because the system is interactive must be added to the list of existing transactions. Otherwise, the volume of transactions for normal and peak periods is not determinable.

The **second procedure** is concerned with the highest utilization of one or more system components for a varying period of time. Delays, generally occurring during peak periods, are caused by transactions entering the system at a more rapid rate than they can be processed. They cause an increase in the time required to process a transaction before sending information back to its destination. The systems analyst's job is to ensure that such delays will occur as infrequently as possible.

Once the normal and peak transaction processing loads have been determined, the **third procedure** calls for projecting these volumes into the future as defined by the MIS exploratory survey report. Actually, this procedure is evaluated along with the overall level of throughput, the **fourth procedure.** The performance requirements are often specified as **x** number of transactions per second, minute, or hour, with a certain response or turnaround time. For example, the system is required to process 600 transactions per hour or 10 transactions per minute. If the response time of the system is 20 transactions per minute, the system is capable of handling the throughput. Although the ability to handle the present volume is ample, consideration must be given to processing future peak loads. If they fall into a range of 13 to 14 transactions per minute for normal loads and 15 to 16 transactions per minute for peak loads, the system is still capable of providing the desired throughput response for all anticipated volume levels.

SUMMARY: Determine Requirements for New System

New Policies Consistent with Organization Objectives Refers to the need for policy changes as the systems design work progresses. Any approved policy changes by management should be consistent with organizational objectives.

Output Needs Covers the design of outputs, and in particular, reports that are generated on demand, on schedule, and by exception. Many of these outputs provide the link between the system itself and its ultimate users; therefore, timely output may be critical for managing operations.

Data Base Structure (On-Line and Off-Line Files) Refers to a group of coordinated data files—on-line and off-line—whereby duplicate data are kept to a minimum. On-line data files represent storage of up-to-date, accessible data elements that can be used for the preparation of timely information or reports. Similarly, off-line data files represent storage of data elements that can be used to produce meaningful output.

New Methods and Procedures Relates to developing new methods and procedures that link output and the data base structure to input. They should be tested for practicability, efficiency, and low cost.

Planned Inputs Focuses on the means of capturing data initially in the system. If at all possible, the data should be captured in a machine processable form to reduce system processing costs.

Common Data Representation Refers to capturing input in an acceptable format (data code) so that it can be handled without human intervention in subsequent processing.

Internal Control (System Controls) Refers to developing control points in the new system to ensure that what has been processed agrees with prior processing. Likewise, the final design should not allow one person to have complete control over any one area.

System Performance Requirements Involves ascertaining the number of transactions that the system must handle within a certain time period as well as the capacity of the equipment to handle normal and peak volumes.

Design the New System

Determining the requirements above is concurrent with designing the new system. After all, a system is nothing more than a total of all its component parts. The design of a system involves making decisions about each of its parts—outputs desired, on-line and off-line files (data base) to be maintained, planned inputs, and data processing methods and procedures that link input with output. An integral part of systems design work is answering many questions so that no one item is left unexplored.

For a realistic and feasible approach to the design of a management information system, it is highly recommended that the analyst partition the total job into manageable segments, called modules or building blocks. These segments are then tailored for each particular level of operation and for each specific managerial activity. This is not to say, however, that each of these segments is incompatible with all of the other segments; just the opposite is true. By providing the systems analyst with a conceptually and technically sound blueprint as a basis for action, and by employing imaginative, intelligent, and trusted design techniques, a consistent whole can be synthesized from the segments. Thus, the new system can be viewed as a dynamic system consisting of a number of subsystems in constant motion and interaction. The introduction of change in one subsystem generally results in necessary changes in other subsystems.

Systems design steps The various steps involved in systems design are set forth in figure 7.4. The initial step involves determining the requirements for new policies consistent with organization objectives, output needs, a data base structure (on-line and off-line files), new methods and procedures, planned inputs, a common data representation, and internal control (system controls). These requirements, plus system performance and consideration for the human factor, are used to design alternatives in a modular or building block environment, the second step. The third step centers on preparing system flowcharts that depict the modular relationships detailed above. Their purpose is to appraise

---

1. Determine the following requirements for the new system: new policies consistent with organization objectives, output needs, a data base structure (on-line and off-line files), new methods and procedures, planned inputs, common data representation, and internal control (system controls).
2. Devise many system design alternatives through a modular or building block approach—including detailed inputs, methods, procedures, data files, and outputs—and give special consideration to system performance requirements and human factors.
3. Prepare flowcharts showing the modular relationships for the various alternatives in step 2.
4. Review system design alternatives with appropriate personnel.
5. Select the more promising alternatives with the aid of properly designated personnel.
6. Compare the tangible and intangible benefits of the promising alternatives with the MIS exploratory survey report. Cost factors, volumes, and requirements for equipment and personnel should be carefully analyzed to check the report's validity.
7. Consider alternative system designs which incorporate alternative functional modules, equipment, and techniques not covered in the exploratory survey phase. Many times, promising system approaches do not come to light until the creative phase. If step 8 is not applicable, proceed to step 9.
8. Determine the tangible and intangible benefits for these new alternatives in step 7.
9. With the assistance of organizational personnel, select the final systems design that best meets the study's requirements from among the promising alternatives.
10. Prepare final system flowcharts and/or decision tables for the recommended system design and relate it to all other parts of the management information system, thereby forming the basis for bid invitations to equipment manufacturers (see chapter 16).

---

figure 7.4 Systems design—steps for the design of a functional area (subsystem).

the merits of the modular system alternative and to provide a basis for review with appropriate organization personnel, the fourth step. The first four steps in systems design (figure 7.4) have been delineated in prior sections of the chapter.

In the fifth step, the more promising alternatives are selected with the aid of designated personnel. For promising alternatives, their tangible and intangible benefits are compared with the MIS exploratory survey report to check their validity, the sixth step. For the next two steps (seventh and eighth), the tangible and intangible benefits of alternative designs which have come to light during the creative design phase are considered. In the ninth step, the final systems design that best meets the study's requirements is selected from among the promising MIS alternatives. These methodical steps insure a comprehensive review of all promising system alternatives and allow the creative talents of the systems analyst to be utilized to their fullest.

Once the study group has selected the final design based on the existing tangible and intangible benefits, any significant deviations from the findings of the MIS exploratory survey report must be reported to the executive steering committee. It is the function of these top managers to make a final decision on the feasibility or infeasibility of applying the recommended system under new conditions. This is why the feasibility study is a continuing one, even though the task force has endorsed a proposed system in the exploratory survey report, and is not formally concluded until the equipment is selected, the subject matter for chapter 16.

Document the New System

Documentation is the tenth and final step. Detailed documentation is necessary for submitting bid invitations to equipment manufacturers and preparing program flowcharts. Regarding the first item, documentation is needed for data origination and communications, planned inputs, data base, methods and procedures, output needs, and special requirements of the system. Also included in the bid invitation are system flowcharts showing each area under study, as well as those depicting the interrelationships of the various parts to the entire system. Without this documentation, the MIS task force is vulnerable to forgotten details.

In order to prepare program flowcharts for coding at a later date, it is necessary to develop the appropriate logic. Although block diagrams or program flowcharts can be prepared for a detailed documentation of the new system, decision tables are preferred at this point for several reasons. They are easy to construct and give a more compact presentation than program flowcharts accompanied by a written narrative. They are easy to modify and update. They show more clearly than flowcharts the effects of system changes upon the logic of the overall system; the more complex the logic is, the more appealing are decision tables to the systems analyst. Thus, decision tables show conditions (if) and actions (then) in a clear and logical manner which facilitates programming during system implementation.

CHAPTER SUMMARY:

The design of any system goes hand in hand with the creative ability of the systems analyst. A fundamental of systems design is to devise a large number of alternatives with varying inputs, methods, procedures, and outputs that are compatible with the data base (the on-line and off-line files). The systems analyst relies upon his or her experience and inventive ability to develop promising system alternatives incorporating these important components. Each alternative is analyzed to determine its benefits and impact on the organization. The most promising MIS alternative is then chosen and documented with system flowcharts and/or decision tables. The basis for selection is the successful accomplishment of an organization's objectives and attainment of quantitative and qualitative factors as set forth in the MIS exploratory survey report.

Numerous questions must be answered by the systems analyst before applying the modular subsystem design approach. These are extremely helpful in determining the most suitable requirements of the new system and in probing any activity that touches the new system directly or indirectly. The answers to these questions are utilized to develop the desired system. Intensive probing, hard work, and imaginative talents of the systems analyst combined with the constructive criticism of organizational personnel, is the key to designing promising new systems. To try any other approach will generally not result in the best system design.

Questions

1. Define **systems design.** How does this differ from **systems analysis?**
2. How important are creativity and brainstorming in designing a new system?
3. What is meant by the **modular** or **building block approach?** Explain thoroughly.
4. What are the essential components from an overview standpoint for designing a new system? Enumerate.
5. a. Why should the systems analyst consider reporting requirements first when designing a management information system?
   b. What is the relationship between reporting requirements and the structure of the data base in a management information system?
6. What important factor must be taken into consideration when designing an efficient and economical data base?
7. a. How important is the human factor in the design of a management information system?
   b. How important is internal control in the design of a management information system?
8. What is meant by "system performance requirements in a MIS environment"?
9. What are the typical steps that a systems analyst should follow for the design of a functional area (subsystem)?
10. If a better system is determined during the systems design phase, what may have caused this to happen?

Self-Study Exercise

True-False:

1. _____ The accent of brainstorming is on new and untried approaches to systems design.
2. _____ Decision-making ability is the most important characteristic of a well-designed system.
3. _____ Simplicity of a well-designed system refers to its capability to change in response to new environmental conditions.

4. _____ An important feature of the modular design approach is that duplicated modules can be examined for possible elimination.
5. _____ Generally, there is no logical starting point for systems design.
6. _____ One purpose of reviewing the present system is to eliminate its weaknesses in the new system.
7. _____ When designing a new system, the recommended approach is to start with the planned inputs first.
8. _____ A data base will generally contain a certain amount of duplicate data.
9. _____ By and large, internal control can be eliminated when designing a new system.
10. _____ Equipment considerations should always take priority over human factors.

Fill-In:

1. The design of an _____ system must be more closely coordinated and timed than a batch-processing system.
2. _____ refers to the capability of the system to adapt to changing environmental factors.
3. _____ stresses the capability of the system to provide a straight-line flow from one step to the next.
4. A logical starting point for _____ _____ is working with an activity that dominates the new system.
5. A typical question to be asked by the systems analyst is, "Are _____ recorded in a manner that is compatible with their final use?"
6. In a new system, it is recommended that newly formulated _____ be in agreement with organization objectives.
7. A _____ _____ is a store of up-to-date, accessible data that is useful for answering operational and managerial information requests.
8. _____ _____ requirements involve determining the number of transactions that the computer must accept, process, and return within a certain time period.
9. Systems analysts should always examine the impact of _____ and data schemes on organization personnel.
10. The final step in systems design is preparing system flowcharts and/or decision tables, thereby forming the basis for _____ _____ to equipment manufacturers.

Problems:

The American Company, which is currently switching from an integrated information system (batch processing) to an interactive information system, completed the basic systems design phase many months ago. The problems set forth below center on the second step of the feasibility study, the detailed design of the inventory control subsystem and the finance subsystem using the modular (building block) concept.

1. The inventory subsystem (major module) for The American Company consists of two intermediate modules, namely, the perpetual inventory control module and the physical stock control module. The first intermediate module can be expanded into the following minor modules: raw materials, work in process, and finished goods.
   Using the modular tree diagram below, develop appropriate **basic** modules for the raw materials and work in process modules.

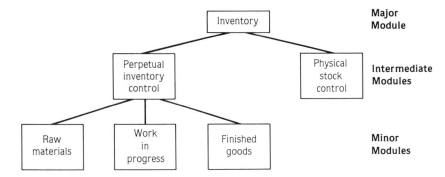

2. The finance subsystem (major module) for The American Company consists of three intermediate modules, namely, cash management, capital budgeting, and sources of funds. The first intermediate module can be expanded into the following minor modules: cash receipts, accounts receivable, cash disbursements, and accounts payable.

Using the modular tree diagram below, develop appropriate **basic** modules for each of the minor modules.

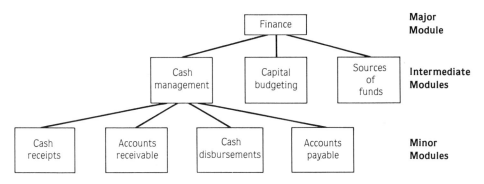

chapter

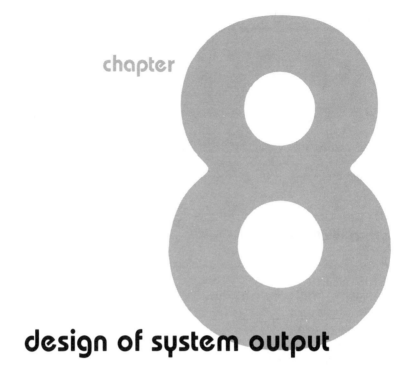

design of system output

OBJECTIVES:

- To set the stage for the design of a new system by starting with output—in particular, managerial and operational reports.
- To set forth important systems design principles that should be incorporated when devising output.
- To relate the human factor to systems design of output, with emphasis on meeting the needs of organization personnel.
- To examine important questions that must be answered before final output can be designed.

IN THIS CHAPTER:

  Introduction to Design of System Output
COMMON TYPES OF SYSTEM OUTPUT
  Printed Output
  Visual Output
  Turnaround Document Output
  Secondary Storage Output
  Microfilm or Microfiche Output
  Audio Response Output
  Plotter Output
INCORPORATE SYSTEM DESIGN PRINCIPLES—OUTPUT
DESIGN OF SYSTEM OUTPUT
  Consider the Human Factor in Designing Output
  Review Information on System Output
  Specify Output Requirements
  Determine the Content of Output
  Design of Output
  Use and Distribution of Output
CHAPTER SUMMARY
QUESTIONS
SELF-STUDY EXERCISE

In the prior chapter, an overview of systems design was presented; in this chapter and succeeding ones, the detailed design of a new system is presented, followed by the design requirements for the order-entry system of the ABC Company. This link of design theory with practice serves to highlight an orderly approach to systems design. Likewise, this presentation will give the reader an opportunity to acquire the proper background for designing a finished product inventory system, an accounts receivable system, or both. Overall, part three is structured to give the reader the required expertise for designing a subsystem of an interactive management information system.

Initially, the focus of the chapter is on those types of system output common to all types of batch and interactive processing systems. After setting forth the underlying principles of system output, the design of summaries, reports, and comparable output is presented. The beginning and and overriding design step is consideration for the human factor, followed by reviewing information on output accumulated to date. This background assists the designer in specifying output requirements, their contents, and the actual location of data items. Finally, the use and distribution of output are discussed.

Introduction to Design of System Output

Output needs, as indicated in the prior chapter, cover more than reports; they include listings, summaries, documents, punched cards, punched paper tapes, magnetic tapes, magnetic disks, magnetic drums, computer display devices, teletype messages, responses from I/O time-sharing terminals, and others. Many of the outputs provide the link between the DP system itself and its ultimate users, the organization's customers and vendors plus organizational personnel at all levels. Information users or departmental managers and representatives, working with the systems analysts, will specify the format, detail desired, the degree of accuracy, and the frequency of the report. System personnel should be able to improve on the output requirements of the user, especially if an interactive management information system is being designed.

Before output requirements are specified, it is advisable to review the common and specialized types of output in order to see current hardware developments in their proper perspective and highlight the need for creative work during this critical design phase. Because of their importance, they are covered in sufficient detail below as a starting point for designing system output. Likewise, system design principles regarding output are set forth before discussing the detailed design of output.

COMMON TYPES OF SYSTEM OUTPUT

Currently, there are many types of system output that can be incorporated into a new system. Because computer hardware and peripheral devices are capable of producing a variety of output quickly, it is necessary that an organization's systems analysts know the capabilities of this equipment; otherwise, the latest technology and developments relative to the forms of output will be missing from the final system design.

Output from computer systems can take a variety of forms. The most common ones are
- printed output
- visual output
- turnaround document output
- secondary storage output
- microfilm or microfiche output
- audio response output
- plotter output

Each one will be explored below.

Printed Output

Of all the various types of output, the most widely used is the printed report, produced on the high-speed printer. Listings, invoices, summaries, as well as management reports, exception reports, and financial statements, to name a few, can be produced. From the typical printer, illustrated in figure 8.1, a wide range of output is available to the systems analyst. This infinite variety of output, as well as the low cost, flexibility, and wide acceptance of printed reports will continue to make the printer the most widely used form of output in the future.

Although speeds can range up to 18,000 lines a minute for the very fastest printers, most computer printers range in the several hundred up to 3000 lines per minute category for printing continuous forms.* Most printers currently available can print up to 144 characters on a single line. Each character can be a number, a letter of the alphabet, or a variety of special characters; it is possible to print both upper and lower case of the alphabet.

Visual Output

With the increasing emphasis on the design of interactive management information systems, the CRT (cathode ray tube) unit is becoming a widely used form of output. A

---

*Continuous forms are those which are connected to one another so that, together, they appear to be one continuous strip of paper.

figure 8.1 A typical high-speed printer, the IBM 3211 High-Speed Printer. (Photograph courtesy of International Business Machines Corporation)

typical system can display thirty lines at one time with 80 characters per line. Data displayed can include numbers, letters of the alphabet, and, in some systems, line drawings.

Another capability of CRT terminals is the entry of data into the computer system. The keyboard can be utilized to enter numeric and alphabetic data into the system for processing. Additionally, some terminals offer the ability to change data through the use of a light pen. As illustrated in figure 8.2, the user can add, delete, or rearrange data that appears on the CRT screen. In turn, these data can be processed by the computer system for some type of visual output.

The most common use of CRT units is inquiry, where a hard copy is not required. Examples include order entry, accounts receivable, and accounts payable inquiries. Other uses are airline and hotel reservation systems and banking transactions. In these interactive modes, an inquiry is made to determine the status of an order or account. Once the information has been displayed, the next inquiry is processed in a similar manner. Answers to inquiries can be displayed faster on a CRT unit than can be printed using some other type of equipment. Thus, speed and timeliness of data make this form of output widely used for interactive processing.

Turnaround Document Output

In order to reduce the input workload at a later date, turnaround documents in the form of punched cards are widely used. To illustrate, when a utility company bills its customers, it asks that an enclosed punched card be returned with payment. In turn, this card is

200 Design of System Output

figure 8.2 A typical CRT unit, the IBM 3278 Display System, with optional Selector Light Pen. (Photograph courtesy of International Business Machines Corporation)

used as data input for recording payment of telephone charges. As shown in figure 8.3, this turnaround punched card becomes input to the system. Other applications include credit card billing and employee time cards. In both cases, these cards serve as input to a computerized system and produce new punched cards for the next time period. Additionally, OCR (optical character recognition) forms can be prepared as output which, at a later date, serve as input to the computer system.

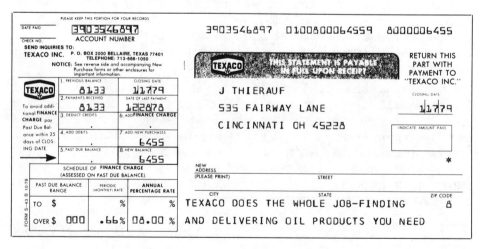

figure 8.3 A typical turnaround document; punched card output ultimately becomes punched card input.

Secondary Storage Output

An increasingly important type of output includes magnetic disk, magnetic drum, magnetic tape, and mass storage. Although these types of output are normally used to store data, they can also be employed to store output. While the information may be printed or displayed as the situation dictates, it may be necessary to store it for use at a later time. For example, a computer inventory updating run may take the beginning balance for each inventory item plus the additions and minus the usage for a certain period to determine new ending balances. These values are then printed for an inventory report as of a certain period. However, the ending balance is stored on secondary storage in order to prepare future inventory reports. Thus, secondary storage can serve as an important form of output for many business applications, especially in an interactive mode to answer real-time inquiries.

Microfilm or Microfiche Output

Because of the large amounts of information that can be generated in the form of printed output (for example, historical records and information reports), many medium- and large-sized organizations have turned to microfilm or microfiche output. The chief advantage of microfilm technology is that printed output can be reduced up to one hundred times or more. By employing microfilm output, it is not only possible to reduce the physical size of output, but also to produce faster output. Computer output can be up to thirty times faster than the typical several hundred lines-a-minute printer.

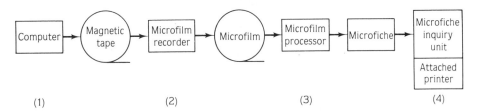

figure 8.4 The typical operation of a microfilm output system.

Inasmuch as computer output is in the form of microfilm, many pages of regular computer printout can be stored as a very small image on a roll, strip, or sheet of film. A most common storage method is called **microfiche,** a sheet of film. A sheet of microfiche ranges in size from three by seven inches to four by six inches, and each sheet can hold from one hundred to three hundred pages. Obviously, microfilm offers important advantages in both processing speed and storage space.

The basic operations of a typical microfilm output system are illustrated in figure 8.4. The magnetic tape output from a computer system (a) provides the input for the microfilm recorder. This unit converts the numeric and alphabetic characters into characters that are displayed on a cathode ray tube for photographing onto microfilm (b). The microfilm is processed (c), and can then be viewed by a microfiche viewer or

Audio Response Output

A newer form of output is the audio response unit. This unit is capable of providing on-line inquiry into the system where output is restricted to short messages. This system consists of a message handling unit (figure 8.5), touch-tone telephone units, and standard telephone lines for an on-line mode. The audio response unit has a vocabulary of up to three hundred words tailored to the user's needs. For example, a bank provides its customers an opportunity to make inquiries about their checking and savings accounts balances. When an inquiry is made to the computer system, the balance is retrieved and output would be as follows: "Customer 579241, balance 14715." A major advantage of a voice response system, then, is the quick verbal response from a computer system by telephone.

figure 8.5 The IBM 7772 Audio Response Unit.
(Photograph courtesy of International Business Machines Corporation)

Plotter Output

Automatic graph plotters, such as the one illustrated in figure 8.6, are utilized where graphic or pictorial presentation of computer data is meaningful and easier to use than extensive numeric or alphabetic listings. They are indispensible when the volume of graphic presentation required is uneconomical or impossible to produce manually. Generally, there is some restriction on format. However, a pictorial representation may include any desired combination of axes, lines, letters, and symbols with an unlimited

figure 8.6 The Cal Comp Model 1012 Plotter. (Photograph courtesy of California Computer Products, Inc.)

choice of scale factors, letter and symbol sizes, and printing angles. Automatic graph plotters, then, offer an effective way of summarizing business data in the form of charts and graphs.

INCORPORATE SYSTEM DESIGN PRINCIPLES—OUTPUT

When specifying outputs for a new system, several important design principles must be considered. First among them is the **Principle of Starting with Output,** that is, an organization's output needs should be considered first before devising appropriate methods and procedures, data base, planned inputs, and effective internal control. The rationale is that the nature of the outputs should determine how they are produced. In essence, it is a waste of time and money to develop processing procedures, data files, and compile input data that do not result in some meaningful output. To process data for its own sake must be clearly avoided.

An equally important design principle is the **Principle of the Acceptability of Reports.** If management and operations personnel are convinced that the reports will assist them in their jobs and result in their promotion, the use of these reports is assured. However, if organization personnel are certain that reports will not benefit them, that they are poor reports, that they do not follow established organization policies, or that there are other legitimate objections, the new reports will be used grudgingly or will be ignored. To overcome this problem, organization personnel should participate in output development.

Although the form and content of output may be accepted by organization personnel, the third design principle, the **Principle of Timely Output,** insures that output will arrive in time for organization personnel to take appropriate action for controlling at least ongoing tasks, not to mention future activities. If it does not produce timely information, the new system is worth very little to users.

The output of an effective system should produce not only timely and pertinent information, but also improve the decision-making ability of its personnel at the lowest possible cost. This is the focus of the fourth design principle, the **Principle of Enhancing the Decision-Making Process.** In meeting information needs, the system should be able to employ techniques that manage resources effectively. A good system is designed to make planning, organizing, directing, and controlling as efficient and automatic as possible.

A very important and new emerging principle is the **Principle of Practicing "Management by Perception."** Fundamentally, high-level management reports should have the capability of perceiving important trends that affect an organization and determining their impact in order to improve future organization performance. This entails employing advanced mathematical models, like corporate planning and venture analysis, so that management can be forward-looking instead of backward-looking. Top management reports are focused on the entire organization as well as its parts, relating its continuing objectives to changing conditions. What was appropriate for yesterday may not be proper today or tomorrow. Thus, a broader and more long-range viewpoint is necessary for developing high-level management reports oriented toward anticipating the future.

Taking the opposite viewpoint, there is a sixth principle, the **Principle of Practicing "Management by Exception."** Management reports should have the capacity to compare what is happening to what should be happening. Detection of potential and actual deviations from planned performance should occur on a timely basis so that corrective action can be taken. This principle holds that the manager should be concerned with analyzing only significant deviations and ignore those items that are "on target." In most cases, corrective action is undertaken at the same level the deviation occurred. However, if the deviation is very significant, it may be necessary to report the results to a higher management level.

The seventh important system design principle focuses on the reports: the **Principle of Eliminating or Reducing Duplication in Reports.** Too often, information is compiled for its own sake and serves no useful purpose for an organization. To avoid this costly problem, managerial and operational reports should serve some useful purpose.

In a somewhat similar manner, there is need for the **Principle of Simplicity in Reports**, an eighth one. Reports should be straightforward and easy to understand: the simpler the report, the more effective it will be in communicating the desired results. This common-sense approach is possible with the cooperation of organization personnel during the output design phase of a system project.

> **SUMMARY: Incorporate System Design Principles—Output**
>
> 1. **Principle of Starting with Output** Good system design dictates that an organization's output needs be considered before developing the data base, methods, procedures, inputs, and internal control. There is little need to go through a lengthy processing cycle if its results are inappropriate.
> 2. **Principle of the Acceptability of Reports** The success of a new system is highly dependent upon the acceptability of its output by organization personnel. For a successful system, the people who will use the reports should participate in their development so as to enhance their acceptability for managing organizational activities.
> 3. **Principle of Timely Output** System output is worth very little unless it arrives in time for organization personnel to take appropriate action for managing ongoing activities.
> 4. **Principle of Enhancing the Decision-Making Process** Output from the various subsystems should enhance the decision-making ability of organization personnel. This approach assists personnel in making more effective decisions.
> 5. **Principle of Practicing "Management by Perception"** Management reports should have the ability to perceive important trends and determine their impact on the organization so as to improve future performance. From this view, reports should be forward-looking as opposed to backward-looking.
> 6. **Principle of Practicing "Management by Exception"** Management reports should have the capacity to compare what is happening to what should be happening. Significant deviations from established plans or performance standards should be brought to the attention of the appropriate management level.
> 7. **Principle of Eliminating or Reducing Duplication in Reports** Duplicate or unnecessary information should be eliminated or kept to a minimum so that costs of producing managerial and operational reports will be minimal.
> 8. **Principle of Simplicity in Reports** Reports should be concise and easy to understand. This is feasible if the systems analyst incorporates simplicity in managerial and operational reports.

DESIGN OF SYSTEM OUTPUT

Now that the various types of system output along with their underlying design principles have been explored, this section will focus on the details of designing system output. Specifically, this design process encompasses the following areas:

- Consider the human factor in the design of output.
- Review output of the present system and information contained in the exploratory survey report to top management.
- Specify the output requirements for the new system.
- Determine the specific content of output that must be produced by the new system.

Design the output, that is, listings, documents, reports, exceptions, and the like.
Specify the use and distribution of the output.

Due to their importance, each of these items is covered in a comprehensive manner below. This material forms the basis for designing system output in the order-entry system of the ABC Company (chapter 10).

Consider the Human Factor in Designing Output

There are many factors to consider when designing output to interface with the human element; foremost is the employment of the computer as a humanizing force in terms of output. Complaints associated with the operation of computers for airline, hotel, and motel reservations service seem to abound these days. Similarly, errors in billings, paychecks, and insurance claims are commonplace.

A typical example would be helpful to understand a humanistic approach to output. The use of computers and associated DP hardware by airline reservation systems is so common that all major airlines take them for granted, and therefore denigrate them any time a problem arises. However, one newer reservation system is United Airline's APOLLO which has a humanized approach to "fringe pieces" of information that help personalize the traveler for the reservation attendant; this "fringe" information includes special dietary needs, wheelchairs, flight attendant aids, and reservations for hotels and rental cars. The most impressive humanized output of this system lists names of family groups including youngsters and elderly passengers who have never flown before. These bits of personalized information are used by the reservation clerk to allay fears and to establish rapport among people who were formerly total strangers. The entire history of the reservation is displayed on a cathode ray tube in seconds to help reduce any complications that might develop as changes are made in flight plans. Thus, this reservation system is able to produce humanized output that pinpoints pertinent information about the specific requirements of any airline passenger.

A most important consideration when designing output is for the systems analyst to make known to appropriate organization personnel what information can be obtained from the system. A statement of what information is available, who has access to this information, and for what purpose the information will be used helps clarify what can be expected from the new system. Within this framework, organization personnel will want to know what provisions have been made for evaluating important information that is retrievable as output. Similarly, they will want to know what provisions have been made to add information which they consider important. If these capabilities are not available to them, managers and operating personnel, as well as customers, may feel that the computer output was not designed with them in mind.

While the foregoing represents the ideal situation, there is a bare minimum that systems analysts should incorporate for satisfying human requirements. When specifying the outputs for a new system, adequate consideration must be given to the Principle of "Management by Exception." The integration of this important concept permits unusual events or knowledge to come to the attention of the appropriate level of management. This principle also insures that events which are under control are processed in the normal manner and that organization personnel can work within established ranges of authority and responsibility. Hence, only extraordinary events which do not fit into regular operations are to be brought to the attention of management at the same or the next higher level.

Other important human factors to include in designing output can be extrapolated from the design principles examined earlier in the chapter. System output should be easy to understand (Principle of Simplicity in Reports); that is, the output which the user receives should be easy to read and easy to follow. Abbreviations and special codes ought to be avoided as much as possible. The output should respond quickly to the user's needs (Principle of Timely Output). If delay in the system's performance cannot be avoided, some reason for the delay should be sent back promptly to the user. In this manner, the user can resort to some other means for producing the desired output. Additionally, the system ought to relieve the user of unnecessary work (Principle of Enhancing the Decision-Making Process); to state it another way, important information for decision making should be the normal output of the new system, thereby reducing the user's workload.

Finally, the system should include provision for correcting its output. If errors occur with sufficient frequency, procedures must be built into the system for detecting them. On the other hand, there is no need for procedures that may be used to correct unanticipated errors in output. To do so would cause the cost of information to exceed its value by a very wide margin.

To illustrate the need for procedures to correct errors, consider the following example. A homeowner received tax notices addressed to the previous owner. The new owner returned these notices with the information that the previous owner no longer resided at that address. When tax notices kept coming, a number of letters were sent to the tax bureau repeating that the previous owner had moved—all to no avail. The reason for this snafu (situation normal, all fouled up) was that the existing procedures stipulated that in cases of unknown residence of a taxpayer. the post office was to attempt to forward mail to a new address if such an address existed. However, the new owner had returned notices to the tax office rather than to the post office. Unfortunately, no one in the DP center knew about the procedure or had any authority to reply to messages, nor was there anyone there with authority to stop sending notices to the last known address of the taxpayer. Thus, an important procedure for correcting errors had never been set up to handle this unfortunate situation. Such an example serves to illustrate the necessity for considering the human factor in systems design, especially for handling errors in output.

Review Information on System Output

After giving due consideration to the human factor, information compiled during the systems analysis phase should be reviewed. This review will highlight current output that may be good, bad, or indifferent. In other words, problems and difficulties with present output should be noted so that they are not incorporated into the new system. In effect, systems analysts will unconsciously duplicate these present shortcomings in the new system if this initial step is bypassed.

Additionally, systems analysts should review the information contained in the exploratory survey report to top management. Special emphasis is placed on the recommended system alternative. Any information that can be extracted about output and its requirements will be helpful in getting started. Thus, any information about output that can be gleaned from material developed to date should be examined for assistance in the actual design steps below.

Specify Output Requirements

Now that current output has been reviewed, the real design work is ready to begin. This entails determining what information is needed as output from the new system to meet the needs of organization personnel. Although the preceding review is important, there is one word of caution. Systems analysts should not merely redesign the output of the previous system, but give primary consideration to the needs of organization personnel and the capabilities of the new system—in particular, its new interactive equipment. In this manner, the creative talents of systems analysts can be employed to their fullest.

When determining the output requirements, there is a need for analysts to work with those people who will be using the informational output. In essence, there is need for a cooperative effort so that managers and operating personnel get what they need from the system. Too often in the past, systems analysts dictated the type and content of summaries, reports, and the like to the users, resulting in less than optimum output. The proper approach, then, is to have both groups work together and determine the form and content of output that will result in more effective planning, organizing, directing, and controlling of business activities.

During the detailed investigation or systems analysis phase, sample forms, documents, and reports were collected. As indicated, these should be reviewed jointly by the systems analysts and the departmental representatives. A good review method is to bring all output samples together and sort them in appropriate categories, so that duplication becomes apparent upon close examination. The experience of many organizations indicates that 10 to 30 percent of all output records do not serve any valid purpose since many are a carryover from earlier days.

Appropriate questions regarding the validity of these reports are found in figure 8.7. Some reports may be retained in their present form, others eliminated, and still others combined. (In general, consolidation of reports that have slightly different purposes into one report results in a standardization of the report's format.) Theoretically, the results of analyzing the present reports should coincide with the user's requirements. However, there will be differences to reconcile in light of the proposed system and user needs before the systems analyst can finalize the output needs. Because of the designer's creative talents—in particular, knowledge about possible output designs and the capabilities of computer equipment—the individual is in a preferred position to recommend a solution that gives the user the desired type of output. Even though the designer does not dictate what informational needs are necessary for each area under investigation, he or she does have the responsibility for designing timely, informative output that meets the needs of organizational personnel at a reasonable cost.

Is the report necessary to plan, organize, direct, or control the activities of the organization?
Is the "management by exception" concept incorporated in reports?
What would be the effect if operating personnel got more or less information than presently?
How would work be affected if the report was received less frequently or not at all?
Is all information contained in the report utilized?
How often is all or part of the information contained in the report utilized after its original use?
Can data on this report be obtained from another source?
How long is the report kept before being discarded?
Is the report concise and easy to understand?
How many people refer to it?
Can the departments (functional subsystems) function without the report?
Are other reports prepared from pertinent data on the report?
Does the use of the data justify the preparation cost of the report?
Is the report flexible enough to meet changes in the organization's operating conditions?
Is the report passed to someone higher or lower in the organization structure?
When and where is the report filed?

figure 8.7   Systems design—questions to test the validity of a report.

Determine the Content of Output

Once the output requirements for the areas under study have been specified in general terms, the next step is to determine the specific contents of each summary, report, and the like. To assist in this detailed phase, comparable reports in the old system can serve as a starting point.

As in the previous step, the analyst must work with management and operating personnel to determine the format as well as the specific content of each report. Specifically, both parties must determine the informational elements; whether the elements are numeric, alphabetic, or alphanumeric; the number of characters for each element; and any special characteristics about these elements. Resolving the content of output is a big step toward designing the final system output.

After a thorough discussion of the output content with organization personnel, the systems analyst drafts a **report analysis sheet** that specifies the information to be contained in the output. As shown in table 8.1 (p. 210), this system output analysis form refers to the preparation of a monthly sales report by products. Additional subheadings could be incorporated into this report. It is always better to have too many columnar headings than too few. This is particularly helpful to new organization personnel who must work with system output.

Design of Output

Now that the output requirements and contents have been determined, the next step in the design process is to devise the output format to suit the needs of the user. This format will be the basis for programming the computer during the system implementation phase. It is important that the systems analyst know the various formats that a computer-generated report can take in either a batch or an interactive processing mode. Fundamentally, most output from a batch processing mode will be printed reports for a certain time period—for example, a weekly inventory report or a monthly departmental statement. In an interactive processing mode, the accent will be on visual output using CRT units for controlling current operations, such as the current status of a customer order and the current balance of a customer account. Because their basic types of output do differ, each will be discussed separately.

table 8.1 Typical system output analysis form.

SYSTEM OUTPUT

Page 1 of 1
Date: July 25, 198_
Type of output: MONTHLY SALES REPORT BY PRODUCTS
Preparer: G. W. Reynolds

Informational elements	Type of characters	Number of characters	Comments
Product number	Numeric	5	
Product name	Alphabetic	15	
Monthly sales units	Numeric	5	
Sales price	Numeric	4	Edit with decimal point
Monthly sales by product	Numeric	8	Edit with decimal point
Total monthly sales for all products	Numeric	9	Edit with decimal point

Comments:
The computer program is to generate the report and column headings. Also, the current date and page number are to be printed at the top of each page.

Batch processing reports A basic type of report produced by the computer in a batch processing mode is the **detailed printed report.** For a typical report of this kind, one line of information is printed for each record processed. The typical report illustrated in figure 8.8 consists of inventory number, description, and inventory figures for the current week. In addition, the computer has generated a column heading, date, and page number. The spacing within the report has been determined by the systems analyst. Two important features of this illustration are simplicity and readability. Hence, organizational personnel should have no difficulty in understanding the report after a casual inspection of its contents.

Another popular type of report is the **detailed exception report** which incorporates the Principle of "Management by Exception." There are two approaches to designing the format of this report. One is producing a report which shows only exception items that are worth reporting to the appropriate management level. For example, referring to figure 8.8, inventory which falls below a predetermined level would be printed as a separate list after running the weekly raw material inventory summary. Whereas, there may be 10,000 inventory items on the weekly summary, there may be only twenty or fifty items on the exception report. This report is forwarded im-

W/E 10/15/8_	WEEKLY RAW MATERIAL INVENTORY SUMMARY				PAGE 01
INVENTORY NO.	DESCRIPTION	BEG. BAL.	ADDITIONS	DELETIONS	END. BAL.
000222	ROCKER ARM	450	050	025	475
000364	BALL BEARING	805	200	400	605
000492	STEEL TEMPLATE	710	060	080	690
000724	STEEL BRACKET	440	000	080	360
000917	COPPER CASTING	210	060	000	270
000970	COPPER PLATE	240	100	050	290
001001	IRON BRACKET	520	000	070	450
001015	IRON CASTING	290	100	050	340
001027	IRON PLATE	175	025	050	150
001031	BRONZE BEARING	550	250	320	480
001045	BRONZE CASTING	605	000	220	385
001056	BRONZE PLATE	110	000	010	100

figure 8.8 Typical detailed printed report—weekly raw material inventory summary.

mediately to the purchasing department for action. It may well be that unless action is taken on many of these items today, the organization may be out of stock for items that are needed in the manufacturing process, resulting in the layoff of factory workers.

The other type of exception report employs the detailed listing approach, starring all important exception items. This approach is widely used for monthly financial statements. As shown in figure 8.9, several items, whose actual cost has exceeded the budgeted amount by ten percent, have been starred for review. Favorable and unfavorable variances are starred. Although the rationale for explaining unfavorable variances is somewhat obvious, the same cannot be said for favorable variances. In the illustration, the favorable variance of $3000 for maintenance and repairs is largely the result of two maintenance personnel having left the month before and not having been replaced. Although this may be a plus factor for this month, it may result in a disaster for this manufacturing department in the months to come: if specialized machinery is not maintained the way it should be, the chance of major repairs and downtime in the future is very high. In fact, this item might be the most significant exception item for the department, even though other items of equal or higher value have been starred for review.

In addition to these detailed reports, there is another type of report that summarizes important information. It is a **summary report,** and is used to accumulate important information in a summarized format. It should be noted that information becomes more summarized as one moves from the lowest to the highest management levels. Unless one area is experiencing difficulty, management at the highest level usually wants only an overview of the organization.

To understand the use of summary reports, one must consider the level of management. As indicated previously, monthly cost reports are processed for use by line supervisors and their superiors, who are responsible for maintaining an even flow throughout the

MONTHLY COST REPORT—DEPARTMENT 4
JULY, 198_

	BUDGET	ACTUAL	FAV. (UNFAV.)
CONTROLLABLE COSTS:			
DIRECT MATERIALS	$150,000	$148,000	$2,000
DIRECT LABOR	205,000	206,000	(1,000)
INDIRECT LABOR	12,000	13,000	(1,000)
OVERTIME	10,000	14,000	(4,000)*
FRINGE BENEFITS	40,500	41,000	(500)
SUPPLIES	20,000	25,000	(5,000)*
MAINTENANCE AND SUPPLIES	20,000	17,000	3,000*
SCRAPPAGE	14,000	18,000	(4,000)*
PERISHABLE TOOLS	3,000	6,000	(3,000)*
UTILITIES	8,000	8,100	(100)
OTHER CONTROLLABLE COSTS	9,500	10,500	(1,000)*
TOTAL CONTROLLABLE COSTS	$492,000	$506,600	($14,600)
NONCONTROLLABLE COSTS:			
RENT	$5,000	$5,000	—
DEPRECIATION AND AMORTIZATION	8,000	8,000	—
INSURANCE	1,000	1,300	($300)
TAXES: REAL ESTATE AND PERSONAL PROPERTY	1,500	1,500	—
OTHER NONCONTROLLABLE COSTS	2,500	2,600	(100)
TOTAL NONCONTROLLABLE COSTS	$18,000	$18,400	($400)
TOTAL MONTHLY COSTS	$510,000	$525,000	($15,000)

*Items have exceeded budget amount by 10 percent.

figure 8.9 Typical detailed exception report—monthly cost report for department 4 line supervisors and their superiors.

manufacturing process. In turn, a summary report of all departments, shown in figure 8.10, would be of utmost importance to the plant manager. Lastly, a summary of factory cost of sales is integrated into the income statement for the current month, figure 8.11. At this last level, the information is of great concern to top management—in particular, the president, since this individual is concerned with whether short-range plans (budgets) have been met or exceeded.

Interactive processing reports In an interactive processing mode, detailed, exception, and summary reports can be produced, just as in a batch processing mode; however, the accent is more on visual than on hard copy output. Specifically, an interactive mode is oriented toward extracting inquiry information from the system to answer questions of interest to both management and operating personnel in controlling present and future operations. Within this framework, a wide variety of inquiry formats can be designed to meet specific operating needs.

To illustrate the use of visual output employing a variety of inquiry formats, the following sales example is presented, including data previously stored on the on-line data base. Inasmuch as the company's forecasted sales are stored on the data base, they allow marketing management, via an inquiry I/O terminal, to compare projected product sales to actual sales activity (updated on a daily basis) for some specific time period. In order to

MONTHLY FACTORY COST OF SALES
JULY, 198_

DEPARTMENT NUMBER	BUDGET	ACTUAL	FAV. (UNFAV.)
1	$200,000	$197,000	$3,000
2	195,000	185,500	9,500
3	300,000	299,000	1,000
4	510,000	525,000	(15,000)
5	840,000	875,000	(35,000)
6	615,000	640,000	(25,000)
7	485,000	490,000	(5,000)
8	475,000	466,000	9,000
9	110,000	113,000	(3,000)
10	170,000	134,500	35,500
TOTAL MONTHLY FACTORY COST	$3,900,000	$3,925,000	($25,000)

figure 8.10 Typical summary report—factory cost of sales for the current month.

INCOME STATEMENT
JULY, 198_

	BUDGET	ACTUAL	FAV. (UNFAV.)
SALES	$6,550,000	$6,650,000	$100,000
COST OF SALES	3,900,000	3,925,000	(25,000)
GROSS PROFIT ON SALES	2,650,000	2,725,000	75,000
MARKETING EXPENSES	1,250,000	1,275,000	(25,000)
GENERAL & ADMIN. EXPENSES	600,000	620,000	(20,000)
NET PROFITS BEFORE FEDERAL INCOME TAXES	$800,000	$830,000	$30,000

figure 8.11 Typical summary report at the highest level—income statement for the current month.

make this information available to the proper marketing personnel, it is necessary to program different inquiry formats for maximum flexibility and ease of inquiry from the interactive system. Systems analysts have decided on a three-part division of inquiry formats: total sales, detailed sales, and exception sales.

"Total sales," the highest level of marketing inquiry, is the overall performance summary that describes the month-to-date or current month, last month, year-to-date, last year-to-date, and other data pictured in figure 8.12. A sample inquiry display for the company's seven basic product groups is shown in figure 8.13. A review of the illustration indicates that the function of this level is to guide further inquiry and analysis in an area which is over or under budget forecasts. If a given sales performance is deficient compared to the budget figures, control can then be transferred to successive levels of detailed reports relative to the area of interest.

The second level of inquiry, "detailed sales," contains many modules of optional levels of detail. Normally, the inquiries requested of this structure are triggered for further

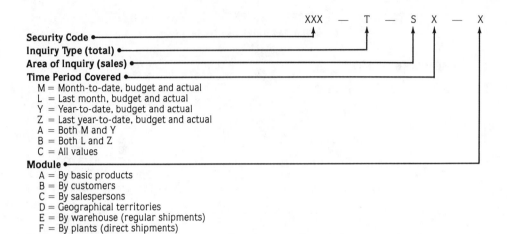

figure 8.12 A typical sales inquiry format in an interactive processing mode. (Robert J. Thierauf, **Systems Analysis and Design of Real-Time Management Information Systems,** © 1975. Reprinted by permission of Prentice-Hall, Inc., Englewood Cliffs, New Jersey.)

analysis by the total sales inquiry. As with the first level, data are available for the same time periods and by the same sales categories. This inquiry format is programmed so that the user can format his or her own report structures since marketing management might be interested in certain sales ratios or sales-to-expense ratios.

The last inquiry level, "exception sales," is more complex and flexible than the first two. The exception inquiry is structured to highlight a certain condition or conditions. A sample inquiry is, "Which products are above 110 percent or below 90 percent of forecasted sales?" Another inquiry example is, "Which salespersons are below their sales quotas?" In essence, this inquiry level provides the user with the capability to ask basic questions about the efficiency or lack of efficiency of the company's marketing effort.

BASIC PRODUCTS	MTD (000) BUD	ACT	LTD (000) BUD	ACT	YTD (000) BUD	ACT	ZTD (000) BUD	ACT
GROUP 1	412	400	394	410	1240	1195	1214	1275
GROUP 2	673	675	711	720	2103	2050	2202	2340
GROUP 3	1415	1450	1375	1379	4292	4095	4252	4242
GROUP 4	2040	2035	2100	2140	5953	5875	6010	5990
GROUP 5	1594	1505	1652	1695	4753	4852	4852	4950
GROUP 6	1141	1145	1133	1155	3502	3550	3493	3510
GROUP 7	1180	1195	1191	1193	3471	3480	3478	3495
TOTAL	8455	8405	8556	8692	25314	25097	25501	25802

Question: What are the budget and actual month-to-date, last month-to-date, year-to-date, and last year-to-date total sales for the company's seven basic product groups?
Code: XXX — T — S C — A (based upon figure 8.12)

figure 8.13 A typical total sales inquiry format in an interactive processing mode (for display on a CRT device). (Robert J. Thierauf, **Systems Analysis and Design of Real-Time Management Information Systems,** © 1975. Reprinted by permission of Prentice-Hall, Inc., Englewood Cliffs, New Jersey.)

This area alone might justify the interactive approach since this information represents new knowledge to management for planning and controlling marketing activities.

Actual design of reports Having set forth the basic types of reports available in a batch or an interactive processing mode, the systems analyst can now specify the exact location of desired information on the report. The individual will normally use a standardized "printer spacing chart" for laying out the report format. This chart will indicate the content and spacing of information and will serve as documentation for future reference, especially when the programming phase is undertaken.

An example of a printer spacing chart is found in figure 8.14 (p. 216). Inspection of this typical printer spacing chart indicates that there are 150 print positions. (Current printer capabilities range from 80 to 150 print positions.) The horizontal lines are numbered from 1 through 50, representing the lines on the report. For the chart illustrated, 6 lines to the inch can be printed.

The proper way to use a printer spacing chart is to select the specific fields for printing and place Xs in the selected positions for printing **variable information.** Variable information is defined as information that will change from one report to another. As depicted in figure 8.15, the inventory number, description, beginning balance, additions, deletions, and ending balance will change as each new line is printed. In addition, **constant information** will be printed on the report. This includes report titles, column headings, and the like. (pp. 218-219)

The illustrated report also indicates where information is to be printed. As shown, "inventory no." is printed in locations 3 to 15, "description" in locations 44 to 54, and so forth. Similarly, variable inventory numbers are to be printed in locations 10 to 15, variable description in locations 30 to 54, and the like. The report is to be triple-spaced; if single spacing were desired, there would be no space between lines. (Other types of spacing are permitted to increase the readability of the report.)

Another important function of the printer spacing chart is to specify **editing** of the printed report. Editing is the process of inserting special symbols, such as the comma, the decimal point, and the dollar sign; four examples of inserting a comma are found in figure 8.15. Additionally, editing includes the suppression of insignificant leading zeros, which is useful when printing dollar amounts that are preceded by one or more leading zeros. Zero suppression is accomplished by placing a zero in the position where the zero suppression is to stop.

Although the preceding presentation centered on a printer spacing chart that utilized a single page, the final report is to be printed on **continuous forms.** After the forms are printed, they can be separated at perforation lines before forwarding to the proper sources. Figure 8.16 (p. 217) exemplifies a continuous form with lined and no preprinted information, whereas figure 8.17 (p.217) is a typical example of preprinted continuous forms.

One last important function of the printer spacing chart is concerned with setting forth the size of the fields to be used for totals. The lowest level is generally referred to as "minor" totals, the next level as "intermediate" totals, and the highest as "major" totals. For any type of report, care must be taken that the proper levels of totals are specified. Otherwise, totals may overlap adjacent fields.

A comparable approach must be undertaken to produce visual output. As indicated in the previous example of sales inquiries, certain fields of information must be called in from

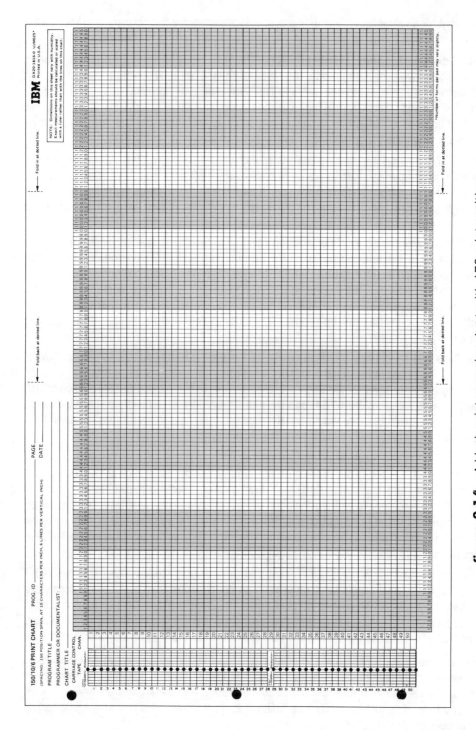

figure 8.14 A blank printer spacing chart with 150 print positions.

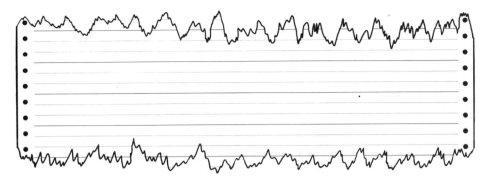

figure 8.16 A standard continuous form.

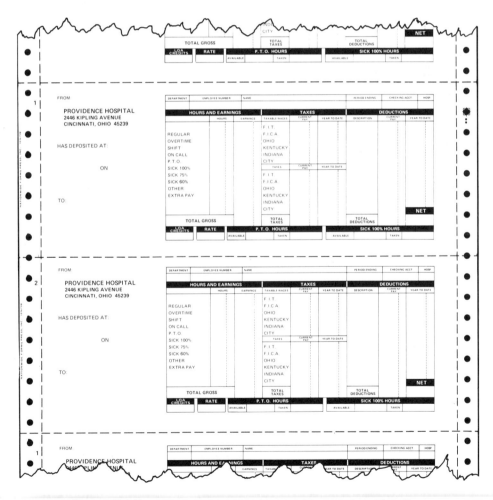

figure 8.17 Typical continuous forms with preprinted headings.

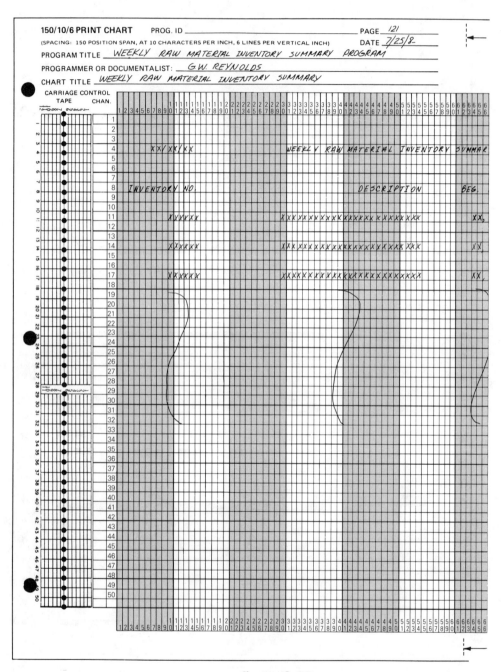

figure 8.15 A typical printer spacing chart

	PAGE XX		
BAL.	ADDITIONS	DELETIONS	END BAL.
XXX	XX,XXX	XX,XXX	XX,XXX
XXX	XX,XXX	XX,XXX	XX,XXX
XXX	XX,XXX	XX,XXX	XX,XXX

for the weekly raw material inventory summary.

the data base in conjunction with an inquiry format (figure 8.12) to produce the desired sales information—total, detailed, and exception. Because this type of output requires the use of a data base, further discussion of relating the data base to output is set forth in the next chapter.

Use and Distribution of Output

Basically, there are two types of output reports: those that serve the internal and external needs of an organization. Internal reports are those that are used by personnel within an organization to assist in the performance of their jobs. Their major requirements are that they be designed for specific tasks and that they conform to desired organization objectives. Examples are inventory summaries, production reports, payroll reports, and accounts payable registers.

In contrast, external reports are those which focus on satisfying demands that originate outside the organization. Whereas internal reports are designed essentially for their usefulness, external reports serve a dual purpose in many cases—usefulness to and satisfaction of customer, government, stockholder, and legal requirements. Typical examples are customer sales invoices, dividend payments, federal income tax returns, and payroll tax forms. External reports, then, are necessary to relate an organization's activities to the real world—in particular, its customers, competitors, vendors, labor unions, banks, governmental agencies, and the community. With the rise of **consumer awareness** and the demand for more **social responsibility** on the part of business organizations, more information must be produced to meet outside demands.

Whether output is either internally or externally oriented, consideration must be given to its distribution. In an interactive processing mode, there is no need to discuss the distribution of output from an inquiry since the user is the sole recipient of output. On the other hand, those batch operations, such as customer sales invoices and payroll checks, which are run by the thousands require some type of output procedures. If there are five copies of a sales invoice, it is essential that a decollator, a device for separating forms, and a burster, a device for separating the continuous forms, be utilized. In an operation where thousands of payroll checks are produced, it is required that an automatic check signer be used after computer processing. Only after consideration has been given to the proper handling of output is it ready for distribution. Depending upon the type of output, distribution can be made either internally or externally.

SUMMARY: Design of System Output

Consider the Human Factor in Designing Output Focuses on those factors that assist organization personnel to accomplish their assigned tasks.

Review Information on System Output Refers to reviewing present output as compiled during the systems analysis phase and information contained in the exploratory survey report to top management on the recommended system alternative.

Specify Output Requirements Relates to determining what information is needed as output from the new system. Important questions relating to output must be answered before finalizing the content of output. It is recommended that organization personnel work with systems analysts for best results.

Determine the Content of Output Results in the determination of the specific contents of each report, summary, and comparable output. Generally, a system output analysis form is prepared for each output.

Design of Output Refers to determining the output format. Reports can be in a batch or an interactive processing mode. Also, they can be detailed, exception, or summary reports.

Use and Distribution of Output Relates to internal and external reporting as well as the methods of completing output for its distribution to various users within and outside the organization.

CHAPTER SUMMARY:

The task of designing effective system output at a reasonable cost is a formidable task for any group of systems analysts. With the wide variety of output—printed, visual, turnaround documents, secondary storage, microfilm, audio response, and plotter—the systems analyst can devise many alternative forms for management and operating personnel. The systems analyst must work with organizational personnel to determine the best type of system output to meet specific needs. Otherwise, the technical aspects of the new system will predominate over all other considerations, in particular, the human factor.

Underlying the output design process are important system principles which must be kept in mind by systems analysts. Many of these principles are reflected in the initial design step. After all information compiled to date on output has been reviewed, the next steps are to specify the output requirements and determine their specific contents. This, in turn, provides a basis for the actual design of the reports. Then the internal or external reports are distributed to the appropriate organization personnel or forwarded to someone outside, respectively. This succession of steps results in a systematic approach to the design of desired system output; other procedures generally lead to less than optimum results. Thorough use of these steps will be made to determine printed and visual output for the ABC Company in chapter 10.

Questions

1. Of the common types of system output, which ones are oriented toward:
   a. batch processing?
   b. interactive processing?

2. How important is secondary storage as a type of output in an interactive management information system?
3. How important is the Principle of Timely Output in an interactive management information system?
4. State two more principles that should be an integral part of designing system output.
5. Which is the most important when designing system output—the technical or the human factor?
6. What is a good approach to testing the validity of system output?
7. How should the systems analyst go about determining the content of system output?
8. How do batch and interactive processing reports differ?
9. Distinguish between "variable" and "constant" information as found in typical printed reports.
10. a. Distinguish among the following: detailed reports, exception reports, and summary reports.
    b. Which are helpful in assisting top management? Middle management? Lower management?

Self-Study Exercise

True-False:

1. _____ The most common type of all system output is the utilization of plotters.
2. _____ Turnaround documents are rarely used as a form of system output.
3. _____ The Principle of "Management by Exception" is not found in interactive management information systems.
4. _____ Present reports should be used as is in the new system.
5. _____ Computer output can be designed in such a way that it is a humanizing force.
6. _____ It is necessary to specify output requirements before determining the content of output.
7. _____ Detailed printed and exception items can be contained in one report.
8. _____ Inquiry formats are used widely in designing output for interactive management information systems.
9. _____ Variable information is defined as that information which does not change for each line item in a report.
10. _____ An example of an internal report is an inventory summary.

Fill-In:

1. _____ needs of a typical organization include more than just printed reports.
2. _____ output is an essential part of an interactive management information system.
3. Since such large amounts of information can be generated in the form of printed output, many organizations are utilizing _____ output.
4. The Principle of the _____ of Reports centers on reports that will assist organizational personnel in fulfilling their work tasks.
5. The Principle of "Management by Perception" is considered to be a _____ looking approach to system output.
6. Eliminating or keeping _____ of information to a minimum is an important consideration when designing new reports.

Design of System Output **223**

7. The inclusion of the _____ factor in output design can increase the chances of a successful system project.
8. Specifying output _____ entail determining what information is needed as output from the new system.
9. A _____ printed report provides for the printing of one line of information for each record processed.
10. Zero-suppression in printed reports is one form of _____.

Problems:

1. The American Company is developing a new computerized finished goods inventory system that will print an inventory summary showing the monthly receipts into inventory and disbursements to the production departments. In addition, beginning and ending balances are shown along with provision for inventory adjustments (additions and subtractions). Not only is the computer program to generate the report and column headings, but also the current date and page number at the top of each page.

   Based upon the foregoing requirements, complete the following system output analysis form for a monthly finished goods inventory summary.

SYSTEM OUTPUT			
Date:		Type of output:	
Preparer:			
Informational elements	Type of characters	Number of characters	Comments
Comments:			

2. Utilizing the data set forth in the completed system output analysis form of The American Company, prepare an appropriate printer spacing chart for a monthly finished goods inventory summary. All information on the report is to be double-spaced and easily readable. Also, the spacing chart is to include a provision for printing appropriate totals for management's use. (Use chart on page 224.)

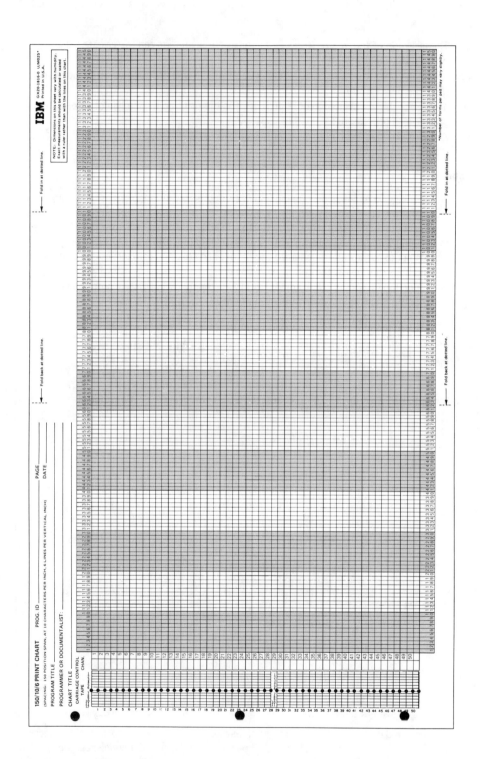

chapter

design of system data files

OBJECTIVES:

- To examine the common types of storage and organization of on-line and off-line files.
- To investigate the important design principles when developing data files.
- To delineate the important data file design criteria for a typical system project.
- To enumerate the specific steps involved in the design of data files in a system project.

IN THIS CHAPTER:

Introduction to Design of Data Files
COMMON TYPES OF FILE STORAGE
   Punched Card Files
   Magnetic Tape Files
   Magnetic Disk Files
   Magnetic Drum Files
   Mass Storage Files
COMMON METHODS OF FILE ORGANIZATION
   Sequential File Organization
   Direct File Organization
   Indexed Sequential File Organization
INCORPORATE SYSTEM DESIGN PRINCIPLES—DATA FILES
DESIGN OF SYSTEM DATA FILES
   Consider the Human Factor in Designing Data Files
   Review Information on System Data Files
   Specify Data File Requirements
   Determine the Content of Data Files
   Design the Data Files
DATA BASE MANAGEMENT SYSTEMS
   Advantages and Disadvantages of DBMS
CHAPTER SUMMARY
QUESTIONS
SELF-STUDY EXERCISE

This chapter presents the underlying principles and procedural steps in designing data files for a new system. Although the desired output has been specified to meet user needs, there is still need to obtain much of this information from some type of data files. In typical systems, these data files will be on line, off line, or a combination of the two. No matter what type of files are used, an effective system requires a logical integration of output with data files. From this view, a great deal of time and effort will be required by systems analysts to specify the proper types of file storage, the most efficient methods of file organization, and the contents of the data files in order to produce the desired output. In chapter 10, this relationship of outputs and data files will be examined through the design of an order-entry system for the ABC Company.

This chapter concentrates on the common types of file storage and the common methods of file organization and highlights the advantages and disadvantages of each. The chapter also explores the underlying design principles for data files. This comprehensive introduction to on-line and off-line data files is helpful in resolving the many alternatives to file design because knowledge of what comprises data files facilitates the determination of the preferred design of an organization's data files. Lastly, the current trend to data base management systems (DBMS) is treated and their advantages and disadvantages are examined.

Introduction to Design of Data Files

As an introduction to the design of data files, it would be helpful to define the term **data file.** In simplistic terms, a data file is a group of related data records. They can be used in an on-line or an off-line mode. There are two broad categories of data files; namely, **master files,** which contain information that reflects the current state of a system, and **transaction files,** which contain individual records of daily activities. To illustrate, the daily inventory transaction file (magnetic tape) which records inventory transactions as they occur throughout the day is used once a day to update the master inventory file (also magnetic tape).

Within these categories, there can be **fixed length records** or **variable length records.** The former refers to a record which always contains the same number of positions while

228 Design of System Data Files

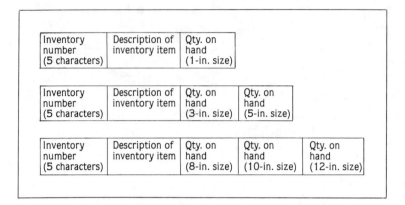

figure 9.1 Typical examples of a variable length record.

the latter refers to a record which may contain a variable number of positions. As illustrated in figure 9.1, variable length records are used when varying amounts of data are available for each record.

COMMON TYPES OF FILE STORAGE

Currently, there are several types of file storage which can be categorized as off-line storage and on-line storage. **Off-line storage** refers to data stored off line, that is, not under the direct control of the computer. The more popular types are punched card and magnetic tape. These data files can be used as input for a computer system to permit the updating and retrieval of information, or they can be used to produce output from the system. Generally, data needed to be referenced periodically, usually weekly or less frequently, are good candidates for off-line storage.

In contrast, **on-line storage** refers to data under the direct control of the computer. Magnetic disk, magnetic drum, and mass storage are the most common types of on-line files. Where there is a need to refer to data frequently, on-line storage is employed. Common examples are inventory updating and accounts receivable retrieval. On-line data files are commonly referred to as an organization's **data base** where information is centrally located in one or several data bases. However, with the advent of distributed processing, data files are presently referred to as **distributed data bases** because many data bases are scattered throughout an organization's operations.

In view of the importance of on-line and off-line files, the more common methods are discussed below.

Punched Card Files

The oldest type of files for batch-oriented computer systems is punched cards, containing either eighty or ninety-six columns of information. The eighty-column card, shown in figure 9.2, is divided into eighty columns numbered from left to right. Each column, in turn, is divided into twelve rows or punching positions. The punching positions, from the top of the card to the bottom, are 12 or Y, 11 or X, and 0 through 9. The top three punching positions of the card—12, 11, and 0—are the **zone punches,** the 0 through 9 are the **digit punches.** It should be noted that the zero-punch position may be either a zone punch or the digit 0. This combined zero-digit approach to punched cards is called the **zoned decimal format.** Reference will be made later in the chapter to this format for storage of on-line data.

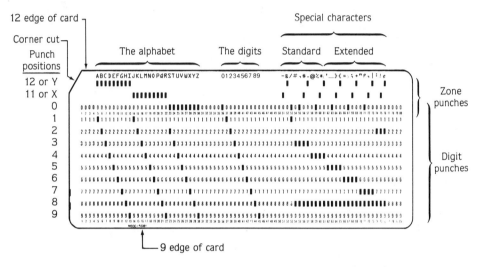

figure 9.2 An 80-column punched card.

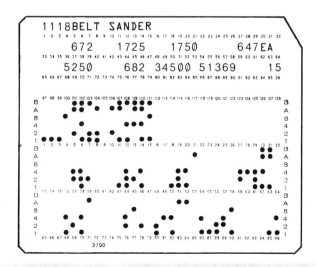

figure 9.3 A 96-column punched card.

The ninety-six-column card can hold sixteen more characters than the eighty-column card (a 20 percent increase) and up to 128 characters can be printed on the face of the card (a 60 percent increase). These differences for a card about one-third the size of the eighty-column card are shown in figure 9.3.

Regarding the card's structure, the punching area is divided into three parts: upper, columns 1 through 32; middle, columns 33 through 64; and lower, columns 65 through 96.

Advantages and disadvantages of punched card files One important advantage of punched card files is that they provide a check of the information printed on the cards. For small volumes of data, punched cards are an economical and efficient form of file storage. On the other hand, a most decided disadvantage is the limited amount of storage capacity. Many times, it is necessary to employ more than one card to store a record of information, such as names and addresses that include ship-to information. Additionally, the processing of punched card input and the punching of card output is slow in comparison to magnetic tape and magnetic disk. From an overall standpoint, there has been and continues to be a trend away from punched cards for off-line storage as well as for computer input.

Magnetic Tape Files

As the size of the data files become larger, magnetic tape files offer an effective means to handle large processing loads. Currently, equipment manufacturers have a variety of magnetic tape units compatible with their own computer lines. They differ basically in two ways: the speed in reading or writing data on tape and the data density (number of bits, digits, or characters per inch) of the tape. Regarding the tape's speed, magnetic tape has a very fast data transfer rate with speeds ranging up to about 1,250,000 characters or bytes per second for either input or output. Many tape units can read tape as they move in either direction.

Regarding the tape's data density, early recording devices placed 200 parallel characters on an inch of tape. Current systems allow 6,250 bits per inch. Latest developments in serial recording provide capability for 80,000 bits per inch. This increase in capacity has been the result of improved magnetic tapes and magnetic heads, and increased sophistication of the methods used to record and recover data.

Magnetic tape drives use magnetic tape, which is made from a very strong and durable plastic. Data are recorded on the tape surface by means of magnetized spots. As indicated in figure 9.4, numbers, letters, and special characters can be recorded by using the 7-bit alphanumeric code. A single reel of tape is usually 0.5 to 1 inch wide and 2400 feet long.

Current magnetic tape units The foregoing characteristics of magnetic tape units apply to most models currently on the market. For example, the IBM 3420 magnetic tape unit (Models 3 and 8) reads and writes data at densities up to 6,250 bits per inch on 0.5-inch magnetic tape. It is under the control of the 3803 control unit where several tape drives can be combined (figure 9.5).

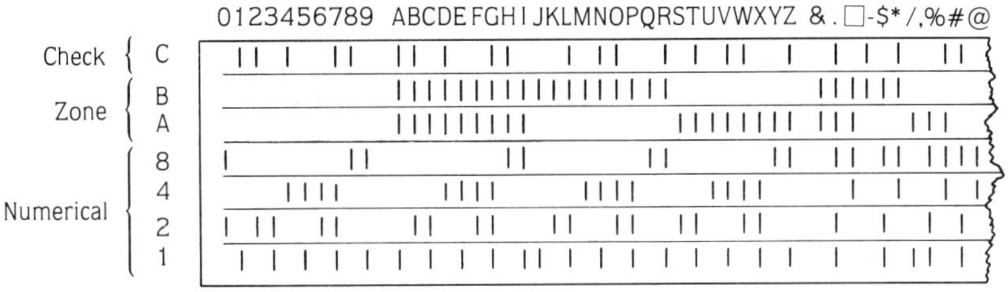

Common Types of File Storage **231**

figure 9.4 A piece of magnetic tape using a seven-bit alphanumeric code.

figure 9.5 The IBM 3803 tape control and four IBM 3420 magnetic tape units. (Photograph courtesy of International Business Machines Corporation)

Blocking of magnetic tape records The traditional method of reading and recording data on magnetic tape is to read or write a single record, followed by an inter-block gap on the tape. As illustrated in figure 9.6, the inter-block gap is a blank space on the tape that is approximately 0.6 inch long. It indicates to the magnetic tape unit that the record has ended. Each time a read or write command is given to the tape unit, only one **logical record** is read or written. It is obvious that in this method there is wasted space at the inter-block gap.

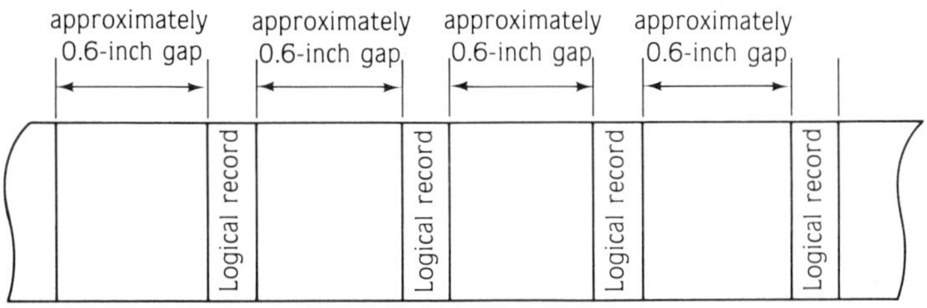

figure 9.6 The traditional way of recording data on magnetic tape.

To overcome wasted space and to increase the speed of reading and writing, it is helpful to block the logical records, that is, to group two or more logical records together and write them on magnetic tape as a **physical record** or **block**. Figure 9.7 illustrates a physical record comprised of three logical records. When compared to the traditional way of reading and writing data, magnetic tape is obviously used more efficiently with the blocking approach.

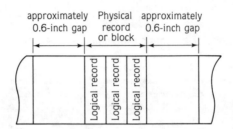

figure 9.7 The method of blocking records, that is, grouping two or more related records together.

There is a word of caution for systems analysts. The larger the block of records, the more main storage is required for the block. If, for example, there are forty 70-character records comprising the physical record, then 2800 positions of main storage are necessary when their contents are transferred in or out. In essence, consideration must be given to how many logical records are to be contained in the physical record in order to have efficient data file processing. Generally, a blocked record should be large as possible, limited only by the amount of main storage that is available for processing input and output records.

Advantages and disadvantages of magnetic tape files As indicated above, the major advantages of magnetic tape files are their fast input/output processing speeds and their capability of storing large amounts of data on a small piece of tape. When compared to punched cards, not only can data be processed faster in terms of input and output, but also a single reel of magnetic tape can hold the equivalent of many thousands of punched cards.

The principal shortcoming of magnetic tape is that data must be stored and processed in a sequential manner. For certain business applications that are processed periodically, this presents no major problem. However, because of the need for timely information in an interactive mode, there is no time to search one or more magnetic tapes for an answer. Thus, this data file method is outmoded for an increasingly large number of applications which require data to be accessed in a very short time frame.

Magnetic Disk Files

The foregoing disadvantage of magnetic tape files is overcome with the utilization of magnetic disk files. The **removable** magnetic disk, similar to a phonograph record, is coated on both sides with a ferrous oxide recording material. Information is recorded as magnetized spots on each side of the disk.

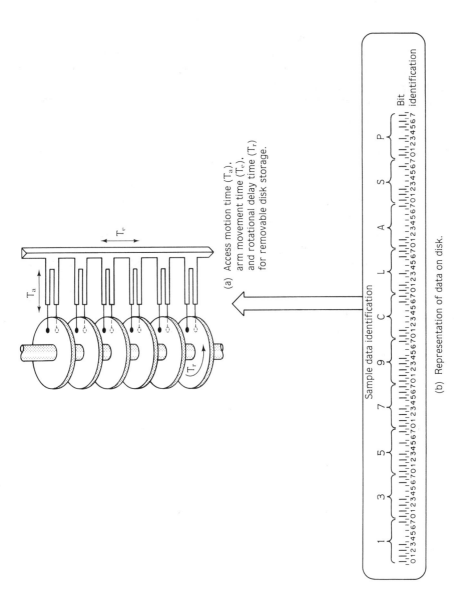

figure 9.8 The typical mechanism of magnetic disk file storage.

For example, six magnetic disks are mounted together as one unit, commonly referred to as a disk pack (figure 9.8a). The disk pack can be easily removed from the magnetic disk unit and be replaced with a new disk pack.

Rather than storing data by columns of characters, it is recorded serially bit-by-bit, eight bits per byte along a track. In figure 9.8b, data are stored serially, using the extended binary coded decimal interchange code (EBCDIC). As shown, enough space is available between each disk to permit access arms to move in and read or record data. An access arm generally has two read/write heads for retrieving or recording on either side of a disk. To read or write data, an access arm must be positioned on the disk over the desired location. The arm moves in or out to locate the correct storage track. If there is only one arm for the entire disk pack, the arm must move out from the stack of disks and up or down to the correct disk (shown as T_e in figure 9.8a). The greater the number of heads, the faster the data may be read or written.

Access to one specific track on a given recording surface is accomplished by the lateral movement of the whole access mechanism from a current track location. The time required for this movement is called **access motion time** (T_a) and is related to the lateral distance the arm moves. In addition to access motion time, there is **rotational delay time** (T_r), the time required for the disk to position the desired record at the selected read/write head. Maximum rotational delay time is slightly more than the time required for one full revolution, average rotational delay time being one-half the maximum. Typical disk speed is 1800 revolutions per minute. The selection of the proper read/write head is performed electronically, simultaneously with access motion time; the selection time is negligible. Total data search time for disk storage, then, includes the access time and rotational delay time.

Data are addressed on the disks by the disk number, the sector on the disk, and the track number on the IBM 2316. The disks, numbered consecutively from bottom to top, are assigned sectors on each side. Generally, they are assigned sector addresses 0 through 4 for the top side and 5 through 9 for the bottom of the disk. There are 200 tracks of recorded data on each side of the disk, having addresses 000 through 199. In some disk files, record lengths may be flexible enough to allow recording variable lengths of data records. Thus, the number of characters available per sector for each track varies with the density of the data stored.

In addition to the removable disk, there are **fixed** magnetic disks which cannot be removed. Designed for environments where data integrity is very critical, these sealed magnetic disk units have a large storage capacity. Since the heads do not move from track to track, as in moving head units, these drives have faster average access times.

Current magnetic disk units The storage capacity of current removable magnetic disk packs varies; some have the capability to store many millions of characters of information. Even though access time is restricted to the revolving action of disks and the arm movements, the average search time for newer equipment (like the IBM 3344 in figure 9.9) is 25 milliseconds (0.025 second), whereas the average rotational delya is 10.1 milliseconds (0.0101 second). Once the data are located by the read/write head, the speed for the data transfer rate can be as high as 885,000 bytes per second.

figure 9.9 The IBM 3344 Direct Access Storage Drive. (Photograph courtesy of International Business Machines Corporation)

figure 9.10 The IBM 3330 Disk Storage Facility with the IBM 3830 Storage Control Unit. (Photograph courtesy of International Business Machines Corporation)

Current multiple magnetic disk units include the IBM 3330 Disk Storage Facility (figure 9.10) which has a maximum storage capacity of 800 million bytes. It features modular construction, allowing users to configure up to four dual-disk modules as their requirements expand.

Advantages and disadvantages of magnetic disk files Important characteristics of the disk is that any item of data is as quickly obtainable as any other, and that it is possible to skip over unwanted data. Another decided advantage over magnetic tape is the ability to process transactions without previously sorting the data. Also, several different but related data may be stored on disk files, thereby allowing a transaction to be processed against these files at the same time. For example, a customer order can be processed on line against the credit check, inventory, accounts receivable, and sales analysis files. In essence, magentic disks are best for random access operations where input data are not arranged in any particular sequence before they are written on the disk. This direct accessibility plus the vast storage capability and relatively fast transfer rate have made them widely used in computer systems.

A principal disadvantage of magnetic disk files is their higher cost over magnetic tape files. In addition, if the magnetic disks are removable, it is necessary to insert the proper disk pack(s) when processing is required. However, for those data files that are updated based upon continuous operations, this is not a disadvantage.

Magnetic Drum Files

An alternative approach to on-line storage capabilities is the use of magnetic drum files. In the early days of computers, magnetic drums were the basic means of primary storage. Today, they are used as on-line storage devices where fast access to secondary storage is required. Basically, a magnetic drum is a cylinder with a magnetized outer surface. Some magnetic drum units are capable of storing several million characters of data. The IBM 2301 Drum Storage (figure 9.11), which utilizes the 2820 Storage Control Unit, can store up to 4,100,000 bytes or 8,200,000 digits. The data transfer rate to and from the central processor may be up to 1,200,000 bytes per second or 2,400,000 digits per second, and the rotational delay to a specific part of the track ranges from 0 to 17.5 milliseconds, averaging 8.6 milliseconds.

Data are normally represented in standard binary coded decimal form, as depicted in figure 9.12. Storage is in the form of invisible tracks around the cylinder, and each track is divided into sections which are, in turn, subdivided into character locations. The number of tracks, sections, and characters depends upon the size of the magnetic drum. As the drum rotates, data are recorded or sensed by a set of read/write heads. These heads are close enough to the surface of the drum to magnetize it and to sense the magnetization on it.

Advantages and disadvantages of magnetic drum files An important advantage of magnetic drum files is the fact that data are always stored on line since the drums are not removable. However, the size of the drum limits the quantity of information that can be stored. For small on-line storage capabilities, magnetic drum files are feasible; for large amounts of on-line storage, magnetic disk is the preferred alternative.

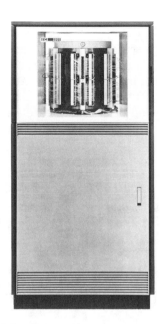

figure 9.11 The IBM 2301 Drum Storage Unit, which utilizes the IBM 2820 Storage Control Unit. (Photograph courtesy of International Business Machines Corporation)

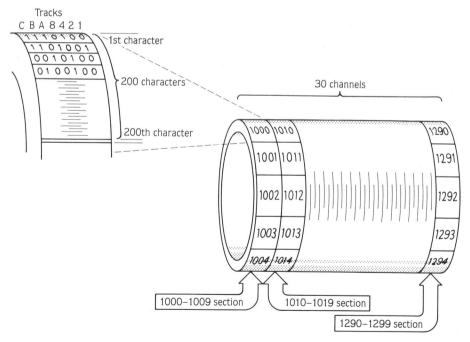

figure 9.12 Schematic diagram for a typical drum storage unit.

Mass Storage Files

Mass storage files provide large secondary storage capabilities. Such storage systems reflect significant improvements in operating efficiency and enhanced data security over conventional computer tape libraries. For example, the CDC 38500 Mass Storage System (figure 9.13) holds 2000 magnetic tape cartridges, each of which has a data storage capacity of 8 million bytes of on-line data, totalling a maximum of 16 billion characters of stored data. System capacity is equivalent to approximately 6400 average tape reels or 200 disk drives.

figure 9.13 The CDC Model 38500 Mass Storage System. (Photograph courtesy of Control Data Corporation.)

The mass storage system may be under the control of up to four computers like the IBM System/370. It also operates through a data-staging process, that is, information is transmitted to intermediate disk storage devices for subsequent computer processing. Overall, such a system has the capability to meet the storage requirements of large computer installations.

Advantages and disadvantages of mass storage files As the name indicates, a primary advantage of this form of storage is the large amount of data that can be stored on line. With the trend toward more interactive MIS and, therefore, more data stored on line, this new form of data file will assume increased importance in medium- and large-sized organizations. At this time, technology is decreasing the cost of mass storage, making this method of storage economically feasible. Thus, the wave of the future may well be this form of storage for interactive processing.

> **SUMMARY: Common Types of File Storage**
>
> **Punched Card Files** Represent the oldest form of file storage for input and output to computer systems. Whether the eighty- or ninety-six-column card is used, their storage capability as well as their reading and punching speeds in a computer system are very limited. In light of these shortcomings, other forms of file storage are recommended over punched card files.
>
> **Magnetic Tape Files** Speed of transfer rates range up to 1,250,000 characters or bytes per second for either input or output. Tape density can be as high as 6,250 bits per inch. It is advisable to **block** records, that is, group together two or more logical records for increased processing speeds and reduced tape storage. For certain business applications which are processed periodically, magnetic tape files offer the most economical method for storage and processing.
>
> **Magnetic Disk Files** Speed of data transfer rates can be as high as 885,000 bytes per second, with an average search time of 25 milliseconds and an average rotational delay of 10.1 milliseconds. Magnetic disk has the ability to process transactions without sorting the data previously, a decided advantage over magnetic tape. This feature of direct accessibility, a vast storage ability, and relatively fast transfer rate have made disk files widely used.
>
> **Magnetic Drum Files** Speed of data transfer rates can be as high as 1,200,000 bytes per second or 2,400,000 digits per second and the rotational delay to a specific section of the track averages 8.6 milliseconds. Magnetic drum, like magnetic disk, is used where fast access to secondary storage is required. However, magnetic drum cannot be physically removed from the storage unit, making its storage capacity more limited than magnetic disk.
>
> **Mass Storage Files** A typical system holds 2000 tape cartridges which provide up to 16 billion characters of stored data—equivalent to about 6400 tape reels or 200 disk drives. This newer form of data storage is employed in medium- and large-sized organizations that require the on-line storage of large amounts of data.

COMMON METHODS OF FILE ORGANIZATION

Having set forth the common types of file storage, it would be helpful to set forth their methods of organization. In certain cases, the method of file organization is dictated by the size of the system to be designed, particularly the small computer or minicomputer system; in other cases, the systems analyst has more latitude in the design of data files in a larger computer system. No matter what type of computer system is used, the most popular types of file organization methods are
    sequential
    direct
    indexed sequential

Sequential File Organization

The sequential file method of organizing relies on the two basic types of files noted earlier in the chapter, master and transaction. Whereas master files contain current information that reflects the status of a system, transaction files contain daily operations of a system. In most business applications, master files are updated by the data contained in transaction files. As illustrated in figure 9.14, a new updated master tape file is produced after making the additions (code A), deletions (code D), and changes (code C). (In this illustration, both the master and transaction files are magnetic tapes.)

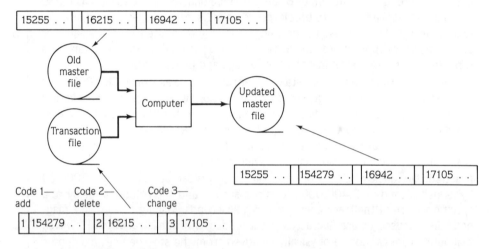

figure 9.14 A typical example of sequential file updating.

The sequential file approach is employed in those situations where there are many transactions to be processed against the master file and the master file is updated frequently. If only a few transactions were processed against the master file on a daily basis, updating would be prohibitively costly and time consuming. Also, the sequential file method requires the writing of a new tape at each updating, as demonstrated in figure 9.14. Thus, the size of the master file and its related activities—additions, deletions, and changes—determine the feasibility of this method of file organization.

Direct File Organization

The direct file organization method is only applicable to direct-access storage devices, such as magnetic disk, magnetic drum, and mass storage. Under this method, an address on the device allows the program to specify an exact location for immediate retrieval of information, so there is no need to process any other record in the file. In figure 9.15, if customer number 324, the key of the record, is represented by cylinder 3, track 2, the fourth record, and customer number 356 is represented by cylinder 3, track 5, the sixth record, then the computer program could be given a command to read the data stored in locations 324 and 356, respectively, for these two customers.

A primary advantage of direct file addressing over the previous method is that any record can be searched without examining the preceding records. This form of file

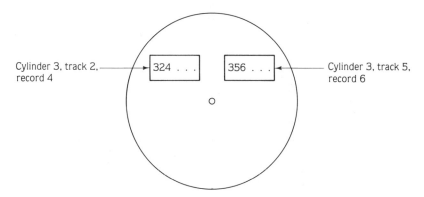

figure 9.15 A typical example of direct file updating.

organization, then, offers one of the fastest forms of inquiry and processing. Likewise, there is no need to create an entire new master file when new data are entered, as is the case with sequential file updating.

Direct file organization, on the other hand, has certain disadvantages. First, there is difficulty of linking the direct address with a key within a record. In the sample illustration, there was no problem; however, in most situations, this type of key-to-address conversion would not work efficiently. For example, many customer numbers are not in use at any given time, resulting in the inefficient use of disk storage. To overcome this problem, it is recommended that some type of arithmetic calculation, i.e., an algorithm, be performed upon the key of a record to generate a usable direct-access address for each record in the file.

Second, although various techniques exist for key-to-address conversion, there can be a problem if the key of the record is long or if it is alphanumeric. Any type of computation technique for key-to-address conversion results in duplicate disk addresses, called **synonyms**. Hence, some provision must be made to handle duplicate addresses. One approach is to record the synonym record at the next available location and **chain** (that is, cross-reference) to the original calculated address by a field indicating the address of the synonym. Generally, 20 percent more space should be allowed than is necessary in a sequential file.

Despite the problems encountered with the direct file organization method—unused storage space, extra programming required, and higher costs in terms of inefficient storage space—it allows faster access to records than is available with sequential files. Where access time is critical and fast retrieval necessary, such as in an interactive management information system, the direct file organization method is used so that on-line records are available for immediate reference.

Indexed Sequential File Organization

The indexed sequential file method of organizing data files provides for both sequential and direct file processing through the use of indexes on direct access devices only. An index is a pointer that directs processing to the location of a record within a file. For most applications, a group of indexes is established so that the method of locating a record in the direct-access file involves a stepping process whereby a high-level index directs the user to a lower-level index which locates the record. As illustrated in figure 9.16, there is a need for an index area on the magnetic disk.

When an indexed file in sequential order is initially set up on a magnetic disk file, all data file records are written on the prime **data area.** Within this area, data can be processed by either sequential or random-access methods, as discussed in the preceding sections. The user can expand the file without reorganizing it by writing records on the **overflow area.** To link the overflow area with the prime data area, as shown in figure 9.16, the records are always presented to the user in a certain sequence, even though they are not actually recorded in that sequence. Thus, the indexed sequential file organization method links all records together sequentially even if one or more records are located in an overflow area.

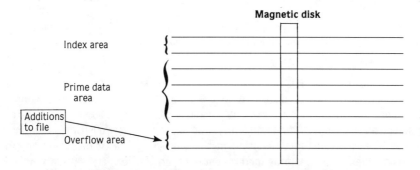

figure 9.16 Addition of records to a typical magnetic disk file.

Like prior methods of data file organization, there are advantages and disadvantages to this one. Although this method allows both sequential and random retrieval, there is a problem of inefficient processing. Since the record locations are referenced by indexes, one or more indexes must be read to locate the desired record. If there is very little use of the overflow area for new records, this method is quite useful in resolving data file organization. On the other hand, in a highly active file with numerous records in the overflow area requiring extensive indexing, this method is not recommended since the pointers which locate records in the overflow area must be read and processed.*

---

*For additional information on methods of file organization, see Edward Yourdon, **Design of On-Line Computer Systems** (Englewood Cliffs, N.J.: Prentice-Hall, Inc., 1972), pp. 261–309.

SUMMARY: Common Methods of File Organization

Sequential File Organization Employs a master file and a transaction file to produce a new master file, as in magnetic tape files. The transaction file consists of those additions, deletions, and changes occurring on a regular basis. The size of the master file and the number of transactions determine the feasibility of this method.

Direct File Organization Uses the location address number of the direct-access device to store information that has been read in with the same number, which is the key of a record. To overcome the problems of key-to-address conversion, it is recommended that some type of arithmetic calculation (algorithm) be performed upon the key of a record to generate a usable direct-access address for each record in the file. Where there is need for fast retrieval, this method is recommended.

Indexed Sequential File Organization Provides for both sequential and direct address file processing through the utilization of indexes to locate records on direct-access devices only. There are three storage areas: index, prime data, and overflow. Where there is very little use of the overflow area for new records, this method is recommended; if numerous records are contained in the overflow area, the processing time of locating records can be excessive.

INCORPORATE SYSTEM DESIGN PRINCIPLES— DATA FILES

Eight design principles are crucial to the design of data files—in particular, the on-line files which constitute an organization's data base. Failure to consider them will result in the deficient design of data files and may result in an unsuccessful system project.

Whether data files are on line or off line, a first and foremost system design principle is the **Principle of Eliminating or Reducing Duplication in Files.** The cause of duplication is that past design approaches structured files for individual applications, requiring the same data to be carried in a number of files. Common keys which provided a means of referencing file data, such as employee number, were mandatory because they were the only method of cross-referencing two or more files. The redundant information, however, included much more information, like employee's name and department number. (figure 9.17a).

The duplication of information in a variety of files is costly because excessive amounts of secondary storage capacity and a greater amount of magnetic tape, disk, or drum manipulation are required. A much more critical aspect, however, concerns updating this

(a)
Payroll file Employee no.	Name	Department	Payroll data
Earnings file Employee no.	Name	Department	Earnings data
Personnel file Employee no.	Name	Department	Personnel data

(b)
Common data base Employee no.	Name	Department	Payroll data	Earnings data	Personnel data

figure 9.17 (a) Redundant data files resulting from the past system approaches versus (b) a common data base approach in the current MIS environment.

multitude of files. It is not difficult to conceive of one file being updated while the status of the same information on a second, third, or fourth file remains unchanged because of an oversight or error in systems design, computer operation, or documentation. To overcome these problems, current approaches to MIS utilize the "single-record" concept, illustrated in figure 9.17b. Under this concept, information which formerly appeared in several data processing files now appears in one common file. Thus, one source—an organization's data base—can provide the same information to many users.

When trying to distinguish between the use of on-line and off-line data files, the **Principle of Suitability of Data Files** comes to the forefront. Only important information should be recorded on magnetic disk, drum, or mass storage files that must be referenced frequently for timely inquiries and reports. In contrast, secondary records that are processed periodically should be maintained on magnetic tape or some other low-cost storage medium. For example, in an interactive MIS environment, all information need not be resident in one or more on-line files. Off-line files can be interrelated with on-line files to permit the automatic updating and retrieval of information. Thus, data that need to be referenced periodically are poor candidates for on-line storage.

A third system design principle that is concerned with data file accuracy is the **Principle of Integrity of Data Files.** To ensure the integrity of data files, there must be a software feature that guarantees their accuracy during processing and updating. Hence, there is need for some type of lockout feature that prevents file data from being accessed during the updating process. Most equipment manufacturers provide such a feature in their data base management systems.

Although data file integrity is maintained, a fourth principle—the **Principle of Security of Data Files**—is needed to allow only authorized users to retrieve certain information. Various control methods can be devised to limit who is allowed to reference or manipulate selected file data. Chapter 14 will examine certain controls, like lockwords and checklists, that can be employed to restrict access to on-line data files.

Inasmuch as organizations are operating in an ever-changing business environment, there is need of a fifth principle, the **Principle of Flexibility of Data Files.** The new system should be so designed that the type and number of on-line and off-line data files can be expanded or contracted. In this manner, data files will be able to meet on-going organizational needs. Otherwise, inflexible or fixed data files generally cause higher data processing costs and longer processing response times.

Compatible with the above is a sixth principle, the **Principle of Variability of Data Files.** To conserve space, especially in voluminous files, there should be provision for adjusting record length to the amount of data. As with the earlier principle, higher DP costs and longer processing times may result for certain applications.

A seventh underlying principle—the **Principle of File Backup and Recovery Procedures**—centers on those recovery procedures that must be an integral part of regular DP operations. File dumps must be produced periodically by the system so that if a file is lost or destroyed, it can be reconstructed from any point in time by processing current transactions against a file dump. In this manner, a file can be restored.

While the foregoing centers on the operational aspects of data files, the last principle—the **Principle of Retention of Data Files**—focuses on the retention of data. Consideration for data retention should be based on management needs and legal requirements. Additionally, the systems analyst should remember that more data should be provided for in current files than is required for present output needs.

Although efficient systems design dictates keeping data to a minimum, this must not hinder the system's ability to meet and satisfy output needs that arise unexpectedly. If output requirements are specified in detail as they should be, the question of which files to maintain and which information to store is no problem. A problem occurs when output requirements are so inadequately specified that it is difficult to design data files to meet the system needs. Thus, designing more files than are needed for the requirements under study assures that many future report requests can be fulfilled. Because there is generally a high cost for constructing and maintaining files of questionable value, the cost of this added information versus value received is the real issued to be resolved.

SUMMARY: Incorporate System Design Principles—Data Files

1. **Principle of Eliminating or Reducing Duplication in Files** Duplicate or unnecessary data should be eliminated or kept to a minimum in on-line and off-line data files to increase economy and efficiency in data file processing and storage. Overall, data files should employ the "single record" concept, that is, information which formerly appeared in multiple files appears in one file to serve all authorized users.

2. **Principle of Suitability of Data Files** For important information that must be referenced frequently for timely inquiries and reports, data can be stored on line using magnetic disk, drum, or mass storage files. On the other hand, secondary records not referenced frequently can be maintained on magnetic tape or other low-cost storage medium.

3. **Principle of Integrity of Data Files** To ensure the integrity of data files, software features can be built into the system to guarantee the accuracy of data files during their processing and updating. This requires some type of lockout device so that file data cannot be destroyed during updating.

4. **Principle of Security of Data Files** This important principle provides for security of files or segments of files. Various control methods can be devised so that only authorized personnel are allowed to reference selected file data.

5. **Principle of Flexibility of Data Files** The system should be designed so that the type and number of data files can be expanded or reduced as the situation warrants. In this manner, on-line/and off-line data files are capable of meeting the organization's needs.

6. **Principle of Variability of Data Files** To conserve space, there should be provision for records of variable length. This type of record is used when differing amounts of data may be available for each record.

7. **Principle of File Backup and Recovery Procedures** This important principle provides for reconstruction of a data file which has been destroyed or damaged beyond use. File dumps are used as backup, and the procedures to restore a file to normal operations are called **recovery procedures**.
8. **Principle of Retention of Data Files** Retention of data files for long periods, whether on-line or off-line, must be considered in light of future management needs, legal considerations, and retention costs.

DESIGN OF SYSTEM DATA FILES

The preceding background on the types of file storage, the methods of file organization, and the principles of data file design serves as a review for the design of system data files. This part of the chapter will examine the designing of the data base and off-line files in terms of these five areas:

Consider the human factor in designing data files.
Review information on system data files, including file design criteria.
Specify data file requirements for the new system.
Determine the specific content of those data files that must be an integral part of the new system.
Design on-line and off-line data files.

Consider the Human Factor in Designing Data Files

Because the accent today is on interactive systems that include on-line data base and off-line data files, **improved data quality** is a major consideration in such systems. Since stored information is continuously updated, relevant updated information is made available for display. At the same time, data management considerations suggest that only required minimum updating be performed on line, postponing certain updates for batch handling. In such interactive systems, the importance of high quality documents cannot be overemphasized. Documents based on timely, accurate, and relevant information from the on-line data base and off-line data files foster a sense of confidence and trust. This, in turn, has a tremendous psychological impact on the user. Thus, the capability to extract quality information for managing an organization's operations accentuates the importance of the human factor in designing data files.

The systems analyst should recognize that he or she is dealing with different users who need different kinds of information from the data files. These considerations should apply to normal day-to-day operations as well as special conditions that might require additional information from the data files. To assist in answering these special requests, the data files should allow for adding critical information. In effect, events that focus on crises ought to be provided with data file procedures to recognize such information and add this information to the data files. In this manner, system files will have the required data to answer important questions at the appropriate level of management. Much of these data items can be used to practice "management by perception" and "management by exception" (as set forth in chapter 8). In effect, the capability to extract meaningful, timely information from the data base goes a long way towards answering a list of "what if" questions facing an organization.

Just as there should be provisions to add data file information, there also should be provision to maintain data quality. This may include correcting data files and collecting information that was disseminated before it was entered in the files. It is recommended that provisions for automatic correction of file information should exist.

In a somewhat similar manner, procedures should exist to give users a choice of how to interact with data files. For example, programs that vary in steps, yet accomplish the same purpose allow for each user to write a program that he or she understands. The capability to make choices may well mark the difference between a humanizing approach to data files and a dehumanizing one.

Additional important human factors to include in the design of data files can be garnered from the design principles delineated in this and the preceding chapters. Data file design should allow the user to understand and control data that is under his or her jurisdiction, should be easy and convenient for the user to control, should be flexible and versatile to meet specific information needs, and should meet response time and timing requirements of the user.

An underlying consideration of the human factor in designing data files is the use of the modular or building block approach. The ability to develop different file modules that are capable of standing alone greatly reduces overall data file complexity. It will also facilitate any modification or updating of data records that may be necessary in the future. Not only does this reduce the possibility of misunderstanding between the system analyst and the user, but also it assists in simplifying the final data file design. Thus, the modular approach is one of the most important human factors in developing organization data files.

Review Information on System Data Files

A logical starting point to data file design, as with output design, is a review of the information compiled to date, including data file information on the present system from the exploratory survey report to top management. The purpose of reviewing the present system is to focus on the problems and difficulties with current data files so that they are not duplicated in the new system.

Closely related to a review of the present data files is an in-depth understanding of the system alternative advocated in the exploratory survey report, particularly emphasizing the file design criteria used to select the recommended system. These include data file activity, data file processing speed, data file capacity, and data file cost. The following discussion of these criteria is from the standpoint of interactive systems that are supplemented by batch processing operations.

Data file activity File activity refers to the number of transactions to be processed against a master file, either a data base or off-line, within a certain time period. In an interactive system, the need for rapid access and fast processing of data files is generally one of the most important considerations. Where activity is high, direct file organization may be necessary. Conversely, if activity is low, an indexed sequential file organization should be considered. However, where there is a low volume of activity over time, the sequential file organization method may be the best approach, resulting in off-line files. Likewise, within a typical batch processing environment, the sequential file method is generally used for large volumes of data, due to its efficiency and low cost.

Data file processing speed   An integral part of data file activity is data file processing speed. Direct-access devices—such as magnetic disk, magnetic drum, and mass storage—offer the fastest file processing speeds for most business applications. Punched cards are the slowest form of file processing.

To determine which type of equipment should be used for speed and economy in data file processing, the required level of data file activity, the type of file organization method, data file costs, the capacity of the equipment to handle peak loads and future growth, and the speed of input/output units linked with the data files must be examined. These factors are applicable to both interactive systems and batch processing systems.

Data file capacity   File capacity refers to the maximum size of on-line and off-line files. Although the size of magnetic disk and magnetic tape files is not limited, there is a limit to the number of magnetic disk and tape drives that can be attached to the CPU for processing. If a certain data file uses either a direct or an indexed sequential file organization method for storage, all disk packs must be mounted on separate disk drives for computer processing. However, if the sequential file organization method is used, only the disk pack or the tape reel that contains the data to be processed must be mounted. Therefore, with very large files, the advantage of sequential files over direct or indexed sequential files should be obvious.

When reviewing data on the recommended system alternative, peak file processing loads must be projected over the next several years. The projection may reveal that the file organization method selected for today's volume will not be appropriate for future file processing loads. The highest consideration must be given to future file processing capacity.

Data file cost   In order to keep within the constraints of the exploratory survey report to top management, the systems analyst must relate the foregoing factors—data file activity, processing speed, and capacity—to their respective cost. Too often it has been necessary to change the file specifications because the capabilities needed for the system cost too much. The final design of the types of file storage and methods of file organization is critical to the cost and effectiveness of the new system.

Specify Data File Requirements

Specifying data file requirements entails determining what types of data file and methods of file organization are needed in the new system to meet the objectives of the system project. The preceding material provides the systems analyst with basic information in undertaking this important design step.

Data base Before the on-line data base can be determined, a number of file design problems should be considered. Generally, the systems analyst should know as much as possible about the various subsystems and the data to be considered in building the data base—types of data, number of data items, and number of characters per item. Consideration must also be given to the number of different application programs to be developed. The systems analyst, then, must balance available on-line storage space against cost and other critical factors.

To assist the systems analyst in structuring the data base, several important questions are detailed in figure 9.18. Of all these, the second one is probably the most difficult to answer.

What is the scope of the data base? What subsystems or application areas should be included in the design?
What data should be retained and for how long? What priorities are applicable to these data in terms of response time required from the system?
How should the various types of data be organized?
What security controls are required to protect sensitive data?
What storage medium should be selected? Is high-speed main memory required or are magnetic disks, tapes, or other storage media suitable alternatives?
What is the optimum method for entering data into and selecting data from the storage medium?
What controls are required for these data to ensure accuracy?
What allowances can be made for growth of records or data sets?

---

figure 9.18 Systems design—questions to assist in structuring a data base.

Another way of viewing these questions for data base design is set forth below. The establishment of certain requirements assists in resolving the data base structure and assures economy in systems design and operation.

Data base requirements should encompass the following:
1. It should permit the establishment of a single area for common information that is usable by all authorized users.
2. It should allow important information to be recorded on magnetic disk, magnetic drum, or mass storage, while secondary records are maintained on magnetic tape or other low-cost storage medium.
3. It should guarantee accuracy of updating by an automatic maintenance feature for all segments of the data file.
4. It should allow variable length records in order to conserve space.
5. It should provide for expanding or reducing the file—both the number of records in the file and the data elements in each record.
6. It should allow for security of files or segments of files.
7. There must be a provision for some type of lockout feature so that certain files or areas of files, or even individual records, cannot be accessed during updating.

Utilizing the questions and requirements will provide systems analysts with a basic framework in specifying file design. The design of the data base will dictate, or in some situations be constrained by, the type of storage devices used, the data base management system chosen to manipulate information in the data base, the degree of security that must be built into the system to protect the data by the types of required responses, and other important considerations. No matter what constraints are placed on the systems analyst, the data base elements should be integrated, as illustrated in figure 9.17. However, when data are too extensive for one unit record, related elements should be grouped logically to minimize file search time, as depicted in figure 9.19.

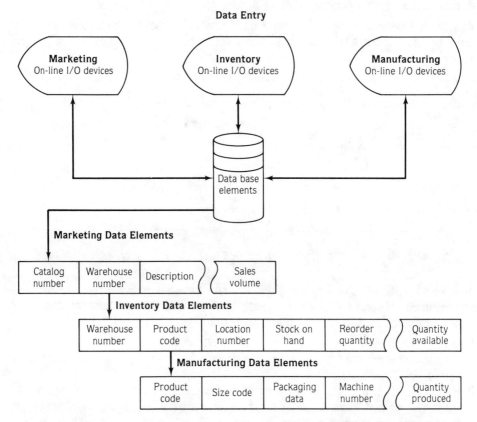

figure 9.19 A data base whose data elements—marketing, inventory, and manufacturing—are integrated. (Robert J. Thierauf, **Systems Analysis and Design of Real-Time Management Information Systems,** © 1975. Reprinted by permission of Prentice-Hall, Inc., Englewood Cliffs, New Jersey.)

Conceptually, the data base is a group of coordinated files of basic data elements in which duplication of data, if any, is kept to a minimum. Each file of individual items within the data base is a **building block** or **module,** thereby permitting constant coordination of functions and data as related to each record. The data base, then, is a store of up-to-date, accessible unit records that are comprehensive enough to serve as a basis for preparation of any required information or report.

Off-line files All information, such as that found in an interactive MIS, need not be resident in on-line files. In addition to data stored on line in magnetic disk, magnetic drum, mass storage, or comparable file devices, there is need for off-line files, such as magnetic tapes, which should be integrated with on-line files to permit the automatic updating and retrieval of information.

Systems analysts should refer to the basic questions found in figure 9.20 to evaluate off-line files. Failure to answer these basic questions may result in duplicating present file deficiencies and difficulties in the new system. Also, these questions serve as a basis for bringing together the factors essential to the design of efficient and economical off-line data files.

Overview of data file requirements The questions contained in figures 9.18 and 9.20 serve a dual purpose. First, they highlight the systems design principles set forth earlier in the text; that is, these questions provide the underlying basis for developing data file design principles. Second, these questions provide the means for specifying data file requirements; specifically, they assist the systems analyst in determining whether data files should be on-line files, off-line files, or a combination of the two. Similarly, these probing questions help the analyst to determine the method or methods of file organization.

What should off-line files contain? Describe the type and kind of records and estimate the amount of each.
How should the off-line files be referenced—by subject, by organization, by geography, by time, or on some other basis?
How should the information contained in the off-line files be structured—by subject or by organization?
What indexing will be necessary for off-line files?
What will be the annual rate of growth for off-line files?
Will any portion of the off-line files be restricted? What portion(s) and to whom?
What methods will be employed to prevent theft, destruction, or unlawful disclosure of off-line files?
Will any unrestricted portion of the off-line files be vital and irreplaceable?

---

figure 9.20 Systems design—questions to test the validity of off-line files.

Even though the main thrust today is on interactive MIS, there is no need to have all file data stored on line, but rather only critical data that is necessary to manage operations. Those file data that lend themselves to a batch processing mode should be processed as such, so that data file costs can be held to those that are set forth in the exploratory survey report to top management. Hence, the interactive system envisioned for the ABC Company has two types of data files—an on-line data base for interactive processing and off-line files for batch processing. Similarly, appropriate methods of file organization will be utilized for operational efficiency and economy.

Determine the Content of Data Files

Once the types of file storage and their methods of organization have been specified, the next step in the design process is to determine the specific contents of the on-line and off-line data files. At this point, it is advisable to involve personnel from the various functional areas to determine all data file content: the records and their data file elements; whether the elements are numeric, alphabetic, or alphanumeric; the number of positions of storage for each element; and any special considerations about these elements. In large organizations, for example, a **data base administrator,** responsible for the layout, allocation, and administration of new data files, would be involved in this phase of the system project.

After the contents of the data files have been finalized, the systems analyst drafts a **data file analysis sheet** that includes the data to be contained in the data base or off-line files. Table 9.1 specifies the data base elements for an accounts receivable file on a cycle billing basis, with appropriate comments about the data elements, secondary storage devices, and method of file organization, as well as the overall approval of such a file. The approval of the data base administrator is needed for this undertaking.

The data file analysis sheet also serves as a permanent record for this step of data file design, helping to avoid future misunderstandings.

table 9.1 Typical system data file analysis form.

SYSTEM DATA FILE

Page 1 of 1
Date: July 30, 198_
Preparer: G. W. Reynolds

Type of record:
ACCOUNTS RECEIVABLE CYCLE BILLING BASIS (Data Base Record)

Record elements	Type of characters	Number of characters	Comments
Name and address	Alphanumeric	45	Edit as three lines
Account number	Numeric	7	Edit with hyphens
Beginning period balance	Numeric	6	Edit with decimal point
Current charges	Numeric	6	Edit with decimal point
Current payments	Numeric	6	Edit with decimal point
Current adjustments	Numeric	6	Edit with decimal point
Current interest charges	Numeric	4	Edit with decimal point
Ending period balance	Numeric	6	Edit with decimal point

Comments:
The data base administrator has approved the record elements, type of characters, and number of characters for direct-access storage (magnetic disk) that utilizes the direct file organization method.

Design the Data Files

After the preceding design steps have been completed, the final step is the actual design of the data files to produce the desired output. Knowing the type of file storage, the

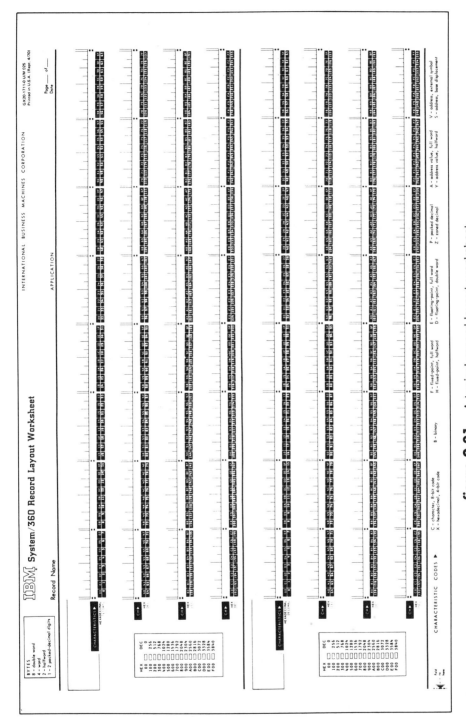

figure 9.21 A typical record layout worksheet.

254   Design of System Data Files

method of file organization, and the contents of the data files enables the systems analyst to specify the detailed file format that will be used for programming in the system implementation phase. Likewise, this information provides a logical basis for applying the modular approach to systems design. This approach will be demonstrated in the next chapter for the required data files of the ABC Company.

In order to bring together the knowledge accumulated on the design of on-line and off-line data files, a record layout worksheet is used. When files are to be contained on magnetic disks and magnetic tapes, the worksheet can be employed (figure 9.21). At the top of the form, space is available for recording the format of two records. Space is provided for a maximum of 256 characters per record at 64 characters per line, with additional space below each position for both hexadecimal and decimal counts. At the bottom of the worksheet, there is a listing of characteristic codes which can be used. This flexibility allows data on magnetic disks and magnetic tapes to be stored in various formats, unconstrained by the character format of data entry into the computer system.

To reduce the storage space of data on magnetic disk or tape when using the record layout worksheet, it is recommended that either a packed decimal format or a binary format (if computer uses EBCDIC)* be employed. For example, if datum 742 is read from a punched card using zoned decimal format, then three bytes of computer storage are needed for this type of format. On the other hand, only two bytes of computer storage are required if the datum is stored in the packed decimal format, as illustrated in figure 9.22. Obviously, considerable space can be saved on magnetic disk or tape.

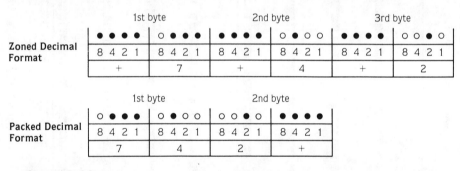

figure 9.22   A zoned decimal format (three bytes) versus a packed decimal format (two bytes).

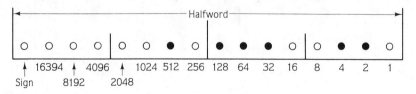

figure 9.23   A binary format (halfword).

---

*Extended Binary Coded Decimal Interchange Code.

	BINARY			PACKED DECIMAL			ZONED DECIMAL	
Numerical value	Binary representation	Byte		Hexadecimal representation	Byte		Hexadecimal representation	Byte
9	0000|1001	1		9F	1		F9	1
99	0110|0011	1		09|9F	2		F9|F9	2
999	0000|0011|1110|0111	2		99|9F	2		F9|F9|F9	3
9999	0010|0111|0000|1111	2		09|99|9F	3		F9|F9|F9|F9	4
99999	0000|0001|1000|0110|1001|1111	3		99|99|9F	3		F9|F9|F9|F9|F9	5
999999	0000|1111|0100|0010|0011|1111	3		09|99|99|9F	4		F9|F9|F9|F9|F9|F9	6
9999999	1001|1000|1001|0110|0111|1111	3		99|99|99|9F	4		F9|F9|F9|F9|F9|F9|F9	7

figure 9.24 A comparison of binary, packed decimal, and zoned decimal formats of from one- to seven-place decimal values.

In a somewhat similar manner, data can be stored more compactly on magnetic disk or tape than on media using the zoned decimal format. Data stored in a binary format is generally stored as either a halfword (sixteen bits), a fullword (thirty-two bits), or a double word (sixty-four bits). Using the same value as before, the binary format for this numeric value is found in figure 9.23.

A comparison of both illustrations indicates that storing data in a binary format is more efficient than either packed decimal or zoned decimal. In turn, the packed decimal format is more efficient than the zoned decimal format. As illustrated in figure 9.24, storage of a seven-digit number requires only three bytes in the binary format, as opposed to four bytes and seven bytes in the packed decimal and zoned decimal formats, respectively.

In addition to the data format, there are other important considerations before finalizing data file design. One is the need to provide for more storage than that required by the data read in; this is particularly necessary when the field in the data file is a total. The reading in of daily sales totals means that the field of the master file record must be larger than that of daily sales; otherwise, an overflow condition will result. Therefore, the systems analyst must determine the highest value of each field for master records to ensure their proper size.

Another important consideration to data file design concerns blocking. When deciding the block size, the amount of computer storage available for the input/output area and the device on which the block is to be stored must be assessed. Although the blocking size can generally be as large as desired for magnetic tape files, the same is not true for direct-access devices, such as magnetic disk. Hence, the general guide for this type storage is to keep the block size equal to or less than the maximum number of bytes that can be stored on a single track. If sufficient storage is not available to store the maximum block size, then the next best blocking size should be considered. In this manner, storage space on direct-access devices is efficiently used, allowing efficient and economical processing of data file records.

SUMMARY: Design of System Data Files

Consider the Human Factor in Designing Data Files Relates to those factors instrumental in helping organization personnel design efficient and economical data files.

Review Information on System Data Files Focuses on reviewing present data files compiled during the systems design phase and information contained in the exploratory survey report to top management. Similarly, consideration is given to data file design criteria—namely, activity, processing speed, capacity, and costs.

Specify Data File Requirements Refers to determining what types of data files and methods of file organization are needed in the new system.

Determine the Contents of Data Files Relates to the determination of the specific contents of the on-line and off-line data files. For the most part, a separate data file analysis form is prepared for each data file. For best results, it is recommended that system personnel work with organization personnel whose files are involved.

Design the Data Files Refers to designing the data files in the form of record layout worksheets, considering not only the data format (zoned decimal, packed decimal, or binary), but also blocking and the size of summary fields.

DATA BASE MANAGEMENT SYSTEMS

With the increase in the number of on-line data base elements available for immediate processing by many users, some method is needed to control the data base. One popular method of maintaining a large, complex data base and managing relationships among the data base elements is to employ a data base management system (DBMS). Among the current data base management systems are ADABAS, DMS, IDMS, IMS, System 2000, and TOTAL. Comparing two of the most widely used systems, the IMS package of IBM is a rather complex approach for controlling a data base, whereas TOTAL of CinCom is relatively straightforward and easy to use.

The reach of a data base management system extends beyond the individual file management systems, such as inventory files and accounts receivable files of the past, to manage an entire data base (or data bases), consisting of accounting, corporate planning, engineering, finance, inventory, marketing, manufacturing, personnel, physical distribution, purchasing, research and development, and other data elements. Furthermore, a DBMS allows **procedure independence** of the data base for a large number of data elements. This means that the programmer does not have to describe the data file in detail, as is the case with procedure-oriented languages like COBOL and FORTRAN, but rather only specify what is to be done in a **data management language.**

The relationship between the application program and the data base in a DBMS environment is shown in figure 9.25 (p. 258). The "data description component" analyzes each requirement of the application program, then transfers control to the "data manipulation component" which retrieves the needed data elements from the data base. In this manner, a data base management system reduces coding time and improves efficiency by allowing the programmer to use a catalog of file definitions for storing and accessing data efficiently. Thus, much of the efficiency of a DBMS is the procedure independence of the data base for producing the desired output report.

Typically, a data base management system (based upon controlled activities as shown in figure 9.25) has the following characteristics:
- operating environment independence—capability to run on many computers with certain types of operating systems.
- user-oriented—provision for an English-like language which enables the user to consider logical entities in place of physical entities, such as hardware and system software.
- data base independence from application programs—use of a data base language and "call" commands from the application program.
- security features—provision for controlling data base access and control over the data base elements.

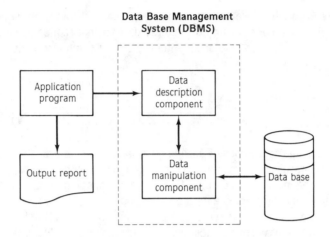

figure 9.25 The relationship between the application program and the data base in a data base management system (DBMS) environment.

Advantages and Disadvantages of DBMS

Inasmuch as future information system design will include a data base management system, the advantages of this arrangement should be considered. First, DBMS is not only effective for generating and maintaining a wide variety of routine management and operating reports, but also adaptable to meeting the new and emerging requirements of management to answer a myriad of "what if" questions. The latter capability means data base management systems will be important aids to managers seeking to explore and understand new relationships among various data elements. Second, data elements can be structured in a manner more suitable to their application, allowing their retrieval with a minimum of effort. Third, DBMS keeps redundancy of data elements to a minimum, since one data file serves many users; as a result of this "single record" concept, a transaction is entered once for all users. The DBMS also allows two or more files to be updated with the entry of a single transaction. Fourth, application programs are independent of the changes in the data base, so that their maintenance is kept to a bare minimum. Fifth, DBMS provides data protection not only for accessing one data base record at a time, but also for preventing data base access by unauthorized personnel.

Although the foregoing provide excellent advantages to DBMS users, the initial investment of system personnel time as well as software and equipment costs can be high; however, increasing experience with data base management systems is reducing these costs. Another disadvantage of DBMS is that incorrect transaction data tend to precipitate additional problems and errors throughout the system. Therefore, various tests should be incorporated to catch errors entering the data base before they are stored. These input tests, such as limit checks and reasonableness tests, will be treated in chapter

14 on system controls. Overall, these disadvantages of DBMS are capable of being overcome to a great degree if the information system is properly designed and managed.

CHAPTER SUMMARY:

The systems analyst's task becomes more complex when system output is integrated with the design of data files. This is particularly true when the various types of file storage, the methods of file organization, and the contents of the data files are considered. All combinations possible either in an interactive or batch processing mode, plus incorporating system design principles, results in an arduous task for any group of systems analysts.

As a first step in the design of system data files, the human factor is viewed from the standpoint of simplifying the design of files that will serve as output for organizational personnel. After information on data files and file design criteria are reviewed, the data file requirements are specified. Specifications are drawn from questions probing the various types of data files to be maintained in the new system. A synthesis of this information allows the systems analyst to determine the contents of data files, which are recorded on system data file analysis forms. Lastly, the data files are designed using record layout worksheets. During this step, consideration must be given to record format, blocking, and size of summary fields. Efficient and economical data files should be the end result of the combined effort of systems analysts and organization personnel, and will be examined in the next chapter for the ABC Company.

Questions

1. Of the common types of file storage, which ones are oriented toward:
   a. batch processing?
   b. interactive processing?
2. a. What are the advantages of magnetic disk over magnetic tape?
   b. What are the advantages of magnetic tape over magnetic disk?
3. What are the shortcomings of magnetic drum files as a principal means of storage? Explain.
4. Contrast the sequential file organization method with the direct file organization method.
5. What is the relationship of the indexed sequential file organization method to the sequential and direct file organization methods.
6. Explain the relationship of the index area to the prime data and overflow areas.
7. How important is the Principle of Eliminating and Reducing Duplication of Files within an interactive MIS environment?
8. What is the purpose of the Principle of Security of Data Files? Explain.
9. What is the relationship of improved data quality to the human factor in designing data files? Explain.

10. How important are the data file design criteria in developing an organization's data base and off-line files?
11. How should the systems analyst go about determining the content of a data file?
12. Of what value is the record layout worksheet to the systems analyst? Explain.
13. What are the important considerations that a systems analyst must take into account when finalizing a data record layout? Explain.
14. How important is it to have organization personnel from the various functional areas involved in finalizing the design of data files? Explain.
15. What are the important characteristics of a data base management system?

Self-Study Exercise

True-False:

1. _____ Currently, punched card file storage is the most popular type of file storage.
2. _____ A major advantage of magnetic tape files is their fast input/output processing speeds.
3. _____ A major advantage of magnetic disk files is that any item of data is as quickly obtainable as any other.
4. _____ The direct file organization method is applicable to direct-access storage devices only.
5. _____ The sequential file organization method combines transaction file data with master file data for an updated master file.
6. _____ An overflow area on magnetic disk represents the additions to a file.
7. _____ The current trend in MIS is toward reducing data file size.
8. _____ Data file processing speed refers to the number of transactions that are to be processed against a master file.
9. _____ Good systems design should include the establishment of a single area for common information to be available to multiple users.
10. _____ A record layout worksheet is one way of designing the detailed aspects of data files.

Fill-In:

1. The current trend in data base design is toward _____ data bases.
2. The grouping together of two or more logical records for writing on magnetic tape is called _____.
3. The principal shortcoming of magnetic tape is that data must be stored and processed in a _____ manner.
4. A principal difficulty with magnetic _____ as compared to magnetic disk is the limited quantity of information that can be stored.
5. _____ _____ files provide large on-line secondary storage capabilities for medium- and large-size computer systems.
6. The _____ _____ organization method relies on master and transaction files for processing.
7. Key-to-address conversion is frequently found in the _____ _____ organization method.

8. The _____ area is used in magnetic disk packs as a means for locating records within a file.
9. Data file _____ refers to the maximum size of on-line and off-line files.
10. A _____ _____ administrator is responsible for the layout, allocation, and administration of new data files.

Problems:

1. The American Company is in the process of developing a new computerized personnel system that will store all pertinent personnel data on line on magnetic disk, including employee name, address, and other data useful in processing the weekly payroll. For this new system, consideration should be given to changing the employee numbering system so that the direct file organization method can be employed. Consideration should also be given to the use of packed decimals to reduce the storage space needed for each record element.

    Based upon the foregoing requirements, complete the following system data file analysis form for an employee master personnel file.

SYSTEM DATA FILE			
Date:		Type of record:	
Preparer:			
Record elements	Type of characters	Number of characters	Comments

Comments:

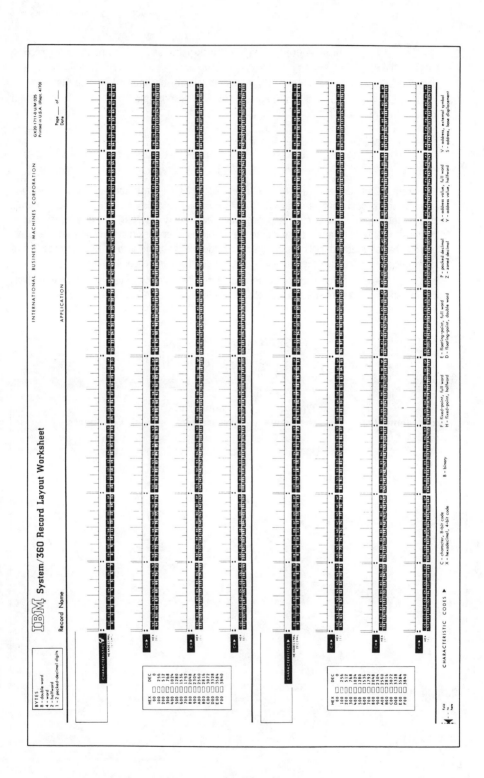

2. Using the data set forth in the completed system data file analysis form of the American Company, prepare an appropriate record layout worksheet for an employee master personnel file. All information on the file is to be logically grouped so that it is easily retrievable for payroll and personnel management uses. Also, where possible, the packed decimal format should be used to reduce storage space.

chapter

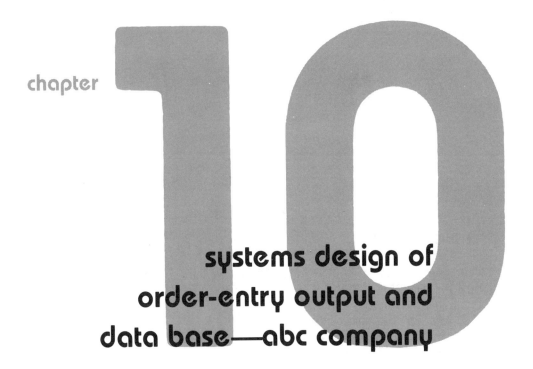

systems design of order-entry output and data base—abc company

OBJECTIVES:
- To relate the modular design concept to the ABC Company's major systems.
- To consider the human factor in specifying output and data base requirements of the new order-entry system.
- To design the output of the new order-entry system.
- To design the data base required to support the new order-entry system.

IN THIS CHAPTER:

    Modular Design of the ABC Company's Major Systems
DESIGN OF ORDER-ENTRY OUTPUT—ABC COMPANY
    Consider the Human Factor in Designing Output
    Specify Output Requirements
    Determine the Content of Output
    Design of Order-Entry Output—Reports
    Use and Distribution of Output
    Overview—Design of Order-Entry Output—ABC Company
DESIGN OF ORDER-ENTRY DATA BASE—ABC COMPANY
    Specify Data Base Requirements
    Determine the Contents of the Data Base
    Design the Order-Entry Data Base
    Overview—Design of Order-Entry Data Base—ABC Company
CHAPTER SUMMARY
QUESTIONS
SELF-STUDY EXERCISE
CASE STUDY 1—FINISHED PRODUCT INVENTORY
CASE STUDY 2—ACCOUNTS RECEIVABLE

In the preceding chapters, an overview of systems design, the design of system output, and the design of system data files were presented. This chapter will follow the design approach previously examined for the order-entry system output and data base of the ABC Company. Similarly, it will build upon the material in chapter 6 that presented the analysis of the present batch order-entry system.

Additionally, the chapter will give further information related to the finished product inventory and accounts receivable case studies. These case study materials will prepare the student for completing the subsystems design. Thus, the sample order-entry system will provide a model for the chapter's case studies.

Modular Design of ABC Company's Major Systems

From the results of the introductory investigation and the systems analysis of the present batch order-entry system, it was clear that customers evaluated the ABC Company not only on quality and price, but also on level of service. As stated in chapter 6, the key objectives of management are

    to consistently provide an order cycle time of five days,
    to provide for a 96-percent of orders shipped complete,
    to produce invoices with an accuracy level of 99 percent.

Inasmuch as management recognizes that the order-entry system can contribute toward the meeting of these goals, many changes must be undertaken.

The new order-entry system is only a part of the total system necessary to meet the customer service requirements. Because it starts when the customer places the order and ends with the printing of the invoice, the order-entry system provides part of the input required for finished product inventory, accounts receivable, sales forecasting, production planning, and several other systems. In effect, this initial module produces input for other important system modules.

The MIS task force recognizes the close relationship between the order-entry system and other major system modules. Since the finished product inventory and accounts receivable systems were specifically addressed during the feasibility study, the MIS task force plans call for the modular design of these three related systems. These systems are to be designed in such a way that the basic shipment information, processed by the order-entry system, will be readily available to the finished product inventory and accounts receivable systems. In figure 10.1, the basic modular design developed for these systems is shown from an overview standpoint. The remainder of the chapter will provide further insight into the output and data files required for these three systems.

As shown in figure 10.1, the daily shipment data and the open order data are the key data required for the finished product inventory system. The finished product inventory system itself is divided into submodules. These submodules produce the today's plant and warehouse inventory status report, projected plant and warehouse inventory status report, recommended plant manufacturing schedules, and recommended warehouse stock replenishment orders report. In a similar manner, the daily shipment data is the key information for the accounts receivable system. This system is also divided into submodules which produce customer statements, an accounts receivable aging schedule report, and an uncollected balances report.

DESIGN OF ORDER-ENTRY OUTPUT—ABC COMPANY

The approach for designing the order-entry sytem output of the ABC Company in the next sections follows basically the same sequence as that found in chapter 8, except that the section on reviewing information of the present system has been dropped. The rationale for doing so is that reference to the present system, where deemed necessary, will be made throughout the other sections. The following design sections to be covered in sufficient depth, then, are these:
 Consider the human factor in designing output.
 Specify output requirements.
 Determine the content of output.
 Design of order-entry output—specifically, reports.
 Determine the use and distribution of output.

The last section is an overview of the design of the new order-entry output for the ABC Company. This final section will coalesce important output that will tie in with other system factors (namely, the data base) presented in this chapter and inputs, procedures, and system controls examined in chapters 13 and 15.

Consider the Human Factor in Designing Output

For the new order-entry system, the MIS task force recognized the importance of the human factor in output design. This aspect was stressed throughout all discussions with the various levels of management. The task force composed a list of design guidelines to insure that the human factor was incorporated into the design of all output reports. These are:
1. Because management is used to reports typed on standard-sized (8-1/2 x 11 inches) paper and because this size is easiest to file, all system reports, to the greatest extent possible, should be on this standard-sized paper, rather than on the standard computer paper size of 14-7/8 by 11 inches.

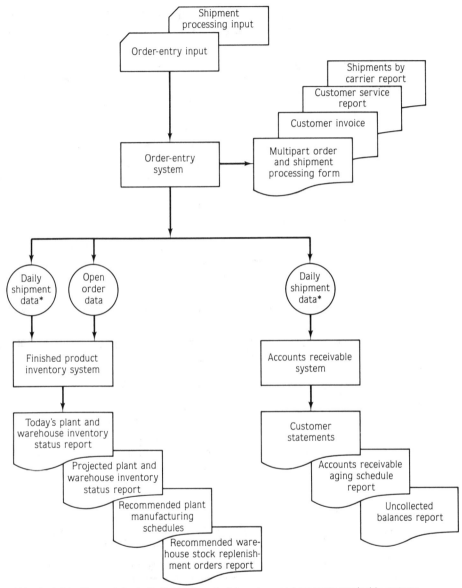

*Identical data file used for both finished product inventory and accounts receivable systems.

figure 10.1 A system flowchart depicting the overall modular design for three major subsystems—order entry, finished product inventory, and accounts receivable—of the ABC Company.

2. No computer jargon should appear on any report, including error listings produced by the system. Only terms well known to the actual system users are to be used.
3. Each report must be clearly identified with a descriptive title that includes the name of the report and the system that produced it.
4. The date the report was prepared must be clearly visible to the reader.
5. The report must be printed in a form that is easy to read and understand. It must read logically from left to right, have descriptive column headings, and allow adequate spacing. In general, reports must have the appearance of a well laid-out manually produced report.

These design considerations can make a big difference between the acceptance and nonacceptance of order-entry output. Failure to include these practical human aspects can cause the best-laid design plans of the MIS task force to be meaningless.

Specify Output Requirements

The MIS task force spent two weeks reinterviewing key individuals associated with the order-entry system to clarify the system outputs they desire. An additional week was spent summarizing and evaluating information from the interviews to insure that all important requirements were identified and separated from requests for output that would be "nice to have," but not essential. The MIS task force identified the following seven outputs as essential to the new order-entry system.

1. **Multipart Order and Shipment Processing Form** Because of the design requirements of the new order-entry system, the five-part order and shipment processing form must be continued. As indicated in chapter 6, one copy of the order and shipment processing form serves as the bill of lading given to the carrier who will make delivery to the customer. Another is used by the company to provide supporting data in case a claim for loss or damages must be filed against the carrier. A third is used by the carrier to acknowledge receipt of freight and as a basis for billing. Thus, the bill of lading must be very carefully worded and accurately prepared according to legal requirements enforced by the Interstate Commerce Commission. Currently, the bill of lading must indicate hazardous materials capable of causing harm to humans or damage to equipment or other freight. Although the ABC Company is not currently indicating hazardous materials on the computer-prepared bill of lading, a requirement of the new system is to print **HM** next to each hazardous material on the bill of lading to conform with the ICC requirement.
2. **Customer Invoice** The new order-entry system will produce the customer invoice directly from information available at the time of shipment processing. The invoice will not be available at the time of shipment. However, this new approach eliminates the use of separate input information for the customer invoice and shipment processing, enables quicker preparation of the invoice and consequent reduction in customer payment time, and insures that information on the customer bill of lading and on the customer invoice will be consistent. Producing the invoice directly from the shipment processing information will greatly decrease the billing errors.
3. **Customer Service Report** A new customer service report must be prepared monthly. The purpose of this report is to improve inventory management and

customer service by providing information on orders containing an item not shipped in the quantity requested or orders not shipped within three days of order receipt. This report will allow management to measure customer service relative to its newly established standards.
4. **Shipments by Carrier Report** The current shipments by carrier report will continue to be produced by the new order-entry system. It will be printed in exactly the same format and provide the same information as in the existing report (see figure 6.11).
5. **Daily Shipment Data** (for Finished Product Inventory System) The daily shipment data must continue to be routed from the order-entry system to the finished product inventory system. These data are created automatically when shipment data are used to update the open order file and to move the open order record to the shipment file.
6. **Open Order Data** (for Finished Product Inventory System) In addition to daily shipment data, it is desired to have the order-entry system pass all open order information to the finished product inventory system. The open order information will enable the finished product inventory system to have the complete inventory status of all items. The future inventory status can be determined by subtracting the open order volume from the current volume. Thus, the finished product inventory system can become a truly forward-looking management control and planning tool.
7. **Daily Shipment Data** (for Accounts Receivable System) Currently, the data required for the accounts receivable system are keypunched and prepared separately from the order-entry system. The new order-entry system must automatically produce a computerized file—a duplicate of requirement 5, above—of input data for the accounts receivable system. This will eliminate the duplicate keypunching currently needed.

Determine the Content of Output

Having determined the information needed as output from the new order-entry system, the next task is to determine the specific contents of each output. Output analysis forms will be completed for each of the seven outputs above to indicate their content in detail and to document the process.

Multipart order and shipment processing form Because the new order and shipment processing form requires only a minor change to accommodate government regulations for hazardous materials, no output analysis form has been prepared. A new column, two characters wide, will be added to the form illustrated in figure 6.10 where **HM** can be printed next to any item which qualifies as hazardous material.

Currently, the heading of the order part of the form includes order number, carrier, shipment date, order date, warehouse, truck or car number, shipment terms, shipment number, consigned to, and destination (city, state, and zip code). Within the body of the form, there are columns for quantity, brand description, unit weight, and short-shipment code/quantity. The columns for the bill of lading part of the form are number of packages, description for bill of lading, and weight. As indicated above, a new column will be added to both parts of the form to allow for printing **HM**.

Customer invoice The customer invoice, as noted on the output analysis form of table 10.1, contains important internal information for determining what products were shipped and billed. Similarly, it contains vital external information for the customer which becomes the basis for payment. Although the contents of the form may seem unusually numerous, critical information for the company and its customers are contained therein; to eliminate any information element from the invoice will generally result in higher costs expended to answer inquiries about the deleted items. Thus, the contents of the customer invoice are critical not only to the success of the new order-entry system, but also to the related accounts receivable module.

Customer service report The customer service report presents summary information about all shipments not shipped within three days of order receipt or for which the quantity shipped is less than the quantity ordered. Key information, which must be shown for each such shipment, is illustrated in table 10.2. Of great interest to marketing

table 10.1 Customer invoice output analysis form for the ABC Company.

SYSTEM OUTPUT			
Page 1 of 1 Date: February 22, 198_ Preparer: G. W. Reynolds		Type of output: CUSTOMER INVOICE	
Informational elements	Type of characters	Number of characters	Comments
Customer name and address (bill to)	Alphanumeric	71	
Customer name and address (ship to)	Alphanumeric	71	
Original order date	Numeric	8	Edit as MM/DD/YY
Shipment date	Numeric	8	Edit as MM/DD/YY
Order number	Numeric	6	
Shipment number	Numeric	6	
Invoice date	Numeric	8	Edit as MM/DD/YY
Invoice number	Numeric	7	
Product code of item ordered	Numeric	4	
Product quantity ordered	Numeric	5	
Product code of item back-ordered	Alphabetic	4	
Product quantity back-ordered	Numeric	5	
Quantity code (indicates cases, drums, or bags)	Alphabetic	1	
Product description	Alphabetic	25	
Quantity shipped	Numeric	5	
Product item unit cost	Numeric	7	Edit with decimal point
Product item extended cost	Numeric	9	Edit with decimal point
Total invoice amount	Numeric	10	Edit with decimal point
Invoice gross amount	Numeric	10	Edit with decimal point
Invoice net amount	Numeric	10	Edit with decimal point
Net amount payment date	Numeric	8	Edit as MM/DD/YY
Total shipping weight	Numeric	7	
Customer account number	Numeric	7	

Comments:
The computer program is to generate the current date and invoice page number.

table 10.2 Customer service report output analysis form for the ABC Company.

SYSTEM OUTPUT			
Page 1 of 1 Date: February 23, 198_ Preparer: G. W. Reynolds		Type of output: CUSTOMER SERVICE REPORT	
Informational elements	Type of characters	Number of characters	Comments
Field warehouse code (from which shipment was made)	Alphabetic	4	
Customer name and address	Alphanumeric	71	
Original order date	Numeric	8	Edit as MM/DD/YY
Shipment date	Numeric	8	Edit as MM/DD/YY
Shipment number	Numeric	6	
Order number	Numeric	6	
Product codes ordered	Numeric	4	
Quantities ordered	Numeric	5	
Product codes shipped	Numeric	4	
Product quantities shipped	Numeric	5	
Reason the shipment failed to meet the customer service level	Alphabetic	30	

Comments:
The computer program is to print the reason for failure to meet the customer service goal. Also, the current date is to be printed at the top of the form.

table 10.3 Daily shipment data output analysis form for the ABC Company.

SYSTEM OUTPUT			
Page 1 of 1 Date: February 25, 198_ Preparer: G. W. Reynolds		Type of output: DAILY SHIPMENT DATA (for Finished Product Inventory)	
Informational elements	Type of characters	Number of characters	Comments
Record identification code	Numeric	3	
Field warehouse code (from which shipment was made)	Alphabetic	4	
Customer account number	Numeric	7	
Product code	Numeric	4	
Quantity shipped	Numeric	5	
Shipment date	Numeric	6	Edit as MM/DD/YY on printed output

Comments:
The above data must be present on the daily shipment file to support the finished product inventory system.

and physical distribution management is the reason the shipment failed to meet the customer service goal.

Shipments by carrier report The shipments by carrier report will continue to be prepared as it is now. In addition to the field warehouse from which the order was shipped and the shipment date, the report contains carrier name, shipment number, trailer number, order number, shipment weight, shipping freight indicator, and freight amount (if not sent collect). A typical report is illustrated in figure 6.11. Hence, no output analysis form has been included.

Daily shipment data (for finished product inventory system) The daily shipment data include the informational elements shown in table 10.3. As shown in figure 10.1 (p. 269), these data are output of the new order-entry system and serve as input for the finished product inventory system. They are stored on a computerized file until needed for processing.

Open order data (for finished product inventory system) The open order data required for the finished product inventory system are found in table 10.4. Like the previous output, these data are output from the new order-entry system and are an integral part of the finished product inventory system. They are stored on computerized files.

Daily shipment data (for accounts receivable system) The daily shipment data needed for the accounts receivable system are illustrated in table 10.5. These data are written automatically to a computerized input file representing each shipment processed.

table 10.4 Open order data output analysis form for the ABC Company.

SYSTEM OUTPUT

Page 1 of 1
Date: February 26, 198_
Preparer: G. W. Reynolds

Type of output: OPEN ORDER DATA
(for Finished Product Inventory)

Informational elements	Type of characters	Number of characters	Comments
Record identification code	Numeric	3	
Field warehouse code (from which shipment was made)	Alphabetic	4	
Customer account number	Numeric	7	
Product Code	Numeric	4	
Quantity ordered	Numeric	5	
Planned shipment date	Numeric	6	Edit as MM/DD/YY on printed output

Comments:
The above data must be present on the daily open order file to support the finished product inventory system.

table 10.5 Daily shipment data output analysis form for the ABC Company.

SYSTEM OUTPUT			
Page 1 of 1 Date: March 1, 198_ Preparer: G. W. Reynolds	Type of output: DAILY SHIPMENT DATA (for Accounts Receivable)		
Informational elements	Type of characters	Number of characters	Comments
Record identification code	Numeric	3	
Field warehouse code (from which shipment was made)	Alphabetic	4	
Customer service exception code	Alphabetic	1	
Order number	Numeric	6	
Order date	Numeric	6	Edit as MM/DD/YY on printed output
Product Code	Numeric	4	
Quantity shipped	Numeric	5	
Product unit price	Numeric	5	
Extended price	Numeric	5	
Shipment date	Numeric	6	Edit as MM/DD/YY on printed output
Shipment number	Numeric	6	
Shipping terms indicator	Alphabetic	1	
Carrier name	Alphabetic	24	
Invoice number	Numeric	7	
Invoice date	Numeric	6	Edit as MM/DD/YY on printed output
Customer account number	Numeric	7	

Comments:
The above data must be present on the daily shipment file to support the accounts receivable system.

Design of Order-Entry Output—Reports

Once the output requirements and their contents have been specified, the next step in the design process is to format the reports. Because our concern centers on the new order-entry system primarily, only the reports from this system will be explored. Those outputs dealing with the finished product inventory and accounts receivable systems will be treated in this chapter's case studies. Since the shipments by carrier report (figure 6.11) has not been changed, only the multipart order and shipment processing form, customer invoice, and customer service report are designed below.

Multipart order and shipment processing form As indicated earlier in the chapter, the new order and shipment processing form requires only a minor change to accommodate government regulations regarding hazardous materials. A new column must be added to the form to allow the printing of **HM** next to any item which qualifies as a hazardous material (figure 10.2).

Customer invoice The new customer invoice form is designed to present clearly all necessary information to the customer. An example of the new customer invoice is

Order Number:		Carrier:	Ship Date:	
Order Date:	Warehouse:	Truck or Car Number:	Ship Terms:	Shipment Number:
Consigned To:				
Destination: City:		State:	Zip Code:	
Quantity	Brand Description		Unit Weight	Short Ship Code/Qty.	HM

	Bill of Lading			
No. of Pkgs.	Description for Bill of Lading		Weight	HM

figure 10.2 New multipart order and shipment processing form for the ABC Company.

depicted in figure 10.3. In addition, all pertinent information needed for internal purposes is contained within this invoice.

Customer service report The new customer service report, as shown in figure 10.4, reflects consideration of the human factor in output design. First, the name of the report and the system which produces the report are clearly stated. Reference to these two items ensure that the users of the report will know the source of the information and will not confuse this report with any current or future report presenting shipment information. Second, the date and time the report was prepared are available as well as the time period for which the information is summarized, so there can be no confusion on the time period for which short-shipments are reported. Even if the report had to be prepared a second time during the month (due to incomplete input, processing errors, and the like), it is possible to tell which report is the most current version.

The retention period reminds the user that the data have a definite period of usefulness; it has been decided that this lifespan is six months for short-shipments. Also, the detail lines of the report are carefully designed to make the report easy to read and to highlight exceptions. There is an asterisk next to each item that was short-shipped or caused the order to be shipped late. The reason the shipment may have failed to meet customer service goals is clearly described.

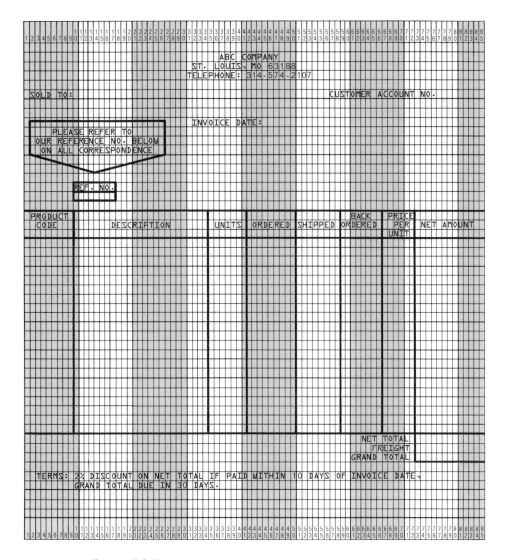

figure 10.3 A new customer invoice for the ABC Company

Use and Distribution of Output

Just as the section above focused on the design of three outputs—namely, the multipart order and shipment processing form, the customer invoice, and the customer service report—the same direction will be taken in this final section on the distribution of new outputs for the order-entry system. These outputs are generated to meet specific internal and external reporting needs.

Multipart order and shipment processing form The five-part order and shipment processing form will continue to be used and distributed in the current manner. By way

```
PREPARED  06/30/8-       ABC COMPANY CUSTOMER SERVICE REPORT            PAGE: 12
  TIME    08:33:25           ORDER-ENTRY SYSTEM            RETAIN UNTIL: 06/30/8-

                       FOR THE PERIOD: 04/01/8- TO 06/30/8-


ORDER        SHIP        ORDER         SHIP
DATE         DATE        NO.           NO.      CUSTOMER NAME & ADDRESS
06/15/8-     06/21/8-    34331         24187    ACME HARDWARE
                                                NEWTON FALLS, OH
                                                44444
             ITEM       ORDERED      SHIPPED            REASON
             1070          10      *     5      A - PRODUCT ALLOTMENT EXCEEDED
             2431         105      *    50      O - PRODUCT OUT OF STOCK
             5482          75            75

ORDER        SHIP        ORDER         SHIP
DATE         DATE        NO.           NO.      CUSTOMER NAME & ADDRESS
06/18/8-     06/20/8-    34352         24199    GENERAL SUPPLY CO.
                                                CINCINNATI, OH
                                                45239
             ITEM       ORDERED      SHIPPED            REASON
             1432          25      *    25      C - CARRIER DELAY
             2050         100      *   100      C - CARRIER DELAY
```

figure 10.4 A new customer service report for the ABC Company.

of review (refer to chapter 6 and figure 6.9), this multipart form is used and distributed as follows:
1. original copy—mailed to customer as confirmation that the order has been received and is being processed.
2. red copy—kept by field warehouse accounting clerk to answer questions from the salesperson or customer; all corrections are made from the yellow copy.
3. yellow copy—used by the stock loader to note items which must be short-shipped due to insufficient inventory on hand; sent to field warehouse accounting clerk for matching with red copy.
4. blue copy—used as file copy by the shipping clerk; all corrections noted on the yellow copy are written on this copy.
5. bill of lading copy—given to the carrier as the shipment bill of lading.

Customer invoice The new customer invoice will be produced in four copies—one sent to the customer, one to the field warehouse accounting clerk, and two copies to the accounts receivable section at headquarters in St. Louis. The field warehouse accounting clerk will use the customer invoice in a "tickler file" to follow-up on all orders for which a backorder was required.

Customer service report The new customer service report will be produced in multicopies. One copy will be distributed to each manager with responsibility for meeting the customer service goals. These managers are: all field warehouse managers (each will receive the portion of the report for his or her warehouse), all manufacturing plant managers (each will receive a complete copy of the report), and the vice-presidents of purchasing and inventory, physical distribution, manufacturing, and marketing.

Overview—Design of Order-Entry Output—ABC Company

The key outputs of the ABC Company's new order-entry system have been identified as follows:
1. The current **multi-part order and shipment processing form** has been upgraded to meet government requirements for identification of hazardous materials.
2. A new **multi-part customer invoice** will be prepared at the time of shipment processing.
3. A new monthly **customer service report** will highlight all shipments which fail to meet the new customer service standard of complete shipping within three days of order receipt.
4. The existing **shipments by carrier report** will continue to be produced daily, as it is now.
5. **Shipment data for the finished product inventory system** will continue to be prepared daily, as it currently is.
6. **Open order data for the finished product inventory system** will be automatically passed from the new order-entry system to the new finished product inventory system. This information will provide management with a projection of the inventory status at future times.
7. Automatic preparation of **daily shipment data for the accounts receivable system** will be produced to eliminate the duplicate keypunching which occurs presently.

Overall, the format of the order and shipment processing form has been fixed, a new customer invoice has been designed, the content and form of the new monthly customer service report has been determined, the shipments by carrier report will be used as is, and the data necessary for input to the finished product inventory and accounts receivable systems have been defined.

SUMMARY: Design of Order-Entry Output—ABC Company

Consider the Human Factor in Designing Output   Relates to those human factors that assist the experienced and inexperienced in understanding order-entry output and using it effectively.

Specify Output Requirements   Refers to determining the order-entry information needed for effective and efficient operation, versus requests for order output that would be nice to have but not essential to daily operations. Outputs essential to the new order-entry system are:
1. multi-part order and shipment processing form
2. customer invoice
3. customer service report
4. shipments by carrier report
5. daily shipment data (for finished product inventory system)

6. open order data (for finished product inventory system)
7. daily shipment data (for accounts receivable system)

Determine the Content of Output Focuses on the specific contents of order-entry output. The output analysis forms contain informational elements and their type and number of characters. Additionally, appropriate comments are specified for each informational element, as well as overall comments about the output.

Design of Order-Entry Output—Reports Refers to the order-entry output formats, that is, where fields of information are to be printed on the report.

Use and Distribution of Output Relates to the uses of order-entry output and where these outputs are to be distributed for more control and efficiency of operations.

DESIGN OF ORDER-ENTRY DATA BASE—ABC COMPANY

Now that the specific order-entry outputs have been devised, the second part of the chapter will focus on designing the data base for the new order-entry system. The design approach corresponds to that detailed in chapter 9, although the sections on considering the human element and reviewing the present system have been dropped. Specifically, the following steps are covered:

Specify data base requirements.
Determine the contents of the data base.
Design the order-entry data base.

As with the order-entry outputs, an overview section on the design of the order-entry data base for the ABC Company concludes this section.

Specify Data Base Requirements

Since a new order-entry data base is the heart of many systems, it provides their key input. It is imperative, then, that the order-entry data base be developed in a manner supportive of important objectives for future systems of the ABC Company. By way of review (chapter 6), these objectives include:
1. Provide better customer service.
2. Improve selling efficiency.
3. Help production more closely match output to customer demand.
4. Provide more timely information for analysis of the company's present and future operations, both at plant and corporate levels.
5. Improve coordinated control of the overall company.

The order-entry data base must be designed to interface effectively with the proposed finished product inventory system by providing information about customers, orders, shipments, and customer service level. Similarly, it must be capable of providing the data needed for the accounts receivable system. Also, the data base must be designed such that the recommendations of the exploratory survey report to top management are implemented—that is, the data base must be designed for use in a distributed processing environment that will employ remote intelligent terminals.

Data base In addition to meeting these general requirements, the MIS task force identified the need for the following files to be part of the new order-entry system data base:

1. An **open order file** contains information on all orders received but not yet shipped. This file will contain all the data elements required for open order information for the finished product inventory system as well as other data elements of potential value.
2. A **shipment file** is created from the open order file. Whenever an open order is shipped, it is removed from the open order file and placed on the shipment file. The contents of the shipment file will be carefully designed to include all the data elements required to support both the accounts receivable and finished product inventory systems. In this manner, a single file can be used as input to both systems.
3. A **customer name and address file** will be created and used to reduce the order input errors. Each customer will be given an account number for use in both order entry and accounts receivable. Use of the account number will insure that the correct name and address is associated with all orders, shipments, and accounting information.
4. A **product code file,** another new file, will assign each product a unique four-digit product code. It will provide the product description and current price, and will be used to identify hazardous materials.

An overview of the new order-entry data base is seen in figure 10.5 (p. 282). Only the files directly concerned with order entry are shown; the files required for finished product inventory and accounts receivable are not shown. The "old" label on the open order file and shipment file designates these as primary input files or as the starting point before today's orders or shipment data are processed. The "old" label distinguishes them from the "new" files, or primary output files, created by processing today's order and shipment data against the "old" files. In addition to the primary input and output files, there are table files, including customer name and address and product code files, which are an integral part of the new order-entry system.

File activity and volatility **File activity** is a measure of the proportion of records affected by an update run and is computed by dividing the number of records updated by the total number of records in the file. If the activity ratio exceeds 40 percent, sequential processing rather than direct access processing is usually more appropriate. **File volatility,** on the other hand, is a measure of the frequency with which a file is updated. Highly volatile files, updated several times per day, indicate that direct access processing is likely more appropriate than sequential processing. Files with low volatility are updated periodically, perhaps weekly or monthly, and sequential processing is generally more appropriate.

The open order file will be updated each night. New orders will be added, and shipment data will be used to delete an order from the open order file and add a record to the shipment file. At any one time, there will be approximately four days of orders on the open order file and ten days of shipments on the shipment file. Thus, about 25 percent of

Systems Design of Order-entry Output and Data Base—ABC Company

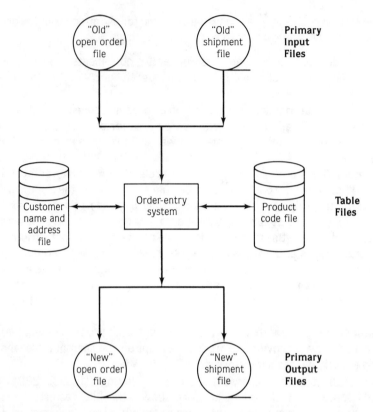

figure 10.5 A system flowchart illustrating the overall design of the new order-entry data base of the ABC Company.

the open order file will be updated with new records each night, and about 10 percent of the shipment file will be updated with new records at the same time.

Initially, the customer name and address file must be created from scratch. About 5000 records will be used to create this file. Thereafter, new records need only be added when new accounts are added. Typically, there are about four or five new customers each day. The customer name and address file must be updated each night with about 0.1 percent new records. In like manner, the product code file must also be created from scratch; there are currently three hundred products, with an anticipated growth of ten products per year. This file will need to be updated infrequently with 1 to 3 percent new records. Price changes, on the other hand, will require updating about twice every week with four to eight changes at each update; this represents a 1 to 3 percent change per update.

File access techniques The open order and shipment files will normally be processed in a sequential fashion, either in order number sequence or shipment number sequence. In a similar manner, the customer name and address file will be accessed as the open order file or shipment file is processed to provide name and address information. Since orders and shipments will be processed in sequence by order number or shipment number, it is highly desirable to have the capability to access directly the customer name and address file by the seven-digit customer account code.

The product code file must be referenced as the open order file is processed during the preparation of the order and shipment processing form. Information, such as product description and hazardous material indicator, is required for each item ordered. Direct access to the product code file via the four-digit product code is ideally suited to this form of processing.

Basic information about the files needed for the order-entry data base is summarized in table 10.6. The activity ratio, volatility, need to refer to a specific record, and average number of records all support the decision of which file access techniques to use. As illustrated, the sequential access file method will be used for open order and shipment files, and the direct access file method will be used for customer name/address and product code files.

table 10.6 File access techniques for the new order-entry system of the ABC Company.

Order-entry data base	Activity ratio when updated	Volatility (update frequency)	Need to refer to specific record on a file	Average number of records	File access technique
Open order file	25%	Daily	No	450	Sequential
Shipment file	10%	Daily	No	1500	Sequential
Customer name and address file	0.1%	Daily	Yes	5000	Direct
Product code file	2%	Weekly	Yes	300	Direct

Determine the Contents of the Data Base

Now that the appropriate data files have been determined for the new order-entry system and their related file access techniques have been specified, the next step in the design process is to determine their detailed contents. The contents of the open order file, the shipment file, and the customer name and address file are delineated in tables 10.7 through 10.9, respectively. The customer name and address file will allow for a "bill to address" as well as a "ship to address" for all customers. Because the open order and shipment files need not contain name and address information, only the seven-digit customer account number is included. The new shipment file will contain information on products and quantities ordered, as well as products and quantities shipped, to facilitate an analysis of customer service. For each product shipped, there is an allowance for a customer service exception code to indicate which item, if any, led to violating customer service standards.

table 10.7 Open order data file analysis form for the ABC Company.

SYSTEM DATA FILE

Page 1 of 1
Date: March 20, 198_
Preparer: G. W. Reynolds

Type of record: OPEN ORDER DATA FILE

Record elements	Type of characters	Number of characters (bytes)	Comments
Record identification code	Numeric	3	
Order number	Numeric	6	
Order date	Numeric	6	
Field warehouse code	Alphabetic	4	
Customer account number	Numeric	7	
Shipment number	Numeric	6	
Shipment date	Numeric	6	
Desired shipment date	Numeric	6	
Products Ordered (up to 10)			
Product code	Numeric	40 (4 × 10)	
Quantity	Numeric	50 (5 × 10)	
Hazardous material indicator	Numeric	10 (1 × 10)	

Comments:
The data base administrator has approved the record elements, type of characters, and length of field in bytes for a sequential access file method. The total number of bytes is 144.

The design of the product code file requires careful study of the company's pricing policies. Generally, each product has two or more price/quantity breaks. This means that cost per unit is constant for order quantities from one unit up to the first price/quantity break, at which point it is decreased; it is then constant until the order size meets or exceeds the second price/quantity break. Therefore, the cost per unit paid by the customer is a function of order size. The price structure for a typical product of the ABC Company is illustrated in figure 10.6.

Through discussions with the vice-president of marketing, the MIS task force uncovered the following facts related to the design of the product code file:

1. The ABC Company currently markets about 300 unique and different products, as well as product variations.

Product Code = 1234

Order size	Cost/Unit for each unit ordered
1 to 49	$3.05
50 to 99	2.95
over 99	2.80

figure 10.6 Typical price structure for the ABC Company.

Design of Order-Entry Data Base—ABC Company

table 10.8 Shipment data file analysis form for the ABC Company.

SYSTEM DATA FILE

Page 1 of 1
Date: March 21, 198_
Preparer: G. W. Reynolds

Type of record: **SHIPMENT DATA FILE**

Record elements	Type of characters	Number of characters (bytes)	Comments
Record identification code	Numeric	3	
Order number	Numeric	6	
Order date	Numeric	6	
Invoice number	Numeric	7	
Invoice date	Numeric	6	
Field warehouse code	Alphabetic	4	
Customer account number	Numeric	7	
Shipment number	Numeric	6	
Shipment date	Numeric	6	
Desired shipment date	Numeric	6	
Shipping terms	Alphabetic	1	
Carrier name	Alphabetic	24	
Trailer number	Numeric	15	
Products ordered (up to 10)			
Product code	Numeric	40 (4 × 10)	
Quantity	Numeric	50 (5 × 10)	
Hazardous material indicator	Numeric	10 (1 × 10)	
Products shipped (up to 10)			
Product code	Numeric	40 (4 × 10)	
Quantity	Numeric	50 (5 × 10)	
Hazardous material indicator	Numeric	10 (1 × 10)	
Customer service exception code	Alphabetic	10 (1 × 10)	
Product unit price	Numeric	50 (5 × 10)	

Comments:
The data base administrator has approved the record elements, type of characters, and length of fields in bytes for a sequential access file method. The total number of bytes is 357.

2. The number of different products marketed will approach 350 in the next five years.
3. Currently, the most complex pricing structure for any product involves three price/quantity breaks.
4. The most expensive product now or in the next five years will not exceed $25 per unit.
5. While price changes occur once or twice per year for each product, the timing and impact of a price change is such that four to eight product prices are changed at the same time. Thus, the product code file would be updated on the average of twice a week with four to eight new prices per update.

Using this information, the MIS task force determined the content of the product code file as shown in table 10.10.

Design the Order-Entry Data Base

The final step in designing the order-entry data base is the actual design of the data files to produce the desired output. Using figure 10.7, the record layout for the product code file, the reader can arrange the data in tables 10.7 through 10.9 back to back in the same

table 10.9 Customer name and address data file analysis form for the ABC Company.

SYSTEM DATA FILE			
Page 1 of 1 Date: March 22, 198_ Preparer: G. W. Reynolds	Type of record: CUSTOMER NAME/ADDRESS DATA FILE		
Record elements	Type of characters	Number of characters (bytes)	Comments
Record identification code	Numeric	3	
Customer account number	Numeric	7	
Customer name	Alphabetic	24	
Customer address (ship to)			
Street	Alphanumeric	26	
City	Alphabetic	14	Edit with comma
State	Alphabetic	2	
Zip code	Numeric	5	
Customer address (bill to)			
Street	Alphanumeric	26	
City	Alphabetic	14	Edit with comma
State	Alphabetic	2	
Zip code	Numeric	5	

Comments:
The data base administrator has approved the record elements, type of characters, and length of fields in bytes for a direct access file method. The total number of bytes is 128.

field format to obtain the record layouts for the open order, shipment, and customer name and address files. Thus, the record layouts for these data files are relatively straightforward.

Product code data file—record layout In order to store numeric data on the product code file for most efficient calculation, shipping weight and the price/quantity break information are stored as packed data, which avoids the need for conversion routines. The shipping weight is stored in five bytes of packed data representing the weight as XXXXXXX.XX pounds per unit. The order size information is stored in four bytes of packed data representing XXXX.XXX units. The cost per unit information is also stored in four bytes of packed data representing XXXXX.XX dollars per unit. Because of the relatively small number of records on this file, total bytes per record is not an important concern. Thus, each product code file record will carry an additional fourteen bytes of currently unused filler. These unused bytes provide the flexibility to add data elements, not currently considered, to the record without a major impact on the systems design. Figure 10.7 (p. 288) shows the record layout for the product code file.

File security The open order and shipment files are critical to the operation of the ABC Company. Loss of either file for even one day would make it impossible to satisfy customer needs and prepare accurate billing information. For this reason, the open order and shipment files will be designed as multigeneration data group files. All order and shipment input data will be saved for at least three working days so that, if anything happens to the current generation of either file, the previous day's input can be

table 10.10 Product code data file analysis form for the ABC Company.

SYSTEM DATA FILE

Page 1 of 1
Date: March 24, 198_
Preparer: G. W. Reynolds

Type of record: **PRODUCT CODE DATA FILE**

Record elements	Type of characters	Number of characters (bytes)	Comments
Record identification code	Numeric	3	
Product code	Numeric	4	
Product description	Alphabetic	25	
Hazardous material indicator	Numeric	1	
Quantity code	Alphabetic	1	
Gross shipping weight per unit (in pounds)	Numeric	5	Edit with decimal point
Effective date of price	Numeric	6	
Number of price breaks	Numeric	1	
Price information for up to five price breaks			
Maximum order size for this price	Numeric	20 (4 × 5)	
Cost per unit within this size	Numeric	20 (4 × 5)	Edit with decimal point
Filler space out to 100 bytes		14	

Comments:
The data base administrator has approved the record elements, type of characters, and length of fields in bytes for a direct access file method. The total number of bytes is 100.

---

reprocessed against the previous generation of these files to recreate the most current generation.

The customer name/address and product code files are fairly static and contain data relatively easy to recreate. However, recreating these files from manually kept records would require duplicate effort in updating both computer files and manual records. Thus, it is appropriate to eliminate manual records and provide backup capability by requiring multiple generations of these two files as well.

Lastly, the open order file, shipment file, and customer name and address file contain information highly valuable to competitors. For this reason, these files require access protection to guard against unauthorized use. However, this is unnecessary for the product code file since it is comprised of information published for the ABC Company's customers.

Overview—Design of Order-Entry Data Base—ABC Company

The key files required to support the new data base for the ABC Company have been defined as:
1. An **open order file** contains information on all orders received but not yet shipped. The file will be updated nightly with new order and shipment data. Also, it will be a

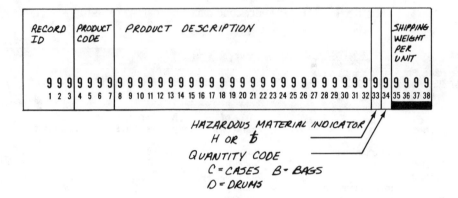

figure 10.7 A record layout worksheet for the product code file of the ABC Company.

multigeneration data group to provide full backup capability and will be processed sequentially by order number.
2. A **shipment file** is created from the open order file. Whenever an order is shipped, it is removed from the open order file and placed on the shipment file. The file will be updated nightly with new shipment data. It, also, will be a multigeneration data group to provide backup capability and will be processed sequentially by shipment number.
3. A new **customer name and address file** contains information on approximately 5000 customers. The file will be updated with only four or five new customers each night. It will be a multigeneration data group in order to avoid keeping duplicate manual records for backup. The key to the file will be the seven-digit customer account number to be used in both the order entry and accounts receivable systems. The file will be accessed directly.
4. A new **product code file** contains product description and price information for all of the company's products. The file will be updated about twice a week—when prices change, at the introduction of new products, or upon elimination of old products. The file will contain about 300 records keyed by the four-digit product code. It will be a multigeneration data group, accessed directly.

SUMMARY: Design of Order-Entry Data Base—ABC Company

Specify Data Base Requirements Refers to specifying the data base (data files) and their related file access techniques for the new order-entry system. The data base essential for this system includes:
1. open order file—primary input (old) and output (new) files
2. shipment file—primary input (old) and output (new) files
3. customer name and address file—reference file
4. product code file—reference file

EFFECTIVE DATE OF PRICE			PRICE INFORMATION											FILLER SPACE OUT TO 100 BYTES FOR FUTURE USE
			BREAK #1		BREAK #2		BREAK #3		BREAK #4		BREAK #5			
MO.	DAY	YR.		ORDER SIZE	COST PER UNIT	ORDER SIZE	COST PER UNIT	ORDER SIZE	COST PER UNIT	ORDER SIZE	COST PER UNIT	ORDER SIZE	COST PER UNIT	
9 9	9 9	9 9	9	9 9 9	9 9 9	9 9 9	9 9 9	9 9 9	9 9 9	9 9 9	9 9 9	9 9 9	9 9 9	9 9 9 9 9
39 40	41 42	43 44	45	46 47 48	49 50 51	52 53 54	55 56 57	58 59 60	61 62 63	64 65 66	67 68 69	70 71 72	73 74 75	76 77 78 79 80–100

↙ NUMBER OF PRICE BREAKS IF LESS THAN 5, THEN ZEROS ARE PLACED IN REMAINING PRICE BREAK FIELDS

figure 10.7 Continued.

Determine the Contents of Data Base Relates to the specific contents of the data files (including order-entry record elements, their type and number of bytes, specific comments about any record element, and overall comments about the data file). These conclusions are recorded on data file analysis forms.

Design the Order-Entry Data Base Refers to designing the order-entry data files, including provisions for file security, in the form of record layout worksheets.

CHAPTER SUMMARY:

Building upon the material presented in chapter 6 regarding the present order-entry system of the ABC Company, this chapter followed the systems design approach recommended in chapters 7 through 9 to develop the new order-entry system output and data base. The format, content, use, and distribution of key outputs were defined. These outputs were the revised multipart order and shipment processing form, the customer invoice, the new customer service report, the shipments by carrier report, and the input files to be created for the finished product inventory and accounts receivable systems. Next, the data base requirements were defined for the open order, shipment, customer name and address, and product code files. The content, format, and file access technique for each data file were specified. It should be noted that the human factor was given due consideration throughout this process.

Questions

1. How was the modular design concept used for the new order-entry system?
2. What are the essential outputs of the ABC Company's order-entry system? Why were these particular ones selected?

Systems Design of Order-entry Output and Data Base—ABC Company

3. a. What is the function of the new customer service report?
   b. What is the function of the shipments by carrier report?
4. What are the basic requirements of the new order-entry system data base?
5. What specific files will be designed for the new order-entry system data base? Why were these particular ones selected?
6. What are the key factors in determining the file access technique for the four data files in the new order-entry system data base?

Self-Study Exercise

True-False:

1. _____ The modular design is not applicable to the major systems of the ABC Company.
2. _____ A major output of the new order-entry system is the open order data file for the accounts receivable system.
3. _____ The multipart shipments by carrier report becomes the bill of lading to the customer.
4. _____ One result of incorporating the human factor in the output design of the new order-entry system is the clarity of printed reports.
5. _____ When specifying output requirements, it is necessary to separate those that are essential from those that are nice to have.
6. _____ Final design of output for the new order-entry system is accomplished by completing output analysis forms.
7. _____ The first step in the design process of the order-entry data base is to determine the contents of the data files.
8. _____ The "old" open order file is an example of the primary output files in the new order-entry system.
9. _____ The product code file consists mainly of customer names and addresses.
10. _____ The shipment file for the new order-entry system will utilize the sequential file access technique.

Fill-In:

1. One of the important inputs for the new order-entry system is the _____ _____ input.
2. An important human factor consideration in the new system is that the output forms be _____ to read and understand.
3. The new customer service report is to be prepared _____.
4. The open order information will enable the finished product inventory system to have the complete _____ status of all items.
5. The output analysis form for the customer service report contains eleven _____ elements.
6. The informational elements of the open order data for the finished product inventory system consist of record identification code, field warehouse code, customer account number, product code, _____ ordered, and planned shipment date.
7. The last step in designing the order-entry output is its use and _____.

8. The customer name and address file for the new order-entry system is an example of a _____ file.
9. The "new" shipment file for the new order-entry system is an example of a primary _____ file.
10. File _____ is a measure of the frequency with which a file is updated.

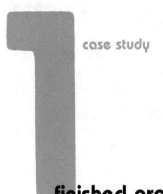

case study

chapter 10

finished product inventory

Modular Design Approach

The primary objective of the new finished product inventory system is to raise the percentage of orders completely filled from 84 to 96 percent. Another is to create the proper inventory balance between low-profit and high-profit items, to result in a net reduction of inventory of from $1 to $2 million. The new finished product inventory system will be designed as three modules in order to accomplish these objectives. Case study1–figure a depicts three inventory modules.

The **first module** will use the plant and warehouse inventory status information as of the start of the current business day and update this information with the current day's shipments to obtain the plant and warehouse inventory status information for the next day. Warehouse receipts representing shipments from plants, returns from customers, and shipments from other warehouses must also be accounted for, since they represent an increase in warehouse inventory. Likewise, the results of production, interplant or plant–warehouse shipments, and customer returns must be added to the plant inventory. This module will produce the following two primary outputs: (1) today's plant and warehouse inventory status report and (2) today's plant and warehouse inventory status file both illustrated in case study 1–figure a.

The **second module** will produce inventory projections that take into account the current inventory status, current open orders, planned manufacturing schedules, and reorder lead times. The current open orders will be subtracted from the current inventory status at the warehouses from which the open orders will be filled. The planned manufacturing schedules will be used to increase the plant inventory. The inventory projections will be made for one, two, three, and four lead times. The lead time represents the average time from placement of a warehouse stock replenishment order to receipt of the replenishment order at the warehouse, assuming no delay caused by an out-of-stock condition at the resupply plant. The lead time is typically three to six days depending on the product being reordered, the warehouse from which it is reordered, and the plant from which the product is shipped. A given warehouse is always resupplied a given product from the same plant; however, different plants resupply different products. Case study 1–figure b illustrates the typical warehouse resupply pattern. This module's primary outputs are: (1) the projected plant and warehouse inventory status report and (2) the projected plant and warehouse inventory status file.

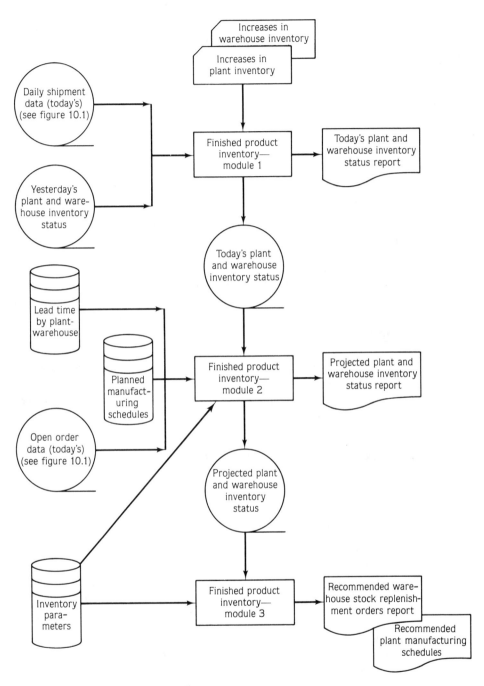

case study 1 figure a A modular approach to the finished product inventory system of the ABC Company.

294 Systems Design of Order-entry Output and Data Base—ABC Company

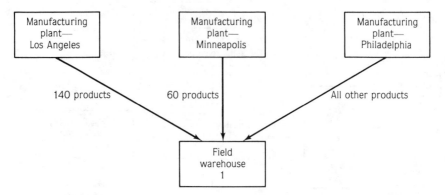

case study 1   A typical field warehouse resupply pattern for the ABC
figure b   Company.

The **third module** uses the projected plant and warehouse inventory status file and a file of inventory parameters to prepare two reports: (1) recommended plant manufacturing schedules and (2) recommended warehouse stock replenishment orders. The inventory parameters file contains the reorder point, economic reorder quantity, and average usage during one lead time for each product at each warehouse, and the hazardous material indicator.

Because both customer demand and lead time for a product varies by warehouse location, the inventory parameters for a product also vary by warehouse location. This module will be used by the central inventory control group to make the decisions necessary to place stock replenishment orders for manufacturing plants, so that they, in turn, can ship specific items to the field warehouses they supply.

Assignment: Design of Finished Product Inventory System Output

1. Prepare a system output analysis form, using case study 1–table a (p. 296) as a sample, specifying the informational elements and their length to be shown on each of the following reports:
   a. **Today's plant and warehouse inventory status report**   This report, shown as output from module 1 in case study 1–figure a, presents the current inventory status of each product at each plant and warehouse. The plants and warehouses are the columnar headings listed across the top of each page of the report. The product codes are listed in ascending sequence in the extreme left column. The row entries opposite a product code represent the inventory of that particular product at the plants and warehouses.
   b. **Projected plant and warehouse inventory status report**   This report, shown as output from module 2 in case study 1–figure a, presents the projected inventory status of each product at each plant and warehouse. A separate report is prepared for each plant and warehouse listing the projected inventory status for each product normally stocked at that plant or warehouse. Inventory projections for one, two, three, and four lead times must be shown for each product. The average withdrawal during one lead time must also be shown. The reorder point and reorder quantity must be shown as well as the lead time for warehouse inventory reports. For plant inventory reports, the average manufacturing run size must be shown.
2. Prepare a printer spacing chart, using case study 1–figure c as a sample, for each of the reports above.

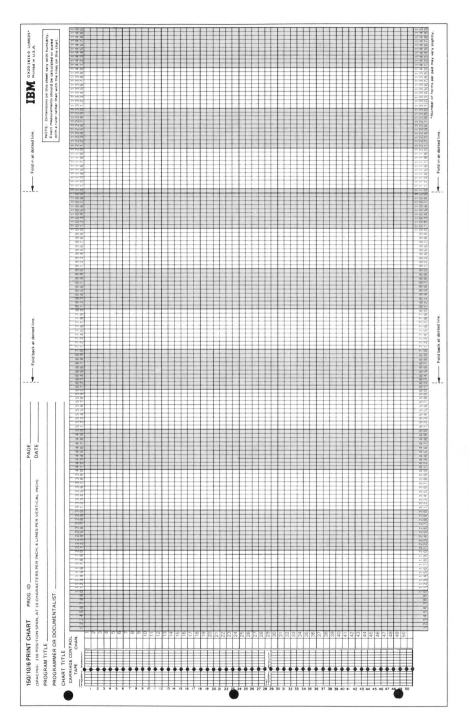

case study 1—figure c A printer spacing chart for the ABC Company.

cast study 1-table a System output analysis form for the ABC Company.

SYSTEM OUTPUT			
Date:	**Type of output:**		
Preparer:			
Informational elements	Type of characters	Number of characters	Comments

Comments:

case study 1-table b System data file analysis form for the ABC Company.

SYSTEM DATA FILE			
Date:	**Type of record:**		
Preparer:			
Record elements	Type of characters	Number of characters (bytes)	Comments

Comments:

Assignment: Design of Finished Product Inventory System Data Base

3. Prepare a system data file analysis form, using case study 1–table b as a sample, for the following finished product inventory files:
   a. Lead time by plant—warehouse
   b. Inventory parameters

   For the foregoing, in addition to determining the size in bytes of each record element for these data files, calculate the total disk storage in bytes required to store each of these files.
4. Prepare a record layout worksheet, using case study 1–figure d (p. 298) as a sample, for data files above.

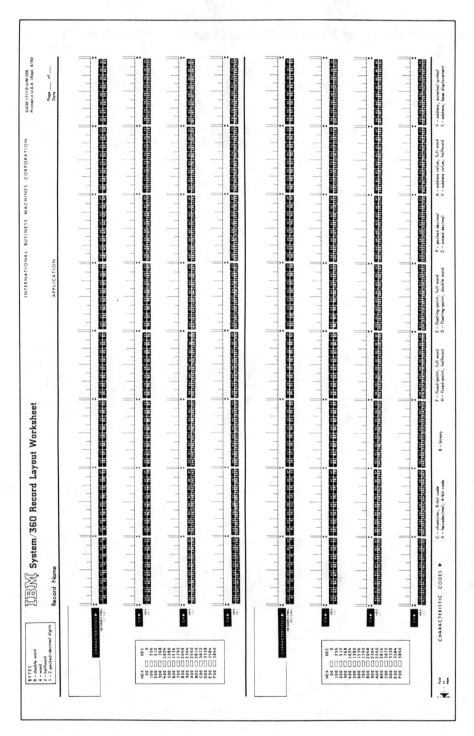

case study 1—figure d A record layout worksheet for the ABC Company.

chapter 10

case study

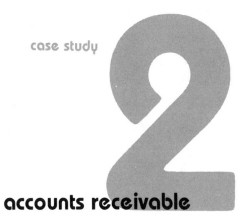

accounts receivable

Modular Design Approach

The primary objective of the new accounts receivable system is to improve the accuracy of the record keeping of customer accounts. This objective is to be met by supplying input data directly from the order-entry system to the accounts receivable system. Currently, input for the accounts receivable system is prepared separately from input for the order-entry system. Not only will duplication of effort be eliminated, but also the order-entry system and the accounts receivable system will be in agreement on the status of each order (shipped or unshipped) and the actual quantities of products shipped. With the billing error rate reduced from 7 to 1 percent by improvements in the order-entry system, there will be a corresponding improvement in the accuracy of customer accounts.

Another objective of the new accounts receivable system is to produce accurate and reliable information in order to help monitor credit and collection activities. Report accuracy is made possible by improving the accuracy of the individual customer accounts.

To accomplish these objectives, the new accounts receivable system will be designed in two modules, as depicted in case study 2–figure a. The **first module** will use the accounts receivable master file as of the start of the current business day and update this information with today's receipt of customer payments and daily shipment data passed directly from the new order-entry system. This module will be run daily. The result of this processing will be the current, up-to-date accounts receivable master file. Customer accounts receivable statements will also be printed according to the following schedule:

1. Customer account numbers from 0000001 to 2000000 will be printed on the 5th working day of each month.
2. Customer account numbers from 2000001 to 4000000 will be printed on the 10th working day of each month.
3. Customer account numbers from 4000001 to 6500000 will be printed on the 15th working day of each month.
4. Customer account numbers greater than 6500000 will be printed on the 20th working day of each month.

The **second module** will use today's accounts receivable master file—updated with today's receipts of customer payments and daily shipment data passed from the order-

300 Systems Design of Order-entry Output and Data Base—ABC Company

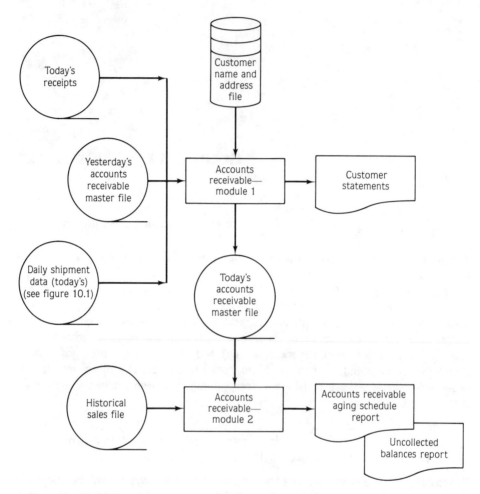

case study 2 A modular approach to the accounts receivable system of the
figure a   ABC Company.

entry system—as its primary input. The accounts receivable master file will be used to prepare the accounts receivable aging schedule report and the uncollected balances as a percentage of original sales report. This module will be run monthly to provide managerial information to monitor the accounts receivable collection process.

Assignment: Design of Accounts Receivable System Output

1. Prepare a system output analysis form, using case study 1–table a as a sample, specifying the informational elements and their length to be shown on each of the following reports.
    a. **Customer statement** The ABC Company currently uses an open-item receivables accounting method. When cash is received, the remittance is applied to specific open invoices so that the discount earned (or forfeited) can be determined. Hence, each

customer account must state the current status of each open invoice, indicating if the invoice is current, thirty to sixty days old, sixty to ninety days old, or over ninety days old. In addition, a summary of invoices paid since the last statement must be prepared; the summary must show discount earned or forfeited on each paid invoice. It is important that dates of all invoices and payments be shown, but individual items on each invoice need not be shown.

    b. **Uncollected balances report** The uncollected balances, as a percent of original sales report, will be prepared monthly for the credit manager's use in monitoring the accounts receivable function. The report presents sales and accounts receivable information for each of the three most current months. For each month, three figures will be shown: (1) the sales during the month, (2) the receivables outstanding at the end of the current three-month period, and (3) the receivables outstanding expressed as a percentage of the respective original sales that gave rise to those receivables (computed by dividing item 2 by item 1, then multiplying by 100 percent). The ratio of receivable balances to original sales will provide a much more accurate indicator of the actual collection experience. This ratio will not be distorted by the month-to-month variation in sales.

2. Prepare a printer spacing chart, using case study 1–figure c as a sample, for each of the reports above.

Assignment: Design of Accounts Receivable System Data Base

3. Prepare a system data file analysis form, using case study 1–table b as a sample, for the following accounts receivable files:
    a. **Customer Name and Address File** Assume there are 4000 customers on the file.
    b. **Historical Sales File** This file contains two years of shipment information at the individual product code level for each invoice. Refer to table 10.1 for information needed to calculate file size. For the foregoing, in addition to determining the size in bytes of each record element for these files, calculate the total disk storage in bytes required to store each of these files.
4. Prepare a record layout worksheet, using case study 1–figure d as a sample, for each of the data files above.

chapter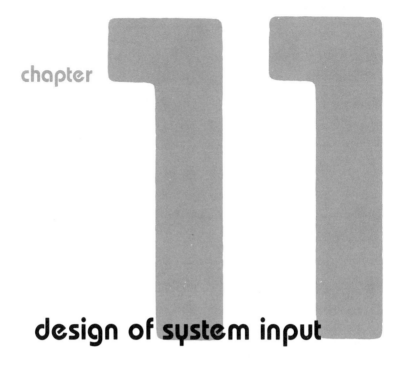

design of system input

OBJECTIVES:

- To explore the common methods of system input found in interactive and batch systems.
- To set forth important system principles to consider when designing input.
- To enumerate human factors that should be evaluated before finalizing the design of system input.
- To set forth a methodical approach to the design of system input.

IN THIS CHAPTER:

    Introduction to Design of System Input
COMMON METHODS OF SYSTEM INPUT
    Card Punches and Verifiers
    Key-to-Tape Systems
    Key-to-Disk Systems
    Key-to-Floppy-Disk Systems
    Distributed Data-Entry Systems
    Input Terminals
    Other Input Methods
INCORPORATE SYSTEM DESIGN PRINCIPLES—INPUT
DESIGN OF SYSTEM INPUT
    Consider the Human Factor in Designing Input
    Review Information on System Input
    Specify Input Requirements—Relate Output Through Files and Procedures to Input
    Determine Specific System Input
    Determine the Method(s) of System Input
    Design of System Input
CHAPTER SUMMARY
QUESTIONS
SELF-STUDY EXERCISE

The previous chapters have focused on system outputs and data files in the design of a new system. Within this chapter, the accent is on system inputs—the raw data captured and utilized by the system to produce the desired output. Each input is not "an island" unto itself; rather, inputs are devised to be compatible, as much as possible, with their final use and the interrelating parts of the DP system. Although system inputs start the flow of data within a system, they are linked directly with data files, methods, procedures, and especially with outputs.

The format of this chapter is similar to that of previous chapters on systems design. Initially, the common methods of system input are explored, with emphasis on the current trend in data-entry equipment to keep human intervention to a minimum. After setting forth the important underlying design principles of system input, the steps in designing input are enumerated, highlighting the relationship of input to files, methods, procedures, and outputs. In addition, the importance of organization personnel and analysts working together to resolve the problems of devising efficient and economical input is discussed. From this broad perspective, an optimum system is designed that will meet user needs.

Introduction to Design of System Input

As with the prior elements in the system design process, inputs must be planned carefully by the systems analyst because of the time, effort, and cost involved in converting raw data to usable information. Proposed inputs must be examined and evaluated from many viewpoints so that the best decisions are made. Inputs may be handled in a more efficient manner on an individual or batch basis. Generally, it is best to capture the data initially onto a machine processable medium. The accuracy requirements of input data may call for few or extensive editing methods. Inputs, in some cases, can be processed on a random sequence basis, whereas in others, a particular sequence is needed before processing. Time constraints on inputs and variations in input volume will affect many design decisions. Thus, many approaches will be presented in the chapter for designing system input. It is up to the systems analysts, assisted by organization personnel, to devise a system to meet user needs from the input stage on up to the final output stage.

Another important consideration for input design is the provision for retention of raw data, which provide a basis for preparing reports and answering questions not presently formulated. Here, caution must be exercised because of the high cost storage, as well as eventual retrieving and processing the data for meaningful output. The criterion used for storage of file information is also applicable to inputs; that is, the potential value from the stored data should be greater than the related cost of storing, retrieving, and processing the data.

COMMON METHODS OF SYSTEM INPUT

Several methods are available for capturing data from originating documents and sources. Data-entry devices include: keypunching and key verifying with a card-punch machine, keypunching to magnetic tape or disk, and keyboard data-entry recording. Of these methods, the keypunch method has been the most widely used. However, data punching directly onto magnetic disk is rapidly being used to replace keypunch machines because it saves time and expense. Optical character readers are yet other methods of system input.

Because of the importance of various methods that can be used as system input, the major characteristics and limitations of the more common ones are examined below. These methods include:
- card punches and verifiers
- key-to-tape systems
- key-to-disk systems
- key-to-floppy-disk systems
- distributed data-entry systems
- input terminals
- other input methods

Card Punches and Verifiers

As input for small computer systems, various types of card punches and verifiers have been used for batch processing. Likewise, punched cards serve as input for larger systems where interactive processing is not economically feasible. In the presentation below, the more common types of card punches and verifiers are explored.

Card punch In the past, the most common method of converting source data into punched cards was the use of the card punch or key-punch machine. Basically, the operator of this machine reads a source document and transcribes the data into punched holes by depressing the appropriate keys on the keyboard. In order for the operator to perform an efficient job during key punching, the key punch feeds, positions, and ejects cards automatically. The card punch stores different card formats and operates in the range of ten to fifteen strokes per second. The IBM 29 Card Punch is shown in figure 11.1.

Key verifier The most widely used verifying method for the card punch is the card verifier, such as the IBM 59 Card Verifier (figure 11.2). The verifier looks similar to the

Common Methods of System Input **307**

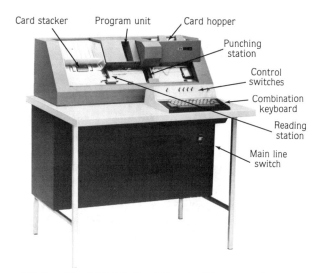

figure 11.1 The IBM 29 Card Punch. (Photograph courtesy of International Business Machines Corporation)

card punch, but does not punch holes. As the second operator or verifier duplicates the original card punching process by reading from the same source document and depressing the same keys (as the first key punch operator did), the card verifier compares each depressed key with the punched hole already in the card by a sensing mechanism which has twelve pins rather than twelve punching dies. If the entire card is correct, then a notch is put in the extreme right-hand portion of the card between 0 and 1 horizontal punching positions. However, if there is an error, the keyboard locks immediately and a red light comes on. The operator has two more chances to obtain agreement between the verifying and original keypunching; otherwise, the top of the card is automatically notched in the column of the error, so that the card punch operator can prepare a new, correct card. An error can occur at the key punch or at the key verifier.

When cards are removed from the card stacker, all correct cards (notched on the right-hand side) can be easily identified. The others can be removed, corrected by repunching new cards, and returned to the group after verifying.

Data recorder An addition to eighty-column card punching is the IBM 129 Card Data Recorder, also capable of verifying cards. Resembling the IBM 29 Card Punch, it has a memory that serves as a buffer before the cards are punched. This approach means that the operator can key data continuously while another card is being punched and stacked. Since corrections can be made before a card is punched, the entire card does not have to be repunched because of a single mistake. In addition, the 129's memory will store up to six different card formats, enabling the operator to change from one format to another without interrupting the work flow. Options include an "accumulate" feature that will total selected card fields and a count of key strokes and cards.

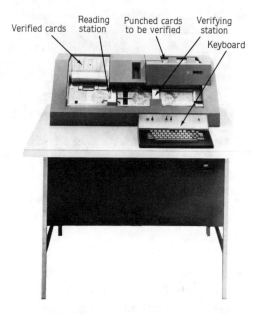

figure 11.2 The IBM 59 Card Verifier. (Photograph courtesy of International Business Machines Corporation)

figure 11.3 The IBM 5496 Data Recorder. (Photograph courtesy of International Business Machines Corporation)

The IBM 5496 Data Recorder (figure 11.3) is the basic punching device for the ninety-six-column card. (Burroughs and Honeywell also manufacture similar equipment.) Its distinguishing characteristics are buffered storage, four control program levels, and the ability to punch directly from data written on the face of the card. Buffered storage stores the complete contents of a card image before any punching or printing takes place. This feature permits the operator to correct any known errors before the card is punched and printed, which increases the productivity of the operator by about 10 percent. The second characteristic allows up to four programs to be loaded into the Data Recorder by reading prepunched program cards. Four program-level function keys are found on the keyboard. Programming provides control of field lengths, automatic skipping, automatic duplicating, upper/lower shifting functions, and file or word erase operations. In addition, verifying cards is a standard function of the IBM 5496 Data Recorder.

Punched cards, despite their past popularity, are limited by the number of columns that can be punched. They are also relatively bulky to store when compared to magnetic disk or tape. Also, they are more susceptible to errors since they are subject to mishandling by operations personnel. In view of the foregoing difficulties, punched cards are normally used when input volume is relatively small.

Key-to-Tape Systems

Although the punching devices above are capable of handling large volumes of input, a considerable amount of time and expense is involved. During the past several years, manufacturers have developed input equipment to remedy these deficiencies. Among the various data-entry devices to record data directly onto magnetic tape is the Honeywell Keytape (figure 11.4). Some key-to-tape stand-alone devices use cassettes or cartridges as an intermediate medium, but generally data are stored on half-inch computer-compatible tape.

During data entry with a typical key-to-tape method, a complete record is key-entered and stored before it is released to the magnetic tape. This feature simplifies and increases error correction. Errors sensed by the operator while keying data from the originating document can be corrected immediately by back spacing in memory and keying in the correction. The same correction method is applicable to verification. Studies indicate that an operator's productivity is increased by approximately one-third over regular punching methods. These improvements in operator efficiency primarily result from quieter operation, quicker setup, and simpler verification.

Although there are many advantages for the key-to-tape system, there are some disadvantages. Among these are higher equipment costs and the inability to access, manipulate, and read individual records. Of all the disadvantages, perhaps the most important is the fact that, for most applications, tapes must be merged together and sorted before further processing can occur. The additional merge step lengthens input processing time.

figure 11.4 The Honeywell Keytape, a stand-alone unit. (Photograph courtesy of Honeywell Information Systems)

Key-to-Disk Systems

To overcome the problems of gathering magnetic tapes from the many input devices and processing them on the computer for a merge operation, equipment manufacturers have developed key-to-disk entry systems, one of which is the Key Processing System developed by the Computer Machinery Corporation (figure 11.5) that utilizes a minicomputer.

A minicomputer-controlled key-to-disk system is more efficient to operate than keypunches or key-to-tape, stand-alone machines. As data are entered through the keyboard, they are processed by the system's minicomputer and stored on a magnetic disk in locations appropriate to the keystation of original entry. Once recorded data are keyed and verified, completed batches are transferred automatically from the disk onto a single reel of magnetic tape. This tape reel is the input for any computer batch processing run.

Figure 11.6 compares the input data processing methods of key-to-disk, keypunch, and key-to-tape (stand-alone) systems. Based on this illustration, it is obvious that fewer steps are required with the key-to-disk system, resulting in economy of operation through increased productivity. Also, this input method allows for monitoring of operators since the processor can accumulate performance statistics, such as the number of keystrokes per hour and the number of jobs completed. On the other hand, its major

Common Methods of System Input **311**

figure 11.5 The Key-to-Disk System. (Photograph courtesy of Computer Machinery Corporation)

shortcomings include a higher initial outlay for the keystation units and the necessity to increment the system up to four units at a time.

Key-to-Floppy-Disk Systems

In addition to key-to-disk devices, there are also key-to-"floppy"-disk units where data are recorded on only one side of the diskette. Illustrated in figure 11.7 is the IBM 3741 single-station unit with a display (at left) for preparing data input and computer programs. The output from data-entry devices, called a **diskette,** is fed into the IBM System 370 series computer through a 3540 diskette peripheral unit which simulates a card reader/punch. There are also "flippy floppys" which allow both sides of the disk to be used.

Distributed Data-Entry Systems

The current trend in data entry is toward distributed data-entry systems whereby single or multiple data-entry units are distributed at various company locations. As illustrated

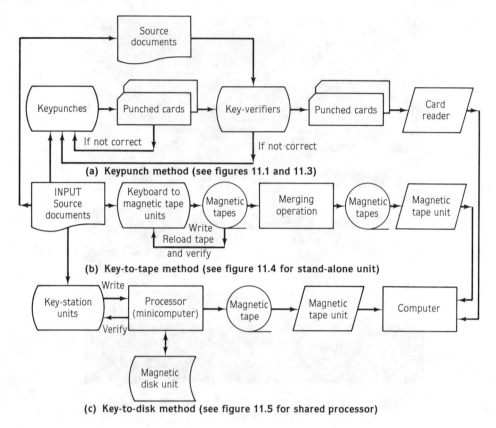

figure 11.6 Three different methods of preparing data for computer entry: (a) keypunch, (b) key-to-tape, and (c) key-to-disk.

in figure 11.8, data-entry units are used to capture input data on either magnetic tape or disk, which are then communicated on-line to input devices of the computer system. To speed up the process, the data from the key-entry units are communicated directly to the computer for on-line processing. A slower alternative, as illustrated, is to transport the tapes and disks to the computer center for processing. Thus, a distributed data-entry approach brings flexibility to an organization's data-entry operations.

Input Terminals

Another important method of system input is the utilization of input/output terminals for capturing and transmitting data to a computer system. There are two basic types: terminals with only keyboard input and terminals with keyboard input displayed on a cathode ray tube (CRT). Input terminals can be either on-site, or remotely located from the computer. In the latter mode, they can communicate with the central processing unit via standard communication lines like telephone lines and provide both input and output capabilities. Figure 11.9 illustrates a typical CRT terminal.

Common Methods of System Input **313**

figure 11.7 The IBM 3741, a single-station data-entry unit. (Photograph courtesy of International Business Machines Corporation)

An important recent development is the **intelligent terminal** (figure 11.10). It has certain processing capabilities, such as the editing and validating of data. Due to its processing capabilities, it relieves the CPU of preliminary processing required for validating input data. Thus, the systems analyst must weigh the increased cost of intelligent terminals against the reduced computer processing time.

Inasmuch as the current trend is toward interactive systems, there is a great need for terminals that can interact with the computer system on-site or from a remote location. Although the initial capital outlay is higher than for traditional input methods, they allow

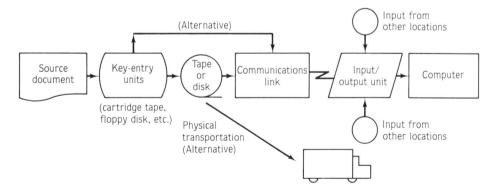

figure 11.8 The current method of preparing input data for computer processing in a distributed data-entry mode.

314 Design of System Input

figure 11.9 The IBM 3278 Cathode Ray Tube (CRT) terminal. (Photograph courtesy of International Business Machines Corporation)

the user to bypass many input procedures and communicate directly with the computer; from an overall standpoint, the total operating costs for an interactive mode may well be lower. Thus, possibly lower operating costs, more timely information, and the capability to control current and future operations more effectively may well prove that terminals are the best input method for an organization.

Other Input Methods

The foregoing input methods are the more common ones found in systems today. In addition, there are other ways of capturing input data. They include:
1. **optical character readers** provide a way of reading source documents via optical fonts. It is possible to read pencil marks on preprinted forms, a wide variety of type fonts, and hand-printed characters.
2. **point-of-sale devices** provide a convenient way for retail stores to capture input data. In turn, these data are converted into a machine processable form for computer processing.
3. **data collection systems** allow the collection of input data, either on-line or off-line, for controlling manufacturing activities. Data collection devices are located throughout the production area.

Hence, a wide range of input methods are available. For best design results, system personnel should be knowledgeable about the latest developments in computer and peripheral equipment.

Common Methods of System Input **315**

figure 11.10 The UNIVAC Universal Terminal System 400 operates as an intelligent remote terminal station or as part of a cluster of stations. (Photograph courtesy of Sperry UNIVAC)

SUMMARY: Common Methods of System Input

Card Punches and Verifiers Were the most common methods of converting source data into punched cards and verifying these same data. Due to their limited storage and file handling problems, newer forms of data entry are generally utilized.

Key-to-Tape Systems Allow a complete record to be key entered and verified before it is released to magnetic tape. Since each unit is a stand-alone unit, a disadvantage is that tapes must be gathered and merged before computer processing can occur.

Key-to-Disk Systems Permit data to be recorded and verified on magnetic disk before transferring completed batches to magnetic tape for computer processing. Of the current input methods of data entry, this one is currently experiencing wide acceptance because there are fewer steps than with prior input methods.

Key-to-Floppy-Disk Systems Allow data to be recorded on diskettes rather than on magnetic disks; this, too, is experiencing wide acceptance.

Distributed Data-Entry Systems Center on single or multiple data-entry units distributed at various company locations. Since input data are captured on either tape or disk, they are communicated in an on-line or off-line mode as input to the computer system.

Input Terminals Make use of either keyboard terminals or CRT terminals that can be used on-site or can be remotely located. The latest development is the **intelligent terminal** which is capable of performing certain editing and validating functions. The chief advantage of input/output terminals is the capability to bypass many of the input processing procedures of traditional methods, thereby providing a direct link with the computer in an interactive mode.

Other Input Methods Include optical character readers, point-of-sale devices, and data collection systems. These methods complete the wide range of methods available.

INCORPORATE SYSTEM DESIGN PRINCIPLES—INPUT

Earlier chapters discussed design principles related to system output and data files. Similarly, several are applicable to system input. Omission of these important design principles may jeopardize the entire system project.

A first principle for consideration, the **Principle of the Human Aspect of Data Input**, treats the human side of data input. Although the computer is capable of processing and transmitting data accurately, people are fallible. An individual has a difficult time reading a long string of numbers, such as account numbers or inventory numbers, and an even worse time reading special symbols and characters. In those cases where input personnel have a difficult time reading data, input design should give preference to this human consideration over technical ones to keep the number of input errors caused by data-entry personnel to a minimum.

Closely related is the **Principle of the Acceptability of Inputs.** If the methods of inputting data are acceptable to organization personnel, the likelihood of a successful system is much greater. To ensure acceptability, it is highly recommended that data-entry personnel work closely with systems analysts to finalize the new system. To ignore the skills and capabilities of present personnel who will be operating newer data-entry equipment is also a mistake. Thus, the success of the new system may well hinge on the acceptability of new system input and equipment.

For an efficient and economical system, there is need for a third design principle, the **Principle of Integration of Inputs and Outputs.** Fundamentally, this means that output from one functional area should not stop there, if it is logically related to other areas. Rather, it should serve as input for those other functional areas, resulting in an integrated approach. Functional areas, for example, should be under the direction of a single program or a set of programs if there is an integration of computer subsystems; otherwise, system complexity grows with an inordinate increase in implementation effort and cost. Hence, this latter view is not recommended, whereas the integration of system inputs and outputs for logically related areas is.

To assist in keeping manual intervention to a minimum, there is need for the **Principle of Common Data Representation.** Good input system design dictates that a common data representation, that is, a common data processing language, be used throughout the system so that communication between various types of DP equipment can be accomplished without manual intervention. Data representation can be in the form of punched cards, punched paper tape, magnetic tape, magnetic disk, or other machine-processable media. In essence, this principle states that processing by some type of machine-processable medium is more accurate, less costly, and faster than using conventional manual methods.

Another important design principle, the **Principle of Eliminating Duplicate Input,** demands that any type of duplication in terms of input be eliminated. Duplication can take several forms: for one, instead of processing multiples of the same data to serve many users, the data are captured at one time; for another example, the need for visual and computer files of the same input data can be eliminated by utilizing an interactive system that allows instantaneous visual display of critical input data. Overall, duplication of data-entry items, input methods, and comparable items should be eliminated for an efficient system.

In order to facilitate the foregoing principles, a sixth principle should be employed in good system design—the **Principle of Simplicity of Input.** Data input formats should be designed simply, thereby assisting in providing clarity of output. Thus, before desired output can be produced to meet specific user needs, input is needed that is comprehensive but, at the same time, simple for ease of operation.

An underlying principle of input as well as other areas of system design is the **Principle of Garbage-In—Garbage-Out.** Its acronym is GIGO. Output from a system cannot be correct if its input is inaccurate and/or incomplete. It does not have the human attribute of correcting itself for imperfect or bad conditions. Furthermore, as organization personnel rely on computers, they tend to trust output as correct, thereby not checking it as they would check human output. Inaccurate and incomplete output can be caused by a host of problems at the input phase.

SUMMARY: Incorporate System Design Principles—Input

1. **Principle of the Human Aspect of Data Input** A logical starting point for system principles of input is one that deals with the human side of data input. Since most errors are found at the input phase where the human element interfaces with originating data, it is essential that systems analysts favor this important element over equipment considerations.
2. **Principle of the Acceptability of Inputs** The success of a new system starts with the acceptability of system inputs. To ensure acceptability of input

methods, organization personnel who work with input should participate in their development.
3. **Principle of Integration of Inputs and Outputs** Integration of outputs from one functional area implies their use as inputs to other areas. Although this approach is desirable, each of these areas should be under the direction of a single program or set of programs so that their control is assured.
4. **Principle of Common Data Representation** Good input system design dictates that a common data representation be established in order to keep subsequent manual tasks to a minimum. Data representation can be in the form of punched card, punched paper tape, magnetic tape, and magnetic disk, among others, that is readable by DP equipment, thereby requiring little or no intervention by organization personnel.
5. **Principle of Eliminating Duplicate Input** Good input system design demands that duplication of data-entry items, input methods, and comparable input items be eliminated. From this vew, one input data item can serve many users.
6. **Principle of Simplicity of Input** Data input formats should be designed for simplicity of input and, in turn, clarity of output. In this manner, there is a link between simplifying inputs and outputs.
7. **Principle of Garbage-In—Garbage-Out** This important design principle means exactly what it says. A system cannot function effectively without accurate and complete data inputs. It does not have the human attribute of correcting imperfect conditions.

DESIGN OF SYSTEM INPUT

With the preceding background on the common methods of system input and their underlying design principles, the main thrust of this section is on the steps involved in designing system input. To be specific, the input design process includes the following steps:

Consider human factors in designing input methods.

Review input of present system and data contained in the exploratory survey report to top management.

Specify input requirements, including the relationship of outputs through files and procedures to required inputs.

Determine the specific content of system input using system input analysis forms.

Determine the methods of system input—that is, data-entry devices, terminals, and comparable equipment.

Design the inputs by determining the source documents, input records, and input codes, among other items.

Each of these areas is covered in a comprehensive manner below to give the reader an insight into the design requirements of system input for the order-entry system of the ABC Company.

Consider the Human Factor in Designing Input

As with the preceding principles underlying good system design, there are a number of important human factors to consider. Foremost among these is service to the user. Too often overlooked is the need in input system design to remember that the user is most important. One of the dangers of turning system design over to system specialists is that they may be more concerned with the efficiency of operations than with the operations the system is supposed to be supporting or controlling. For example, a systems analyst might be more concerned with the type of data-entry equipment than with the utility of its input. This is no indictment of systems analysts **per se,** but rather it is a warning that any system specialist cannot be expected to know everything. For this reason, both managerial and system personnel must work together to finalize efficient input methods that satisfy user needs.

Also included in this teamwork approach to systems design is establishing the proper level of automation. Total automation of input would completely eliminate any need for human intervention; typically, today's input systems are automated to a far lower degree because, if for no other reason, cost and complexity of total automation would be prohibitive. Because decisions based on relatively simple logic are the forte of the computer, logic at a higher level of complexity is still handled more economically by humans. Unless there is a real advantage, there is no reason to automate what people can do better; this is especially true at the input level.

Going beyond the overall factors in designing input, the National Bureau of Standards (NBS) suggests that systems analysts make a special effort to avoid special symbols and characters that can be confused with other symbols; for example, the letters I, O, and Z may be confused with numbers, 1, 0, and 2, respectively. Likewise, it is recommended that similar character types be grouped, such as AB4 and not A4B. Also, coded data should be linked to something meaningful to people, such as M and F, instead of 1 and 2, for male and female. Thus, many errors that originate at the interface of user and data-entry equipment can be overcome.

When designing system input, consideration should be given to the level of noise. Industrial psychologists maintain that noise fatigues keypunch operators, thereby adversely affecting their output. Relatively, the key-to-tape and key-to-disk equipment runs quietly and can, therefore, be located near other offices rather than clustered in one place. Although the keyboard could be made absolutely silent, operators prefer to hear a click with each keystroke as an acknowledgment of keying action by the system. Also, the clicks set up a cadence that can actually add to keystroking rates.

Tape and disk input equipment allows operators to move from record to record at faster speeds since they do not have to wait for cards to be dropped and registered. On card equipment, short operators often have to stand up repeatedly to change drum cards, load blank cards, and remove cards from hoppers. Key-to-tape systems eliminate card handling, and key-to-disk systems go a step further by reducing tape handling to a minimum. Only the supervisor loads and unloads the single tape drive in a disk system. In some key-to-disk systems, the disk unloads automatically every six minutes.

Large-user organizations can afford to try a selection of key-input units made by different manufacturers. One organization, for example, rented three different kinds of key-to-tape equipment for a six-month evaluation. Seven operators took turns operating the units. After the trial, the operators were asked which equipment they liked best and why. Some of the reasons given included "attractive physical keyboard characteristics, touch, layout, display" and "feeling more at ease" with a certain manufacturer's units.

Overall, effective input system design that incorporates the human factor must allow the user to understand the data-entry equipment, and the equipment, in turn, must be relatively easy and convenient for the user to use. Data-entry personnel should experience less fatigue with the new data-entry equipment. The equipment must be flexible and versatile to meet changing input needs and be capable of meeting speed and accuracy requirements.

Review Information on System Input

Although the human factors are most important in system design, a starting point for the input design process is a review of the information compiled during the systems analysis phase. Facts to be reviewed as related to the current system include **what** data are entered, **who** enters it, **where** it is entered, and **when** it is entered. This review highlights basic problems and difficulties with the present system input and distinguishes between areas that are straight forward and those which are complex. In effect, an understanding of the present system's input directs attention to those areas needing improvement for more efficient data entry.

Systems designers should also review the information contained in the exploratory survey report to top management, especially the recommended system alternative. This information is extremely helpful in understanding what the new system inputs are all about. The capability of extracting information developed to date will be of great assistance in the actual design.

Specify Input Requirements—
Relate Output Through Files and Procedures to Input

With the foregoing background, the real design work can commence. This centers on determining what data are needed as input from the new system to produce the desired output by employing various procedures and related files. As mentioned previously, this should not be a redesign of the present system, and prime consideration must be given to user needs.

At this juncture, there is a great need for cooperative effort. Organization personnel and systems analysts must work together to design the desired inputs. Failure to agree upon what the inputs should be not only restricts the effectiveness of system input, but also jeopardizes the entire system project. The rationale is that data input is not "an island unto itself;" rather, inputs become the prime source of data for DP methods and procedures as well as data file storage. After input data have been processed, they become the basis for various types of output. Thus, there is a great need for cooperative efforts of organization and system personnel so that inputs are designed properly. From an overview standpoint, this teamwork is required so that inputs, methods, procedures, data files, and outputs are logically related for an efficient and economical system.

To assist in specifying input requirements, questions can be raised to test the validity of new system input. Typical questions are set forth in figure 11.11. These questions are general, that is, to facilitate more effective input procedures and considerations for the human element. Specific questions can also be asked regarding how data are to be extracted from the source document. Will certain data be read directly into the computer by means of optical character recognition (OCR) equipment? Will most data be keyed directly into the system by input/output terminals located throughout the organization? Will data be keypunched and key-verified for a remote job entry operation, or will input data be considered in light of the above plus other pertinent questions? These questions can be answered by referring to the outputs, methods, procedures, and data files of the system. As mentioned previously, inputs should be compatible with their final use and interrelating parts of the system.

Are the input data necessary to plan, organize, direct, and control organization activities?
Is there a logical flow of inputs to outputs using prescribed procedures and data files?
Can capturing of input data be justified from the standpoint of value received versus cost expended?
Can input data be obtained from another source?
Are input data easy to understand and use?
Are input procedures flexible enough to meet organizational changes?
Is there consideration for the human factor when interacting with input data?
Are input methods and procedures acceptable to organizational personnel who must use them?
Has there been consideration to reduce the noise level of data-entry operations?

---

figure 11.11 Systems design—questions to test the validity of input.

When specifying input requirements, the designer is defining data that are eventually needed to produce the desired output. Inasmuch as the required output has been specified (chapter 8), it is necessary to consult the System Output Analysis forms. They contain the informational elements, type of characters, number of characters, comments, and other necessary information. However, there is no indication on this form of the input data necessary to produce this output. Hence, this phase of system design centers on determining the sources of data as input to the system. The more common ones are:

    data originating from a source document that has essentially the same output
    data originating from a source document that has output different from that document
    data originating from one or more tables in conjunction with other input data to produce output
    data originating from and calculated by a computer program in conjunction with other input data to process output
    data originating from many sources in order to produce the desired output

A popular source of data is that which originates from a single document. For example, for a new customer credit account, a New Customer Name, Address, and Credit Information form would be prepared. The data from the source document would serve as input to data-entry equipment, which, in turn, would be input to a computer program for adding new accounts to direct-access storage devices. Similarly, this computer-updating process would require that new customer names, addresses, and credit information be listed as an output report. In this example, the output content is identical to that of the input, except that the latter is computer-prepared and the former is not.

Although the preceding example centered on directly relating input to output, data contained in single source documents normally undergo some changes before the final report is prepared. For example, sales orders are the source documents for data-entry equipment. When these data are computer-processed, sales invoices are prepared. Many times, the prices on incoming orders are incorrect, items must be backordered, or items must be cancelled (no longer offered for sale). Thus, the final sales invoice may be quite different from the original sales order.

Tables are a third source of input data, an alternative to preparing input data. Often it is necessary to keep information confidential, such as a salesperson's commission rate. To illustrate, a sales commission rate, based upon the type of order, is recorded by the sales order clerk as sales orders are received. When the sales orders are processed on the computer, the type of sales code determines the calculation for sales commissions; this information is recorded on a machine-processable medium and processed periodically to produce a sales commission report. In this example, input related to data contained in a table is used to produce the desired output.

An extension of this approach is to use data originated by a computer program in addition to that from source documents. This approach requires program calculations or employing constants within the program. Because these items are under computer control, they present no problems in terms of system design. However, the need for such calculations and constants should be noted on a system analysis form for consideration during that phase which defines the programming specifications for the computer programmer. Referring to a typical sales invoicing computer program, calculations are made for the price of the products sold (price per unit × number of units sold) and for the total invoice amount (price of the products + freight charges + sales taxes). Similarly, constants are employed, such as the zip code, used to determine the proper city and state to be printed.

Of all the common approaches to specifying input requirements, the fifth (last) is the most difficult. When the data needed to produce some type of output are available neither from a single source document nor a computer program, it is necessary to extract input data from a combination of sources. These include not only source documents, tables, and computer programs, but also master files, transaction files, and infrequently used input sources. To illustrate, the monthly billing of customers in a batch processing mode requires the processing of master and transaction tape files concurrently (as input) along with the need for calculating new ending balances, and checking tables for city and state addressing, sales tax rates, and similar items. In effect, tape files (data input) require the use of programming (program procedures) to produce monthly statements for customers.

In summary, there are many considerations to take into account when specifying input requirements. An understanding of the preceding sources of input data will do much to resolve the design of appropriate input data. In addition, the pertinent questions noted above will assist the systems analyst in devising effective input.

Determine Specific System Input

Now that input requirements have been specified by the systems analyst working with appropriate organization personnel, the next step in the input design process is to specify in detail the content of system input. This means preparing System Input Analysis forms, like the one shown in table 11.1 for sales order-entry. As one might expect, there is some similarity between the information on this form and that on the System Output Analysis form (chapter 8). However, there are differences: first, the "source" (columnar heading) of input must be specified, which refers to the type of source document; second, the "comments" (columnar heading) must state specific requirements of the informational elements.

table 11.1 Typical system input analysis form.

		SYSTEM INPUT		
Page 1 of 1				
Date: July 25, 198_		Type of input:		
Preparer: G. W. Reynolds		SALES ORDER-ENTRY		

Informational elements	Type of characters	Number of characters	Source	Comments
Customer account number	Numeric	7	Customer file	Must be numeric
Customer name, address, city, state, and zip code	Alphanumeric	55	Customer order	
Description of item	Alphanumeric	15	Customer order	
Number of items	Numeric	5	Customer order	Must be numeric
Inventory number	Numeric	5	Inventory file	Must be numeric
Tax code	Numeric	1	Customer order	Must be numeric

Comments:
Customer orders can be received via telephone, mail, salesperson, or Teletype.

In the example, the source of the sales order is the customer. Most input data can be derived from the sales order, except for customer number and product numbers, which are generally derived from files. As noted in the "comments" column, certain fields must be numeric. Techniques of testing for all numeric (or alphabetic) fields are covered in a later chapter.

Determine the Method(s) of System Input

Based upon the determination of what input data are needed and the source of that data, the next step in the input design process is to develop the method(s) of data entry. Specifically, the designer must determine the most efficient method of preparing system input. As demonstrated previously, the common methods of system input are: card punches and verifiers, key-to-tape systems, key-to-disk systems, key-to-floppy-disk systems, distributed data-entry systems, input terminals, and other input methods. Because of the number of methods for capturing input data, it is worthwhile to explore feasible alternatives by asking certain questions.

Typical questions that must be answered include: **who** is involved in data entry; **when** and **where** are data to be captured for economy and efficiency of input operations; is remote data-entry desirable or should there be use of distributed data-entry units; **how often** is it necessary to process the data for timely output; and **what** is the likely volume of input during regular and peak periods? These and other probing questions will help determine which data-entry method will best suit the users' needs. Similarly, consideration must be given to the advantages and disadvantages of common methods of data input. The coalescing of all the pertinent factors will help resolve the need for one or more methods of data entry.

Design of System Input

System input starts the flow of data through a system; it is one of the most important steps in the entire system design process. The analyst must design each input record to be an integral part of the new system by visualizing that integration: how will each input record be handled by operating methods and procedures; what is its relationship to various on- and off-line data files; and, what is its utility in producing the desired outputs. In a few words, the systems analyst will face long periods of intensive creativity and logical thinking to devise the best system possible.

Source document formats Since the source documents are the initial forms on which data are recorded, it is the job of the designer to determine their format. As an organization moves from one system to another, the formats generally need to be changed to accommodate newer data-entry equipment. This is applicable to all source documents, whether **master records** comprising permanent information or **transaction records** reflecting daily business activities.

To have an efficient data-entry operation, it is necessary that certain guidelines be followed for source document formats when transcribing information to an input record through the use of data-entry equipment. The formats should be designed such that:
- data can be read from the source document from left to right and from top to bottom.
- data which are to be duplicated can be grouped together and placed in the left-most positions in the record.
- data which are variable can be grouped together and separated from fields that are to be duplicated or skipped.

data which are numeric or alphabetic can be grouped together.

unused portions of a record which are to be skipped can be placed in the right-most positions of the record.

Using these basic guidelines lends a consistency to input records which increases the productivity of data-entry personnel and decreases the amount of errors.

Input record lengths   The design of the input record length is dependent upon the type of input media. If magnetic tape or magnetic disk is used, there is generally no restriction upon the length of the input record. On the other hand, if data are entered from a CRT device or some other type of remote device, or if the record is to be stored on a punched card, there can be restrictions on the length; for these reasons, more than one physical record may be necessary to contain all the desired input. It is generally recommended that the input record length be kept within the length of the physical record. A typical layout form for a CRT device is illustrated in figure 11.12 (p. 326).

To determine the number of characters for each record, it is advisable to review past records which indicate the number of characters required for the application. This is particularly true for names, addresses, ship-to information, and comparable data that are subject to a wide range of character sizes. Also, consideration must be given to restricting space for data fields. For example, although the date on a report may be **07/25/8_**, requiring eight positions on the output report, the field in the input record would be **07258_**, thereby requiring only six positions of storage. The slashes would be added by the program creating the output report.

Input codes   In addition to discussing source document formats and input record lengths for designing system input, it is helpful to examine input codes. Fundamentally, a **data code** is a short label composed of one or more numbers, alphabetics, or alphanumerics which identify a certain type of data element. To illustrate, code **A** might represent a file addition, code **D** a deletion, and code **C** a change.

Because codes are widely used in DP operations, the systems analyst should have a good understanding of the basic ways of coding data. There are five characteristics of good coding systems:

1. **Easy to use**   This is particularly necessary for new organization personnel or those unfamiliar with DP operations.
2. **Unique**   This means that each element of the code should have only one meaning and that, conversely, each possible meaning be assigned to only one element.
3. **Concise**   Each code should be as short as possible to adequately describe the data item represented.
4. **Expandable**   This allows new data items to be inserted in logical sequence.
5. **Adaptable to DP equipment**   This helps eliminate human intervention and is particularly useful when the code can be machine-stored.

Regarding the basic types of codes, a common method of coding data items is the sequence code wherein numeric values are placed in a sequence to represent different meanings or to indicate different conditions. For example, the codes **T** and **N** could be used to represent taxable items and nontaxable items, respectively. Generally, the most frequently used condition of the data item is represented by the lowest numeric value. This is particularly helpful when programming since the program is

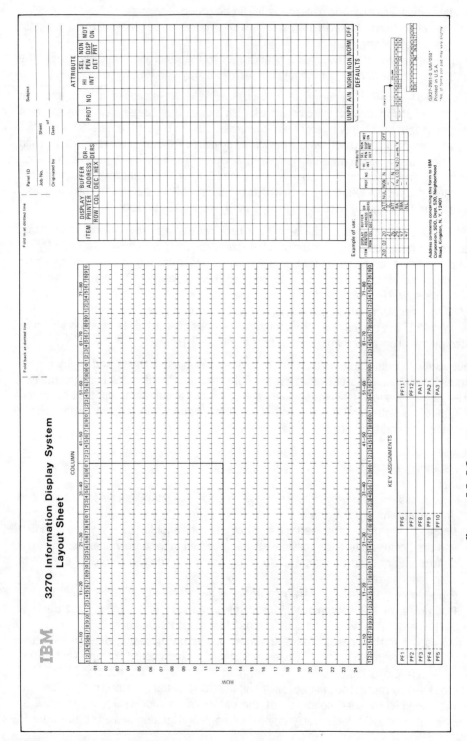

figure 11.12 A typical input layout form for a CRT device.

much more logical and easier to understand if the sequence of comparison is in counting order—1, 2, 3, . . .—instead of some other sequential arrangement.

Another type of widely used code is the **alphabetic code** which permits relating the meaning of a data item with its alphabetic character. An illustration would be the use of M and F, instead of 1 and 2, as the code for male and female. However, caution is needed here for certain data items that must be compared internally by a computer program. In other words, inexperienced DP personnel may find computer program comparisons difficult to understand.

One of the most popular types of codes is one that represents major and minor classifications of data items by succeeding digits within a number; it is sometimes called a **combined sequential code.** To illustrate, the following code is generated

Division code	Product line	Product number
1	0 1	0 5

for division code 1 (there are six divisions), product line 01 (there are fifteen product lines), and product number 05 (there are twenty products in each product line); the code for this specific product, then, is **10105**. Because there are only six divisions and one digit can handle up to nine divisions, only one digit code is necessary. Similarly, two digits can handle up to ninety-nine product lines and product numbers, which is more than ample for this illustrated combined sequential code. In this case or any other, the systems analyst is responsible for devising a coding structure that provides the most efficient approach to processing data items.

SUMMARY: Design of System Input

Consider the Human Factor in Designing Input Underlying the design of system input is consideration for the person who will be involved in data-entry operations. Overall, input operations should be relatively easy and convenient for the user.

Review Information on System Input Reviewing information on present system input and that contained in the exploratory survey report to top management is a logical starting point for devising new system input.

Specify Input Requirements—Relate Output Through Files and Procedures to Input This phase of system input design centers on determining the sources of data as input to the system. In addition, consideration at all times is given to relating input to procedures, files, and finally output.

Determine Specific System Input Accent in this step is on specifying in detail the content of system input. The use of System Input Analysis forms is helpful in accomplishing this task; they also serve as documentation.

Determine the Method(s) of System Input The questions of **who, when, where, what,** and **how often** must be answered in order to determine what method or methods are the best for data entry. In addition, their relative advantages and disadvantages must be evaluated.

Design of System Input Within this final step, the actual design of each input record must be specified. For effective system input design, attention must be paid to source document formats, input record lengths, and input codes.

328 Design of System Input

> **CHAPTER SUMMARY:**
>
> Initially in the chapter, the more common methods of system input were presented with great accent on key-to-disk systems, distributed data-entry systems, and input/output terminals. These newer methods of data entry make use of processors and minicomputers and provide a means of speeding system input. Many of the current principles for designing input underlie these newer forms of data-entry equipment. Hence, card punches and verifiers as a principal means of input are being replaced rapidly by these more convenient and easy to use data-entry devices.
>
> The design of system input is related not only to output reports, but also to methods, procedures, and data files. This broad approach to system input design assures that the "left hand knows what the right hand is doing." Only in this manner can an effective system, such as an interactive management information system, be designed. As will be seen in the chapter on the order-entry system of the ABC Company, the integration of inputs with files, procedures, and outputs is necessary for optimum results.

Questions

1. What is the current trend in system input methods for capturing data from source documents?
2. What are the common methods of input for:
   a. a batch processing system?
   b. an interactive processing system?
3. Explain what is meant by a distributed data-entry system when preparing input data for computer processing.
4. Are input/output terminals more widely used as system input devices than optical character recognition (OCR) devices? Why or why not?
5. Of the many system design principles set forth in the chapter, which one(s) is (are) most important from a management viewpoint?
6. How important is the elimination of noise in data-entry operations?
7. Why should input data during their design be linked to procedures, data files, and outputs? Explain thoroughly.
8. a. What are the most common types of source data?
   b. What are the types of source data that are found in an interactive management information system?
9. What is the purpose of the System Input Analysis form? Explain.
10. Why is it important to distinguish between a master record and a transaction record when finalizing system input design?
11. a. What guidelines should be followed when designing source document formats?
    b. What guidelines should be followed when designing input codes?
12. What is the rationale for designing system input after system output? Explain thoroughly.

Self-Study Exercise

True-False:

1. _____ Card punches continue to be the most popular method of system input.
2. _____ A problem with key-to-tape or key-to-disk systems is the inability to read individual records once they are stored on machine-processable media.
3. _____ Fewer number of input processing steps are needed with key-to-disk systems versus punched card systems.
4. _____ A current trend in data-entry systems is distributed processing systems.
5. _____ The GIGO principle is not applicable to data input.
6. _____ An important consideration when designing input is keeping the user in mind.
7. _____ There is little need for grouping similar character types when transcribing input data.
8. _____ At the input design phase, there is no need to work with organizational personnel on the final system.
9. _____ A popular source of data is that which originates from a single document.
10. _____ The contents of the System Input Analysis form is identical to the System Output Analysis form.

Fill-In:

1. A data processing system cannot supply the desired outputs unless it has read and stored the necessary _____ data.
2. _____ _____ input is susceptible to errors since they are subject to mishandling by operations personnel.
3. Key-to-disk systems do not require any type of _____ operations prior to computer processing.
4. A principle of good input system design dictates that a common _____ _____ be established in order to keep manual tasks to a minimum.
5. Good input system design requirements are not complete until the _____ _____ in data coding and representation is considered.
6. Good input system design demands that _____ of data-entry items, input methods, and comparable items be eliminated.
7. _____ data should be linked to something meaningful with which the user can identify.
8. Throughout the system input phase, there is a great need for _____ between system and organization personnel.
9. To assist in specifying _____ requirements, questions can be raised to test the validity of new system inputs.
10. After determining specific system input, the next step in the system input design process is selecting the _____ of system input.

Problems:

1. The American Company is currently developing a computerized accounts receivable system that will store all current accounts on line. In this manner, all outstanding charges to an account and current payments and adjustments to an account will be available for inquiry by

accounts receivable CRT terminal operators. In the area of payments, checks, money orders, and cash are received through the mail and from salespeople as well as notifications by various banks of their customers. Also, special consideration is given to nontaxable institutions in terms of courtesy discounts and elimination of sales taxes.

Based upon the foregoing requirements, complete the system input analysis form for payments received on account.

SYSTEM INPUT				
Date:		Type of input:		
Preparer:				
Informational elements	Type of characters	Number of characters	Source	Comments

Comments:

2. Using the data set forth in the completed system input analysis form of The American Company, prepare an appropriate input layout form (for a CRT device) that can be utilized in the new computerized accounts receivable system. Make the new format as readable as possible for use by all authorized accounts receivable personnel. Also, consider the handling of nontaxable and taxable transactions so that payments are handled properly.

331

chapter

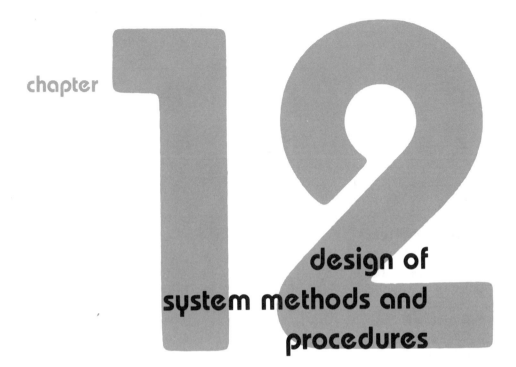

design of system methods and procedures

OBJECTIVES:

- To set forth the basic DP functions that comprise system methods and procedures.
- To explore the underlying design principles of system methods and procedures.
- To discuss the important human factors in designing system methods and procedures.
- To specify in detail the important factors in designing system methods and procedures.

IN THIS CHAPTER:

Introduction to Design of System Methods and Procedures
BASIC DP FUNCTIONS THAT COMPRISE SYSTEM METHODS AND PROCEDURES
    Originating
    Recording
    Classifying
    Manipulating
    Summarizing
    Communicating
INCORPORATE SYSTEM DESIGN PRINCIPLES—METHODS AND PROCEDURES
DESIGN OF SYSTEM METHODS AND PROCEDURES
    Consider the Human Factor in Designing Methods and Procedures
    Review Information on System Methods and Procedures
    Specify Methods and Procedures Requirements
    Determine the Steps in Methods and Procedures
    Devise Back-Up Procedures
CHAPTER SUMMARY
QUESTIONS
SELF-STUDY EXERCISE

After the system input, data files, and output have been designed, the next step in the design process is the determination of what system methods and procedures are needed to produce the desired output from the input. System flowcharts are helpful in relating input to data files and output via system methods and procedures. They are also a means by which to review methods and procedures. Overall, system flowcharts allow the systems analyst to examine the various methods and procedures so that they can be improved upon for efficiency and economy of operations.

Within this chapter, basic data processing functions that comprise the essentials of system methods and procedures are discussed first, followed by the underlying system principles of methods and procedures. As with much of the material on systems design, the steps in specifying appropriate methods and procedures are presented. After considering the human factor of system design and reviewing information on system methods and procedures, the requirements for specifying appropriate methods and procedures are described, and their procedural steps determined. Finally, backup procedures necessary in the event of system failure are noted.

Introduction to Design of System Methods and Procedures

Now that output requirements for the new system have been defined, the systems analyst's attention is focused on the new methods and procedures to produce these output needs. Consideration at this time must also be given to the on-line and off-line files as well as inputs, because it is difficult to isolate methods and procedures from system requirements. This phase requires intensive creativity from the systems analyst; it is a process of logical thinking that involves developing many system design alternatives for a specific area. The new methods and procedures are tested for **practicability, efficiency,** and **low cost.** Attention must be given to those designs that meet the study's objectives as closely as possible. Once the many possibilities have been reduced to reasonable alternatives, the alternative methods and procedures are examined thoroughly for the best one under the existing conditions. Included in these inventive steps is recognition that each functional area or activity is not isolated, but rather a part of the entire system. Thus, each related part must be considered in the final evaluation of the area under study.

Before exploring the underlying principles and the steps involved in the efficient design of system methods and procedures, it would be helpful to define the meaning of methods and procedures and to set forth those data processing methods and procedures which convert input to output. Each is explored below.

System methods and procedures defined As set forth in chapter 5, a **method** is a way of doing something, and a **procedure** is a series of logical steps by which a job is accomplished. Whereas a method proscribes how an action is undertaken, a procedure specifies what action is required, who is required to act, and when the action is to be undertaken. Overall, a method is smaller in scope than a procedure; therefore, in this chapter, emphasis will be placed on procedures, since we are concerned with linking inputs and data files with outputs, and on designing procedures that efficiently and economically accomplish desired data processing tasks.

BASIC DP FUNCTIONS
THAT COMPRISE SYSTEM METHODS AND PROCEDURES

In order to produce the desired results from a system, basic data processing functions, consisting of methods and procedures, are needed to link inputs and data files to outputs. They are:
1. originating
2. recording
3. classifying
4. manipulating
5. summarizing
6. communicating

Relating several to prior chapters, **input** (chapter 11) refers to originating, recording, and classifying, whereas **output** (chapter 8) is linked to summarizing and communicating. In addition, **data files** (chapter 9) are related to all items. These relationships are depicted in figure 12.1.

Because the first three items, the last two items, and data files have been covered in some depth in prior chapters, the main thrust in the next section will be on manipulating data using appropriate methods and procedures. First, however, an overview of all basic data processing functions will establish the logical development from input to output.

Originating

The basic input for any system originates on a variety of business forms, or source documents. Data are initially captured at the time a transaction occurs. As noted in the previous chapter, some common input documents are payroll time cards, sales orders for billing customers, receipt forms for merchandise received, and material requisitions. Input documents are related not only to accounting and finance, but also to marketing, manufacturing, purchasing, engineering, personnel, research and development, and other functions, depending upon the nature of an organization's products and/or

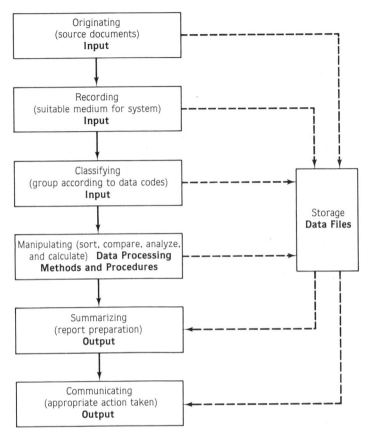

figure 12.1 The basic data processing functions—that is, methods and procedures—for relating input and data files to output.

services. However, these originating forms have two points in common: they form the basis for further data processing, and they verify that a transaction has occurred.

Since source documents become the basis of input to the system, it is necessary that the data be as clear and accurate as possible. Another important consideration relates to the sequence of data on the input form: data should be arranged in the same order as that desired in the transcription, which will be the natural order required in subsequent processing procedures.

Recording

Having captured the data accurately on source documents, it is generally necessary to record them onto another medium. This may necessitate conversion from a verbal or written form to punched card, paper tape, magnetic tape, or magnetic disk. The conversion may involve making a manual entry, entering data through the keyboard of a CRT terminal, using a teletyprwriter terminal, or employing some other input device. In data collection systems, data are captured by the input of cards and badges on which fixed data have been prerecorded in a machine-acceptable language. Variable data are entered by turning dials, knobs, or other kinds of manual settings. Still other recording methods—cash registers, imprinters, and data communication devices—are available for capturing the original data. No matter how the conversion is accomplished, data from the source document are recorded in a form that is acceptable to DP equipment.

During the recording process, every effort should be made to verify the accuracy of critical data. Some form of internal control (internal check and internal accounting control) over input data should be built into the recording function. Not only should individuals check on one another so that no one person has complete control over one operation (internal check), but also totals accumulated throughout the recording process should be checked against those tabulated previously in order to ensure accuracy (internal accounting control).

In addition to the methods and procedures utilized for verification, duplicating and editing are important parts of the data recording function. **Duplicating** is the process of reproducing the same machine-processable data. This may be necessitated by the internal needs of the system or by the need to use the data for two or more functions simultaneously. On the other hand, **editing** is defined as the process of selecting important data for recording and discarding irrelevant data. It can also mean the checking of data to ensure that input records contain valid data, that is, data which are accurate and authorized. Because data can originate from a variety of sources, editing principally refers to the verification of these input data, the second meaning discussed. Thus, appropriate methods and procedures for verifying recorded input data, an integral part of the design of system controls, are discussed in chapter 14.

Classifying

After having been recorded in a machine processable form, data may need to be classified. Classifying is the process of identifying one or more common characteristics on which to base classification. They might be product line, geographic location, sales territory, division, department, price range, or type of cost. It should be noted that the appropriate classification is usually anticipated, that is, classifications are generally determined before the recording process begins.

An essential part of classifying is coding. As noted in the previous chapter, **coding** is the conversion of data to some symbolic form in order to save space, time, and effort during the recording process. A code may consist of numbers, letters, or an appropriate combination. When coding uses a series of numbers, the data are automatically classified according to some logical sequence. For example, numbers 1 through 12 represent the months of the year and customer numbers are used in place of names. Overall, the coding schemes employed should facilitate classifying procedures.

Manipulating

Even after they have been properly recorded and classified, input data are generally not in a form for final use. There is a need to manipulate the data before they can be summarized, reported, and communicated to the proper individuals. The basic data manipulating functions are **sorting, comparing, analyzing,** and **calculating.**

Sorting procedures Sorting is the process of arranging data into some desired order and is dependent upon a key or field contained in each item. It may be performed manually, mechanically, or electronically with computers. The sorting task is determined by the length of the fields and the number of items to be sorted. Like other data processing procedures, sorting is simplified by expressing certain data in coded form.

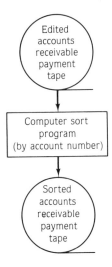

figure 12.2 System flowchart for a computer sort of accounts receivable payment tape.

For the typical batch processing system, it is necessary to sort input data into some meaningful sequence before updating can occur. In figure 12.2, the system flowchart illustrates the computerized processing procedures that take place with a sort program. Specifically, the edited accounts receivable payment tape (transaction data) is sorted by account number, then processed against the master accounts receivable tape file. It should be noted that the systems analyst need only define the accounts receivable file to be sorted and specify the sequence in which the data are to be sorted, since utility sort programs are normally available with the software of most computer systems.

While sorting is usually an integral part of a batch processing system, the same cannot be said for an interactive system. After data have been recorded and classified for entry into the system, the operator can interact on line (using an input/output device) since direct access files, such as magnetic disk and drum, are used. Thus, sorting procedures are minimized or eliminated in an interactive mode.

Updating procedures The degree of **comparing** and **analyzing** is a function of the equipment utilized. The order relationship and relative value of data, respectively, are capable of being determined before making the necessary calculations. **Calculating** is the process of performing arithmetic operations which convert data into a final form for

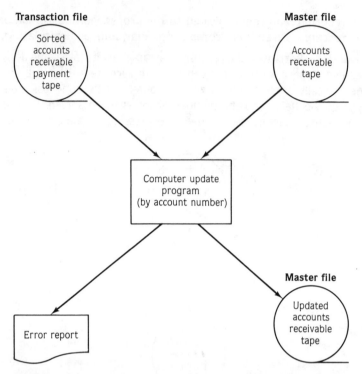

figure 12.3 A system flowchart for updating the accounts receivable master file, which is stored on sequential magnetic tape.

summarizing. Thus, data must generally be manipulated in some meaningful manner before proceeding to the next step in the data processing cycle.

Another way of viewing the foregoing functions of comparing, analyzing, and calculating is to think of updating procedures. In a batch processing mode, after the transaction data have been sorted into the proper sequence, the transaction file is processed against the master file for the purpose of updating the master file.

Referring to the prior example of accounts receivable, the computer update program reads a sorted accounts receivable payment file (a sequential transaction file) and matches it with the records in the accounts receivable file (a sequential master file). Since the two files are arranged in the same sequence by account number, errors can be reported on a printed report, indicating payments for which no master records exist. Also, as a result of the updating process, an updated accounts receivable master file is created, containing master records which were changed and not changed. Records for which a payment was processed without a master record will not be contained on the new magnetic tape file. An overview of this updating process is shown in figure 12.3.

Whereas these processing procedures are based on the sequential file organization method, files can also be organized on either a direct file basis or an indexed sequential file basis (chapter 9). These two types of files can be updated randomly, that is, a transaction record can update any item in a master file regardless of the order of the master record. Although the updating procedures relate to additions, deletions, and adjustments of master records, an entire new master file is not created for each updating run, nor do transaction records have to be sorted prior to the update. Hence, when a random update takes place, the master record to be updated is read into computer storage, updated, and rewritten back onto the master file in the same location.

To illustrate the random batched updating process, consider the preceding accounts receivable example. The transaction payment record and the corresponding master file record (matched by account number) are read into the computer's memory. The master accounts receivable record is then changed and rewritten back onto the master file in the same location. As indicated previously, a new master file is not created, as is necessary for a sequential magnetic tape updating. This accounts receivable updating process is illustrated in figure 12.4.

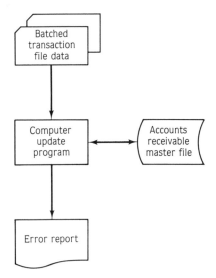

figure 12.4 A system flowchart for updating the accounts receivable master file, which is stored on random magnetic disk (batch basis).

Other updating approaches are necessary for an interactive system. Just as with a batched environment, there are additions, deletions, and adjustments. However, transactions are not batched or sorted. Generally, an update takes place as the transaction is sent from the CRT or typewriter terminal to the computer's memory. Because the update occurs simultaneously, this type of updating is preferable if time is of the essence.

Referring to the preceding accounts receivable example, payments on accounts receivable are entered by the terminal operator as they are received. An operator, for example, could enter a transaction to update the last record in the master accounts receivable file, then a record in the middle, then one in the front, and so forth. Since each payment is entered separately, there is no possibility of batching and sorting the transactions to allow for sequential updating. Hence, most master files found in interactive management

342 Design of System Methods and Procedures

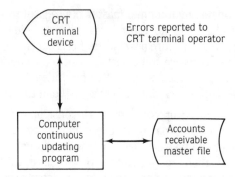

figure 12.5 A system flowchart for updating the accounts receivable master file, which is stored on random magnetic disk (interactive basis).

information systems employ random or direct access files, such as magnetic disk, magnetic drum, or mass storage. The system flowchart of utilizing interactive updating procedures is found in figure 12.5.

Although the foregoing emphasis was on updating one file, two or more files can be updated at a time in either a batch or an interactive mode. In the previous example, it would be possible to update the customer name and address file at the same time as the accounts receivable file.

Summarizing

Summaries provide management with information about the results of business operations at periodic intervals. Fundamentally, summarizing is the process of totaling numerical data. In this respect, totaling is dependent on the sorting process; summarizing goes one step further by providing aggregate values for comparison with other values.

The net result of the basic data processing functions—originating, recording, classifying, manipulating, and summarizing—is processed information, or output. The output medium may be a typewritten or printed form, punched cards, punched paper tape, magnetic tape, or magnetic disk. The most important part of output is accurate and timely reports for an organization's operations. The output is contingent upon the design of the system and the corresponding needs of both management and nonmanagement personnel.

Communicating

Of all the basic DP functions, the communicating function is the most important from a management viewpoint. It is directed toward the assembly and transmission of information throughout an organization so that appropriate action can be taken. Generally, the distribution of information is in the form of reports to the user. However, communicating can mean the transmitting of information in the form of data communications. It is now possible to transmit all pertinent information between a wide variety of input/output devices and the central system by means of telegraph circuits, telephone circuits, or microwave networks. This method allows communication between close and distant points.

An important feature of the communicating function is data storage. Data are generally stored so that they can be easily retrieved. Likewise, a well-designed system stores the information and associates it with previously stored information in order to provide a frame of reference for its next step, making a decision. Results of the decision are automatically transmitted to an output device so that appropriate action can be taken when and where needed.

SUMMARY: Basic DP Functions that Comprise System Methods and Procedures

Originating Refers to the capture of data on a source document at the time a transaction occurs. This is the initial phase of **input**.

Recording Relates to the methods and procedures of converting data from the source document into a form acceptable to DP equipment in the next sequence of operations. This is an integral part of **input**.

Classifying Refers to the process of identifying one or more common characteristics usable for grouping data. This is the last phase of **input**.

Manipulating Emphasizes the methods and procedures for sorting, comparing, analyzing, and calculating data in some meaningful manner before proceeding to the next phase. This is the essential part of **methods and procedures** in a typical DP system.

Summarizing Focuses on report preparation by highlighting important information. This is an important part of **output**.

Communicating Relates to the dissemination of information produced by the system to the respective user for taking appropriate action. This is the most important aspect of **output**. A feature of the communicating function is storage of data upon completion of the foregoing basic DP functions. This is an essential part of **data files**.

INCORPORATE SYSTEM DESIGN PRINCIPLES— METHODS AND PROCEDURES

When determining the specific methods and procedures for a new system, several important design principles must be considered. Although it may not be possible to include every one, it is highly recommended that as many as possible be included in the design of system methods and procedures.

Foremost among important system design principles is the **Principle of Modularity of Methods and Procedures.** Modularity of system design refers to a way of breaking down a function into its component parts; conversely, it is also a way of grouping methods and procedures that logically fit together. The capability of bringing together different procedural modules that are capable of standing alone greatly reduces the complexity of systems design. It also facilitates any modification of a system that may be necessary in the future.

A second important design principle is the **Principle of Automated Methods and Procedures.** Systems design may require that methods and procedures of a certain function be broken apart, then recombined for more effective automation of the function. From this viewpoint, methods and procedures are subservient to the function being automated. It may well be that the recombined methods and procedures cut across established organizational jurisdictions; if so, there may be need to change the organizational structure for automating the function for optimum efficiency and economy of operation.

Closely related to the preceding principle is the **Principle of Economy of Methods and Procedures.** Good system design dictates that methods and procedures be economical; that is, the cost of accomplishing the desired tasks is low, relative to the importance of that task. To achieve economy, data should be captured as near to the source as possible and then flow through the system automatically, without human intervention. Hence, there is minimal need for methods and procedures to accomplish desired processing tasks.

A logical successor to the previous principle is the fourth one, the **Principle of Simplicity of Methods and Procedures.** Fundamentally, simplicity can be implemented by providing a straight-line flow from one procedure to another, thereby avoiding needless backtracking. Similarly, it can be effected by reducing the need for coordination and communication. Simplified methods and procedures can be better understood and more easily followed than complex ones; to state it another way, methods and procedures should be convenient and easy to use.

A fifth principle for designing an effective system is the **Principle of Flexibility of Methods and Procedures.** Well-designed methods and procedures are capable of adapting to changing conditions inside and outside the organization. They provide for expanding or contracting business operations when and where necessary. Without the ability to absorb, the overall system, and perhaps the organization, may fail to meet its assigned goals.

The need for timely information to manage current operations and to react to changing conditions is facilitated by the **Principle of Speed of Methods and Procedures.** Good system design dictates that methods and procedures respond quickly to the user's needs. In this manner, specific time requirements can be met at the various stages in the DP flow, particularly at the output stage in the form of timely information.

In addition to the preceding principles that focus on the essential characteristics of methods and procedures, there is a seventh principle—the **Principle of Reliable Methods and Procedures.** Reliability refers to effective internal control procedures, that is, establishing numerous control points for checking on the accuracy of data. An integral part of this reliability process is the **Principle of Provision for Corrections.** As incorrect data are detected, correcting methods and procedures are initiated before processing continues. In this manner, the reliability of system methods and procedures is maintained, and variances from established norms and standards are detected and corrected before processing continues.

SUMMARY: Incorporate System Design Principles—Methods and Procedures

1. **Principle of Modularity of Methods and Procedures** Effective system design dictates that methods and procedures be modular. The ability to bring together different modules that are capable of standing alone reduces overall system complexity and facilitates any procedural modification or system updating that may be necessary in the future.
2. **Principle of Automated Methods and Procedures** Good system design may disclose that certain methods and procedures be separated, then recombined for automation of a certain function. From this view, methods and procedures are subservient to the larger task that is to be automated.
3. **Principle of Economy of Methods and Procedures** Essential parts of an effective system are methods and procedures that are economical to use; that is, their costs are relatively low in accomplishing the desired processing tasks.
4. **Principle of Simplicity of Methods and Procedures** Efficient methods and procedures should be relatively simple and convenient to use. They should relieve the user of unnecessary work and backtracking.
5. **Principle of Flexibility of Methods and Procedures** Well-designed methods and procedures should be flexible to meet changing conditions experienced by the organization.
6. **Principle of Speed of Methods and Procedures** Good system design requires that methods and procedures respond quickly to the user's needs so that specific time requirements can be met at various points in the data processing flow.
7. **Principle of Reliable Methods and Procedures** An integral part of good system design is the need for methods and procedures that meet certain established reliability requirements—in particular, those for system control points, established to determine the accuracy of data flow.
8. **Principle of Provision for Corrections** Effective system methods and procedures should provide for making corrections when there are deviations from established norms.

DESIGN OF SYSTEM METHODS AND PROCEDURES

The preceding material on the basic DP functions that comprise methods and procedures serves as a basis for designing effective system methods and procedures. The design process for relating input to output and data files is as follows:

Consider the human factor in designing methods and procedures.

Review methods and procedures of the present system and data contained in the exploratory survey report to top management.

Specify methods and procedures requirements as they relate from input through data files (if necessary) to output.

Determine the steps involved in system methods and procedures.

Specify those methods and procedures needed as backup in case of system failure.

Each of the foregoing areas are discussed below. This detailed investigation of designing system methods and procedures will be helpful in the next chapter where order-entry methods and procedures are designed for the ABC Company.

Consider the Human Factor in Designing Methods and Procedures

In the past, there was little regard for the human factor in the design of system methods and procedures, and more for the technical aspects of data processing, in particular, the equipment itself. However, there has been a change for the better whereby the emphasis is placed first on the human factor, and later on equipment considerations.

In the early stage of a system project, a company goes over a checklist with an equipment manufacturer; the checklist covers such items as product environment, operator interaction, frequently accessed controls, routinely serviced parts, maintenance philosophy, and operator convenience features. This checklist leads to a set of design criteria against which solutions can be measured. Some specific considerations for DP equipment might include:

- relative placement of keyboard, controls, and display, including the angle of the keyboard with respect to the operator.
- comfortable seating; ample knee and leg room.
- size, color, and typography of graphics for maximum legibility.
- shielding of display to minimize reflections.
- movable keyboard, storage of personal belongings, and modesty panels.

The size and shape of DP equipment are not dictated by the components within it, as it used to be, but rather, by the fingers which must operate the keyboards and the eyes which must read the displays. Thus, equipment considerations must take a back seat to the human ones.

Going one step further, the equipment itself must be designed such that methods and procedures assist the operator in doing an efficient job. As an example, human interface with key-to-disk systems is maximized by a minicomputer set up to "talk" with operators. In this case, the operator can respond to the system rather than initiate actions. Software guides the operator through different modes, keeps the person informed of station status, and tells when an error has been made, its type, and how to correct it. Data can be entered in a logical sequence and the minicomputer can reformat these data for the central computer.

To illustrate the kind of interaction possible, a key-to-disk system may have a "HELP" key which, when depressed, lists alternative actions on a CRT screen. Subsequent dialogue might go something like the following:

Operator: (Selects) START BATCH
System responds: Batch name?
Operator: (Types) PAYROLL
System responds: Must this batch be verified before output to tape?
Operator: (Types) YES
System responds: What are the names of record formats?
Operator: (Types) WEEKLY, HOURLY

In essence, the system has programmed procedures to assist the operator; it acts like a personal instructor when help is needed. Because of the ease of this mode of instruction, the operator's confidence builds, and he or she becomes productive in a short time.

The foregoing has focused on methods and procedures for assisting equipment operators, but there is need to look beyond the user–machine interface. Specifically, new system methods and procedures should also be concerned with interfacing with organizational personnel where no or very little equipment is involved. In such cases, effective system design requires that operational procedures and requirements do not exceed human capabilities; otherwise, workers will experience fatigue, which results in lower productivity. Additionally, effective methods and procedures should recognize the human need for deriving satisfaction from one's work.

In reference to the last point, the systems analyst should consider the following points when designing appropriate methods and procedures:

> The individual will have a meaningful series of tasks or activities that will significantly utilize his or her capabilities. Hence, there is a "complete job" for the individual to perform.
>
> The individual will have as much decision-making responsibility as possible in executing the assigned tasks and activities.
>
> The individual will be provided with direct performance feedback instead of channeling information through his or her superior.
>
> The individual will be assured of pride from his or her tasks and activities.

Based upon these considerations, the individual will be able to satisfy his or her needs on the job. The human need for a satisfying work experience will be provided within the framework of efficient and economical methods and procedures.

Overall, the design of system methods and procedures is related to the division of tasks between the individual and the DP equipment. Currently, the problem is resolved by allocating to the equipment all jobs it can seemingly do better than the individual. These include repetitive procedures, operations that require speed of computations, tasks involving a well-defined data search, and jobs that call for a precise adherence to instructions. As DP equipment takes over more information processing tasks, the organization personnel will need to be responsible for the remainder. Although both human and machine areas of specialization/participation can be expected to emerge and continually change, there will always be a need for system methods and procedures that are **simple** and **economical** for the individual to use, that are **fast** and **reliable**, that are **flexible** and **versatile**, and that can be **understood** and **controlled**.

Underlying the foregoing human factors is the **Principle of Humanizing System Design** which states that the design of a system should be consistent with applicable human factors since people are responsible for the effectiveness of a system. The term **human factors** includes all those personality traits that consciously or unconsciously shape the actions and reactions of the people who must interact with the system. Also, it includes the same traits in the systems analyst as they may affect his or her ability to design an efficient and successful system.

Review Information on System Methods and Procedures

As noted in earlier chapters, a logical starting point for designing a new system is a review of present methods and procedures, as well as of material contained in the exploratory survey report to top management about the recommended system. (An understanding of the material on system input, data files, and output generally includes a review of methods and procedures.) The rationale is that system methods and procedures are the means of relating input through data files to output. By this time, the systems analyst generally has a good understanding of what methods and procedures are necessary for an efficiently designed system.

Because of the importance of the human factor in designing methods and procedures, it is highly recommended that organizational personnel participate with system personnel in this design phase.

Is the procedure really essential to the organization's operations?
What would happen if one or more steps in the procedure were eliminated?
Can the procedure be improved to realize more fully the organization's objectives?
Is it possible to simplify the procedure through modification of existing organizational policies, departmental structures, practices of other departments, or similar considerations?
Is there too much handling of the document in the procedure?
Does the procedure route the document through too many of the organization's personnel and departments?
Can the procedure be performed in a faster and a more economical manner?
Does the procedure make a contribution to the quality or flow of the work?
Is the cost of the procedure greater than its value to the organization?
Are all of the forms used in the procedure necessary? Can the forms be combined?
Does duplication of work exist in the procedure?
Are the steps in a logical sequence for the greatest efficiency in the procedure?
Are there parts of the procedure that functionally belong to another activity?

---

figure 12.6 Systems design—questions to test the validity of methods and procedures.

Specify Methods and Procedures Requirements

The specification of methods and procedures requirements relates the entire data processing flow from input to output by specifying what data processing methods and procedures (input) link the originating, recording, and classifying functions with the manipulating function (processing) and, in turn, with the summarizing and communicating functions (output). In some cases, the methods and procedures will be manual; in others, they will be fully automated; in yet other cases, they will be somewhere in the middle.

Design of System Methods and Procedures **349**

In order for the systems analyst to validate the newly designed methods and procedures, certain questions must be asked. These are delineated in figure 12.6 and are comparable to the ones developed for testing the validity of a report.

Probably the best method to measure the efficiency of the new methods and procedures is to create an information flow from the input point through data files to the final output stage. Any one of the documentation techniques presented in chapter 2 can be utilized; which one(s) is best depends on the area under investigation. The systems analyst should use whichever flowcharting techniques work best.

To assist the analyst in creating an information flow, it is helpful to consult the System Output Analysis, System Data Files Analysis, and System Input Analysis forms (chapters 8, 9, and 11, respectively). Details on these forms enable the systems analyst to specify detailed methods and procedures that relate input elements to their final output. This phase of systems design brings together the essential elements of the system such that output from one stage in the data processing flow becomes the input for the next. This continuous input/output DP flow ultimately results in the desired output.

Determine the Steps in Methods and Procedures

Now that methods and procedures have been specified, the next step is to determine the detailed steps. A system procedures analysis form can assist the systems analyst (table 12.1, p. 350).

In the example, sales order processing procedures in an interactive mode are listed for the company's salespeople, including sales order operators. These operating procedures can be supplemented by additional detail. Specifically, they can include operating procedures relating to the following items:
   customer order number (from the customer's purchase order)
   ship-to number
   date of order
   mail, telephone, or salesperson
   tax code (the customer's tax type)
   salesperson number
   number of items ordered of each product and catalog number
   job control number

Hence, typical system procedures for day-to-day operations are very detailed in order to cover all operating contingencies. Failure to be thorough may cause confusion at the implementation phase as well as during daily operations.

To assist the systems analyst in specifying exactly what system methods and procedures should be undertaken by organization personnel, reference should be made to system input analysis, data file analysis, and output analysis forms. Knowledge of what data enter the system, how data are stored, and what information will be produced as output are helpful in specifying exactly what methods and procedures are required to convert input data to useful information. In the example, background knowledge about who initiates the order, what it contains, what data files are used, and who receives the processed sales order not only simplifies the design process of appropriate methods and

table 12.1 Typical system procedures analysis form.

SYSTEM PROCEDURES

Page 1 of 1
Date: July 25, 198_
Preparer: G. W. Reynolds

Type of procedure:
INTERACTIVE SALES ORDER PROCESSING

Location affected: All sales offices

SPECIFIC INSTRUCTIONS:

Salespeople Prepare customer sales orders and forward them to closest sales office for processing by sales order operators.

Sales order operators Use CRT terminals for interactive order processing. Initially, the CRT terminal displays a request for customer identification. Enter customer number unless the transaction is for a new customer, in which case enter customer name, address, and other information. If the number is unknown for an established customer, enter the name and the system will display a list of "sound alike" customers; select one. The interactive system then displays successive requests for the entry of data pertaining to the order. A list of questions is projected on the operator's screen. New questions are flashed on the screen in a page-like sequence, and can cover up to 30 items on a single order. Sets of questions continue until the operator enters an **L** for "last." Each keyed-in answer appears on the screen opposite the question. Then, the computer summarizes each series of answers and "plays back" the information for visual review. There is only one edit check for each item on order, and the operator verifies that an entry is correct by keying **OK** (okay) or **NG** (no good); after an **NG**, the operator redisplays the particular item in question and enters changes. An editing program stored in the computer checks the answers against pre-established facts and flashes question marks on the screen for answers unacceptable to the program. Examples of unacceptable answers: if the operator fails to include the number of items ordered, the program will flash a question mark; if the number of items ordered is present, but the catalog number is not, the program will flash a question mark. When all the questions on an order have been answered, the computer checks the customer's credit and either approves the order for filling or, if there is a credit problem, stops order processing. A message is automatically relayed to the credit section of the accounting department on an I/O terminal unit for appropriate action.

---

procedures, but also increases the likelihood that all pertinent processing steps have been included in the design process.

As indicated in an earlier part of the chapter, system flowcharts and program flowcharts should be prepared as necessary during this stage. If all pertinent inputs, data files, and outputs can be tied together with a series of processing steps (which are system methods and procedures), the systems analyst can feel confident that the design process is complete. Likewise, since organization personnel from the functional areas under study are involved, flowcharts assist in illustrating what methods and procedures are required. They also serve to clarify processing steps that may be new and different from those of the present system. Overall, the use of flowcharts is a convenient way of illustrating the various processing steps that link input to output.

One final word of caution is necessary when determining specific methods and procedures. There is need to integrate newly designed processing procedures with other, related procedures; that is, output resulting from a group of processing procedures may serve as input for another set of procedures. From this view, output and input processing procedures for related functional areas should be integrated so that the final system design is complete and efficient to operate. Lack of concern for integration of inputs and outputs may result in higher processing costs than those initially estimated.

Devise Back-Up Procedures

Although the foregoing approach will result in devising effective methods and procedures, there is still need to consider backup procedures in case of system failure. Recovery procedures should be established in case of accidental loss or destruction of transaction or master data. In most batch systems, because transaction data are batched and processed in stages, data can generally be reconstructed by backtracking to the prior stage and reprocessing from that point. However, in interactive systems, recreating transaction data may present difficulties unless separate files of transaction data are maintained.

As illustrated in figure 12.7, one method to recreate current transaction data is to create a transaction file at the same time the master file is updated. If, at a later date, the master file is found to be unusable, the transactions which previously took place can be re-recorded on the master file. Failure to create a current transaction file makes reprocessing a formidable task in an interactive mode. In many cases, the job would not be feasible due to time constraints.

Going beyond the problem of working with **transaction data,** there are also problems associated with maintaining **master files.** Generally, if a master file has been updated many times before becoming unusable, it may be extremely difficult to restore it to its current status. A back-up file can ensure the security and integrity of a master file.

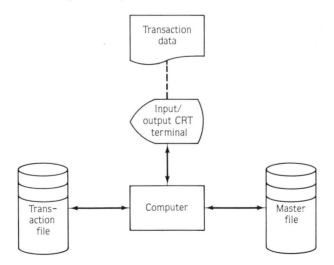

figure 12.7 A typical example of a back-up file, specifically, a transaction file, in an interactive processing mode.

352 Design of System Methods and Procedures

Fundamentally, a back-up file is either an old master file, which can be updated to the current level, or a copy of the current master file.

The updating of sequential files presents no problems to the systems analyst. Because a new master file is created when the updating takes place, the old master file—plus transaction data which update the new master file—are retained, thereby serving as effective back-up. However, the updating process is not that straightforward for direct or indexed files, because the updated master records are written back on the magnetic disk, drum, or other medium in the same location as the previous record. In view of this problem, a master back-up file is created periodically (once a day or once a week) from the information in the updated master file, as shown in figure 12.8. Most master back-up files are kept on magnetic tape because it is less expensive than other recording media.

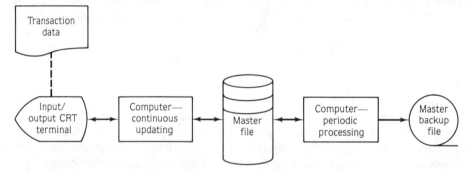

figure 12.8 A typical example of the process of creating a back-up master file in an interactive processing mode.

SUMMARY: Design of System Methods and Procedures

Consider the Human Factor in Designing Methods and Procedures Relates to assisting organization personnel in the performance of their jobs so that their work experience is satisfying.

Review Information on System Methods and Procedures Refers to reviewing methods and procedures currently employed and those contained in the exploratory survey report to top management on the recommended system alternative.

Specify Methods and Procedures Requirements Relates to specifying methods and procedures that link system inputs with outputs and data files (where necessary) such that the entire data processing flow is considered.

Determine the Steps in Methods and Procedures Results in the determination of the specific contents of methods and procedures. Generally, a system procedures analysis form is prepared for each major processing function. Additionally, flowcharts are used to illustrate the various processing steps that link input data to desired output.

Devise Back-Up Procedures Refers to devising back-up procedures to restore transaction data and master files to normal operation in case of system failure.

CHAPTER SUMMARY:

The main thrust of this chapter is the design of system methods and procedures that relate input data to the desired output. In this DP flow, data files may be used as a means of capturing data for future output or may, themselves, be a form of output. Regardless of the links of inputs, data files, and outputs with methods and procedures, a useful way of depicting these interrelationships is by flowcharts. In this manner, system and organization personnel can review the many processing steps and comment on their effectiveness. In essence, the validity of system methods and procedures can be ascertained by a comprehensive review of what they propose to accomplish. To be effective, system methods and procedures must be simple and economical to use, fast and reliable, flexible and versatile, and understood and controllable.

Questions

1. a. Distinguish between system methods and procedures.
   b. Why is the accent of systems design on procedures?
2. What is the relationship of system methods and procedures to the basic DP functions? Explain.
3. What are the essential elements of system methods and procedures?
4. What is meant by updating procedures in the data processing flow?
5. a. Do data processing procedures, before entering a computer system, differ much between a batch processing system and an interactive processing system? Why or why not?
   b. Give two examples to prove your point.
6. What is the relationship of data files to the basic DP functions that comprise system methods and procedures? Explain.
7. How much importance should a systems analyst give to incorporating system design principles as presented in the chapter in the final system design? Explain thoroughly.
8. What is the relationship of organizational personnel needs to newly designed methods and procedures? Explain thoroughly.
9. Why is it necessary to relate methods and procedures to system input, data files, and output when designing a new system?
10. How important are back-up procedures in the design of a batch processing system versus an interactive processing system?

Self-Study Exercise

True-False:

1. _____ A procedure is smaller in scope than a method.
2. _____ The first basic DP function that comprises methods and procedures is recording.
3. _____ In one sense, editing can be defined as the process of selecting important data and discarding irrelevant data.

4. _____ The manipulating function centers on updating procedures.
5. _____ Comparing is the process of arranging data into some desired order.
6. _____ In a batch processing system, there is a great emphasis on updating transaction files.
7. _____ The Principle of Economy of Methods and Procedures refers to devising system methods and procedures that are relatively low in cost for accomplishing the desired tasks.
8. _____ The human need for a satisfying work experience should not be considered when designing efficient and economical methods and procedures.
9. _____ The major focus of system methods and procedures is to relate input data to final output that is meaningful to the user.
10. _____ Program flowcharts are useful in creating an information flow from input data to output reports.

Fill-In:

1. _____ is the process of identifying one or more common characteristics that are usable for grouping data.
2. Another way of viewing comparing, analyzing, and calculating is to think in terms of _____ procedures.
3. _____ data processing procedures are undertaken to update master files and produce desired outputs.
4. From a management viewpoint, the _____ function is the most important.
5. Good system design dictates that certain functions be separated, then recombined in a different manner for more effective _____ methods and procedures.
6. Good system design dictates that methods and procedures be _____; that is, their costs are low in accomplishing desired tasks.
7. Good system design requires that methods and procedures be _____ in responding to user's needs.
8. Effective system methods and procedures should include provision for making _____ when there are deviations from established norms.
9. An essential part of specifying methods and procedures requirements centers on the entire _____ _____ flow.
10. _____ _____ are an effective means for relating input data to files and output reports.

Problems:

1. The American Company, which is operating an integrated information system (batch processing), is currently reviewing its customer billing system. The purpose of the review is to determine whether or not computer equipment is being utilized to its fullest. The computer equipment configuration includes a central processing unit and attached console typewriter, a card reader, a card punch, a printer, and five magnetic tape units.

   The system flowchart for the customer billing program is given below. Based upon this flowchart, develop a new system flowchart that will maximize the utilization of the computer's equipment. Assume that shipped finished goods inventory items can be stored in memory (as temporary storage) for reducing the master finished goods inventory file.

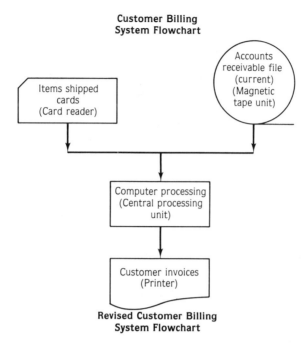

Customer Billing System Flowchart

Revised Customer Billing System Flowchart

2. The American Company, which is operating an integrated information system (batch processing), is currently reviewing its finished goods inventory system. Like the previous problem, the purpose of the review is to determine whether or not computer equipment is being employed to its fullest. The computer equipment configuration includes a central processing unit and attached console typewriter, a card reader, a card punch, a printer, and five magnetic tape units.

The system flowchart for the finished goods inventory system is on page 356. Based upon it, develop a new system flowchart that will maximize the utilization of the computer's equipment.

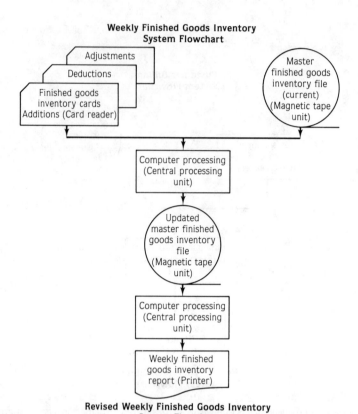

Revised Weekly Finished Goods Inventory System Flowchart

chapter

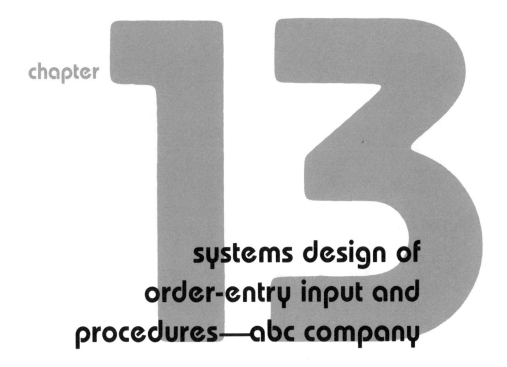

systems design of order-entry input and procedures—abc company

OBJECTIVES:

- To build upon the outputs and the data base set forth in Chapter 10 for the new order-entry system of the ABC Company.
- To consider the human factor and apply basic design principles for specifying inputs and procedures of the new order-entry system.
- To design the content and format of the new order-entry system inputs.
- To determine the required procedures of the new order-entry system to produce the desired output.

IN THIS CHAPTER:

DESIGN OF ORDER-ENTRY INPUT—ABC COMPANY
    Consider the Human Factor in Designing Order-Entry Input
    Incorporate Underlying Principles of Order-Entry Input
    Specify Input Requirements—Relate Output through Files and Procedures to Input
    Determine Specific System Input
    Determine the Method of Order Entry
    Design of Order-Entry Input
    Overview—Design of Order-Entry Input —ABC Company
DESIGN OF ORDER-ENTRY PROCEDURES—ABC COMPANY
    Consider the Human Factor in Designing Order-Entry Procedures
    Incorporate Underlying Principles of Order-Entry Procedures
    Specify Order-Entry Methods and Procedures
    Determine the Steps in Order-Entry Procedures
    Devise Order-Entry Back-Up Procedures
    Overview—Design of Order-Entry Procedures—ABC Company
CHAPTER SUMMARY
QUESTIONS
SELF-STUDY EXERCISE
CASE STUDY 1—FINISHED PRODUCT INVENTORY
CASE STUDY 2—ACCOUNTS RECEIVABLE

In the previous two chapters, the basic concepts for the design of inputs, methods, and procedures were presented. This chapter will develop further the design of the ABC Company's order-entry system by focusing on these concepts. Design principles presented previously will again be followed, illustrated by practical applications.

The chapter will present additional information related to the text's case studies—the finished product inventory and accounts receivable systems of the ABC Company. The new order-entry system will provide a model for designing the inputs, methods, and procedures of these two systems.

DESIGN OF ORDER-ENTRY INPUT—ABC COMPANY

The approach outlined below for designing the order-entry input of the ABC Company is comparable to that set forth in chapter 11, except that the section on reviewing information of the present system has been dropped. (Because information that is pertinent to the present system is brought in where necessary, there is little need for this section.) Also, a beginning section has been added that highlights the important principles that underlie the new system. The design sections, then, to be covered in sufficient depth below, are:
1. Consider the human factor in designing order-entry input.
2. Incorporate underlying principles of order-entry input.
3. Specify input requirements—relate output through files and procedures to input.
4. Determine specific system input.
5. Determine the method of data entry.
6. Design of order-entry input.

Consider the Human Factor in Designing Order-Entry Input

The primary design principle that the MIS task force agreed to follow in the input design process is the **Principle of the Human Aspect of Data Input.** The human factor is to be considered over technical aspects associated with efficiency of processing and economy of design. The MIS task force is extremely sensitive to the fact that order processing personnel at the field warehouses are the key to providing accurate and complete information and, thereby, the key to a successful order-entry system.

The new system essentially eliminates the tedious and error-prone process of transcribing order information from a sales order form and replaces this process with direct telephone communication between the salesperson and a trained order-entry clerk. Furthermore, the order-entry clerk will interact with an intelligent CRT terminal to enter accurate and complete order information directly from the salesperson onto a computer-accessible file. The variability of speaking directly with different salespeople and interacting with an intelligent CRT terminal will greatly enhance the job content and quality of performance for order-entry personnel.

Incorporate Underlying Principles of Order-Entry Input

To ensure an expertly designed system, the MIS task force has decided to reference the design input principles set forth in chapter 11. Such an approach relates theory to practice and provides a means of simplifying input for those who will be interacting with the system. In the discussion below, these input principles are related to the system being designed.

Principle of acceptability of inputs   To ensure the acceptability of the system inputs, the MIS task force will work closely with the order-entry clerks. They will, to the greatest extent possible, design the system inputs so that order-entry clerks will be very comfortable using the new system. When designing the new inputs, they will try to make the inputs similar to the content and format of the old system, with changes only where necessary or strongly desirable; in this way, the amount of time required for the retraining of the clerks will be reduced, and more importantly, the clerks will more readily accept the new system.

In all discussions with the order-entry clerks, the positive aspects of the system, as it relates to these company personnel, will be stressed. The primary ones are: they will have more interesting jobs; they will do a better job—input data will be more accurate and complete; they will process orders more quickly; and, they will experience fewer errors.

Principle of integration of inputs and outputs   The outputs from the new order-entry system will be designed to provide input for the new finished product inventory and accounts receivable systems. Information captured by the order-entry system about orders and shipments is essential to the successful operation of the finished product inventory and accounts receivable systems.* The key to insuring that all three systems contain accurate and complete information is to (1) edit all inputs carefully, (2) provide for interactive correction of errors so that order-entry clerks can correct inaccurate or incomplete data while on the telephone with the salesperson, and (3) pass accurate and complete order and shipment data through the order-entry system directly to the finished product inventory and accounts receivable systems.

---

*Figure 10.1 depicts the information flow among these three systems.

Principle of common data representation A common data representation will be established to minimize subsequent manual tasks. Magnetic tape and magnetic disk will be used as the primary means of data storage and retrieval, thereby requiring little intervention by organization personnel.

Principle of eliminating duplicate input The new system will be designed such that key inputs about orders and shipments need be captured only once. Information like order number, order date, customer account number, products ordered, quantity ordered, and the like will be input at the time of order placement. This information about the order need not be entered again either for further processing in the order-entry system or the finished product inventory system. Similarly, at the time of shipment processing, information like shipment number, shipment date, carrier name, products shipped, and quantity shipped will be input. This information about the shipment need not be entered again for further processing in the order-entry, finished product inventory, or accounts receivable system.

Principle of simplicity of input All system input will be designed with clarity and simplicity foremost in mind. This will make the input more acceptable to the user, reduce the time needed to train order-entry clerks, and reduce input errors. Simplicity of input will be coupled with simplicity of output so that both are clear, easy to understand, and more usable.

Principle of garbage-in—garbage-out Every precaution will be taken to insure that all system input is accurate and complete. Further design effort will be spent to insure that the data are processed accurately. The systems design will be geared to result in a system that produces accurate and complete output.

Specify Input Requirements— Relate Output through Files and Procedures to Input

Now that the basic principles to be followed have been specified, the order-entry system input design work can begin. This step specifies the input requirements of the new system to produce the desired output by employing specified procedures and various data files.

The MIS task force had previously defined the key data elements to be stored on four data files—open order, shipment, customer name and address, and product code—depicted in tables 10.7 through 10.10. Comparable information on system outputs (tables 10.1 through 10.5) was also discussed, in particular, the multipart order and shipment processing form, the new customer invoice, the new monthly customer service report, the daily shipments by carrier report, the open order and shipment data for the finished product inventory system, and the shipment data for the accounts receivable system. In figure 13.1, this information is summarized.

Each file in figure 13.1 has been identified with a **number;** each output file and report is identified by a **Roman numeral;** and each input is identified by a **letter.** The letter or number in a particular row and column identifies that source of that data element. For

DATA ELEMENTS	No. of bytes	DATA FILES: Open order (1)	DATA FILES: Shipment (2)	DATA FILES: Customer name & address (3)	DATA FILES: Product code (4)	OUTPUTS: Order & shipment processing form (I)	OUTPUTS: Customer invoice (II)	OUTPUTS: Customer service report (III)	OUTPUTS: Shipments by carrier (IV)	INPUTS: Order-entry input (A)	INPUTS: Shipment processing input (B)
Order number	6	(A)	(B)				(2)	(2)	(2)	✓	✓
Order date	6	(A)	(1)				(2)	(2)		✓	✓
Invoice number	7		(B)				(2)				
Invoice date	6		(B)				(2)				
Field warehouse code	4	(A)	(B)					(2)	(2)	✓	✓
Customer account number	7	(A)	(B)	✓			(3)			✓	
Customer name	24			✓			(3)	(3)			
Customer address—'Ship to'	47			✓			(3)	(3)			
Customer address—'Bill to'	47			✓			(3)				
Shipment number	6	(A)	(B)				(2)	(2)	(2)		✓
Shipment date	6	(A)	(B)				(2)	(2)	(2)		✓
Desired shipment date	6	(A)	(1)							✓	
Shipping terms indicator	1		(B)						(2)		✓
Carrier name	24		(B)						(2)		✓
Trailer number	15		(B)						(2)		✓
Products ordered (up to 10)		(A)	(1)				(2)	(2)		✓	
Product code	4	(A)	(1)		✓		(2)	(2)		✓	
Quantity	5	(A)	(1)				(2)	(2)		✓	
Unit price	3				✓		(2)				
Hazardous material indicator	1	(4)	(4)		✓		(2)				
Extended unit cost	6						C				
Quantity Code	1				✓		(4)				
Description	25				✓		(4)				
Products shipped (up to 10)			(B)				(2)	(2)			✓
Product code	4		(B)		✓		(2)	(2)			✓
Quantity	5		(B)				(2)	(2)			✓
Unit price	3		✓		✓		(2)				
Hazardous material indicator	1		(4)		✓		(2)				
Extended unit cost	6						C				
Quantity Code	1				✓		(4)				
Description	25				✓		(4)				
Customer service exception code	1		(B)								✓
Total invoice amount	7						CALC				
Invoice gross amount	7						CALC				
Invoice net amount	7						CALC				
Total shipment weight	7						CALC		(2)		

figure 13.1 The input–file–output analysis worksheet for the new order-entry system of the ABC Company.

example, an order number is present on the open order file **(1)** and the input source is the order-entry input **(A)**. An order number is also present on the shipment file **(2)** and the input source here is the shipment processing input **(B)**. An order number appears on the customer invoice **(II)**, the customer service report **(III)**, and the shipments by carrier report **(IV)**. The order number in all these reports is obtained from the shipment file **(2)**. However, many of the elements of the shipment file are obtained from the open order file. This is to be expected since the open order record is first created. Then at shipment processing time, additional data are added to create a shipment record.

In the customer name and address file **(3)** and product code file **(4)** columns, check marks represent the presence of a particular data element. The source of the data element is not specified here. However, it is easy to picture a set of input for adding information to these two files. Under inputs, the check marks indicate that this data element is captured and entered into the system from this input source. Some of the data elements are neither input nor come from another file, but rather are calculated. These data elements are identified by **CALC**.

In addition to relating files to outputs, it is also required at this time to relate files and outputs to inputs. Hence, compatible inputs for the new order-entry system consist of **order-entry input (A)** and **shipment processing input (B)**. As shown in figure 13.1, the check marks in these two columns indicate the presence of a particular data element on these two input sources.

Determine Specific System Input

Having specified the input requirements, the next task of the MIS group is to determine in detail the content of the order-entry system input. Based upon the information contained in figure 13.1, appropriate system input analysis forms can be prepared. These are shown in tables 13.1 and 13.2 for order-entry input and shipment processing input, respectively. As noted earlier in the chapter, the order-entry input is received via telephone from the salesperson for entry by the order-entry operator. On the other hand, the shipment processing input is initiated by the shipping clerk at the field warehouse after goods have been shipped to the customer. Both of these areas will be explained in more detail later in the chapter.

Determine the Method of Order Entry

As originally noted in chapter 6, a very convincing argument has been presented for the utilization of distributed processing input to the new order-entry system. The MIS task force has carefully reviewed the basic conclusions that led them to this decision. They are:
1. Built-in delays associated with the current centralized processing of all order-entry input make it impossible to reduce the average order cycle time or to improve the accuracy of the input. Distributed processing of input could overcome these limitations.
2. Use of remote intelligent CRT terminals is by far the most inexpensive distributed processing alternative for the ABC Company.

table 13.1 Order-entry system input analysis form for the ABC Company.

	SYSTEM INPUT		
Page 1 of 1 Date: April 2, 198_ Preparer: G. W. Reynolds			Type of input: ORDER ENTRY

Informational elements	Type of characters	Number of characters (bytes)	Source
Order number	Numeric	6	Assigned by order-entry clerk
Order date	Numeric	6	Today's date
Field warehouse code	Alphabetic	4	Assigned by order-entry clerk
Customer account number	Numeric	7	Customer name and address file
Desired shipment date	Numeric	6	Customer order
Products ordered (up to 10)			
Product code	Numeric	40	Product code file
Quantity	Numeric	50	Customer order

Comments:
Order-entry input is received via an established WATS telephone network between the salesperson and the order-entry clerk. The total number of characters (bytes) is 119 plus 3 characters (bytes) for a record identification code totaling 122 characters (bytes).

table 13.2 Shipment processing system input analysis form for the ABC Company.

	SYSTEM INPUT			
Page 1 of 1 Date: April 3, 198_ Preparer: G. W. Reynolds			Type of input: SHIPMENT PROCESSING	

Informational elements	Type of characters	Number of characters (bytes)	Source	Comments
Order number	Numeric	6	See below	
Order date	Numeric	6	See below	
Field warehouse code	Alphabetic	4	See below	
Shipment number	Numeric	6	See below	
Shipment date	Numeric	6	See below	
Shipping terms indicator	Alphabetic	1	See below	
Carrier name	Alphabetic	24	See below	
Trailer number	Numeric	15	See below	
Products shipped (up to 10)				
Product code	Numeric	40	See below	
Quantity	Numeric	50	See below	
Customer service exception code	Alphabetic	1	See below	

Comments:
The source of this information is the yellow copy of the multipart order and shipment form at the field warehouse. The total number of characters (bytes) is 159 plus 3 characters (bytes) for a record identification code totalling 162 characters (bytes).

3. Remote intelligent CRT terminals provide the capability to continue order-entry processing even if the central computer in St. Louis is down.
4. The information processing requirements of the field warehouses are insufficient to warrant the installation of computer processors and data storage capability at each field warehouse.

After further review and analysis of these basic conclusions with other members of the ABC Company, the MIS task force held discussions with several intelligent terminal vendors to confirm their basic findings and to obtain detailed information on costs, quality, and service. The best overall vendor was selected, and a two-week evaluation test was mutually agreed upon. Specific equipment criteria are normally used to select the equipment manufacturer. (These factors are covered in some depth in chapter 16.)

For two weeks, the salespeople in the Los Angeles area used a prototype system for order entry and shipment processing that is very similar to the system the MIS task force is prepared to recommend. They used a WATS line to call in all orders to a branch office of the vendor. An ABC Company order-entry clerk was temporarily stationed there and an order-entry clerk was trained for entering the information via the intelligent CRT terminal using programs and procedures provided by the equipment vendor. At the end of each day, a program was used to extract the order information and create a summary report for each salesperson. Each salesperson telephoned at the end of the day to confirm the accuracy of the order information. At the end of two weeks, all criteria of a successful test had been met. These are:

1. The order-entry clerk was easily trained.
2. The job content of the order-entry clerk was greatly enhanced, making the job more interesting.
3. The intelligent CRT terminal was easy to use and highly reliable; it lost no time during the two weeks.
4. Order accuracy exceeded 99 percent.
5. Projected total order-entry processing cost per order was less than $16.63 (the original goal set for the new system).
6. The salespeople were highly enthusiastic about the use of the WATS system to reduce the average order cycle time.

Based upon this experience, the MIS task force asked for and received top management's approval to place a protective order with the vendor for the necessary equipment and to proceed with planning for the installation of the new equipment at each field warehouse.

The equipment to be installed at each field warehouse is an intelligent terminal with a CRT screen for the display of data entered. The terminal will provide for flexibility of input formats by having its own easily defined and constructed formats as needed. It will have display editing capabilities, including (1) insert a new line of information, (2) delete a line of information, (3) insert a character of data, and (4) delete a character of data. The CRT screen itself will allow for up to twenty-four lines of eighty characters each to be entered or displayed simultaneously. There will be storage capacity of 96 thousand bytes of memory to hold both data entered by the operator and tabular information to edit input.

For data entry, the keyboard will consist of the standard typewriter keyboard plus sixteen program function keys. The program function keys can be used to perform special predetermined operations for facilitating data input and display. The analyst can define the operations to be performed: for example, the analyst may specify that any time program function key 1 (PFK1) is depressed, the format for entering the order-entry input is displayed on the screen; similarly, PFK2 could be used to display the shipment processing input format.

Design of Order-Entry Input

Now that the specific order-entry inputs and method of input have been determined, the final step in the input design process involves the design of the specific input that will initiate the flow of order-entry data. Not only must consideration be given to the human element, but also to how each order-entry input will be handled by operating procedures, its relationship to data files, and its usefulness in producing the desired outputs. Overall, there is a great need to visualize the foregoing relationships so that the most logical order-entry input is devised by the MIS task force.

Source document formats The basic input mode for the new order-entry system employs data entry via a CRT display device. Predefined data input formats are stored to assist data-entry clerks in the accurate and complete input of all information needed for order entry and shipment processing. These predefined data input formats can be recalled and displayed on the intelligent CRT device by depressing PFK1 for order-entry input or PFK2 for shipment processing.

As shown in figure 13.2, the data input format is used to assist the clerk in entering **order-entry input.** (Note that the hardware selected limits the size of the display to twenty-four rows by eighty columns.) The underscores represent blanks that must be filled in before order entry is complete. The order-entry clerk completes the input based on information from the salesperson. The account number entered by the clerk is edited for validity, that is, it is located on a table of valid account numbers constructed from the customer name and address file. At this point, the CRT display looks like the sample order illustrated in figure 13.3.

The order-entry clerk next depresses PFK3 to complete information about the order. The product code for each item is found in the product code file and the product description, units, and prices are retrieved. The total cost and total shipment weight of the order are calculated. Finally, the CRT display of the sample order is shown (figure 13.4).

The procedure for **shipment processing input** is similar. Figure 13.5 illustrates the data input format for this second input. The shipment processing clerk (shipping clerk) completes the input based on the information from the yellow copy of the shipment processing form used by the warehouse loader when order picking the items. (This procedure will be explained later in the chapter.) In contrast, figure 13.6 shows what the CRT screen looks like after the information is typed for a sample shipment whose order was illustrated in figure 13.3.

The shipping clerk next depresses PFK4 to complete information about the shipment. The product code for each item is found in the product code file, and the product description, units, and prices are retrieved. The total cost and total shipping weight of the order are calculated. Finally, the CRT display format is shown in figure 13.7.

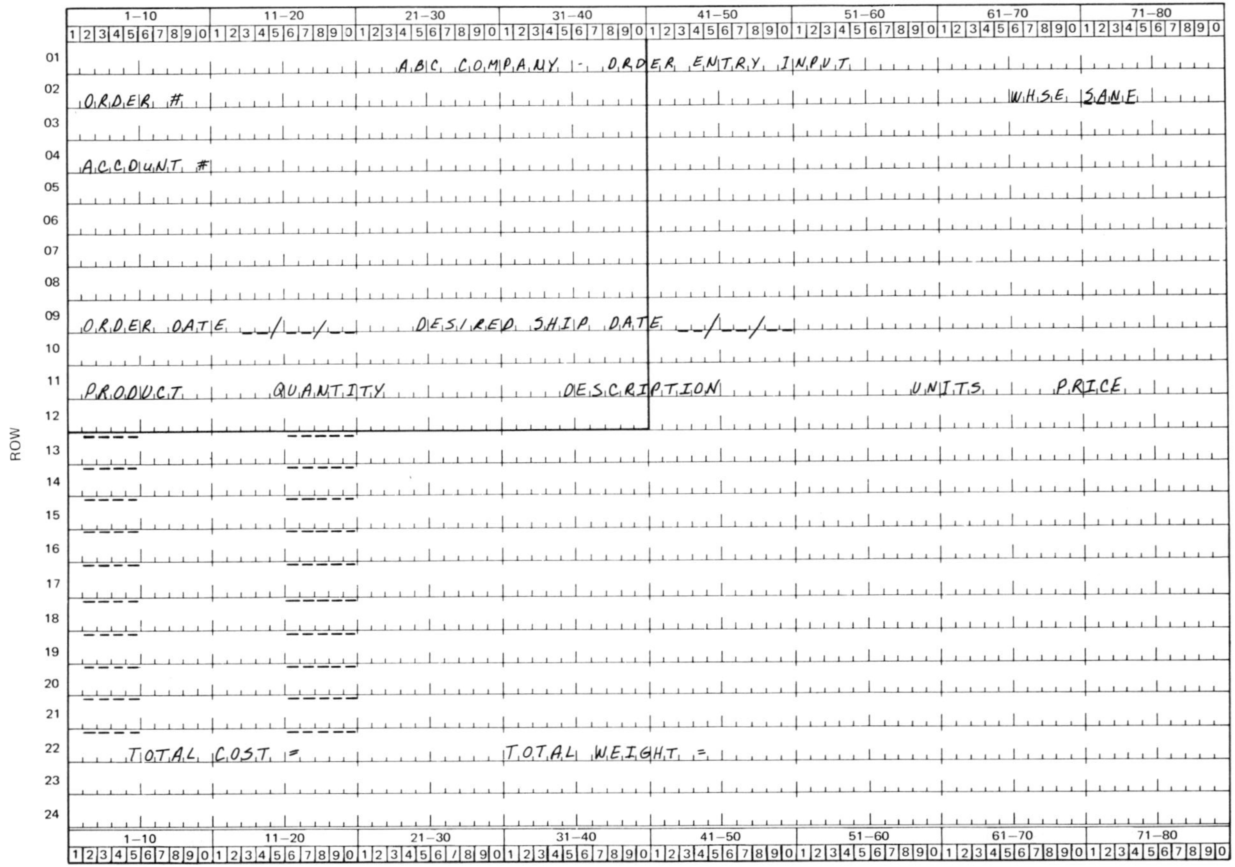

figure 13.2 The order-entry input screen format, showing a sample order for the ABC Company.

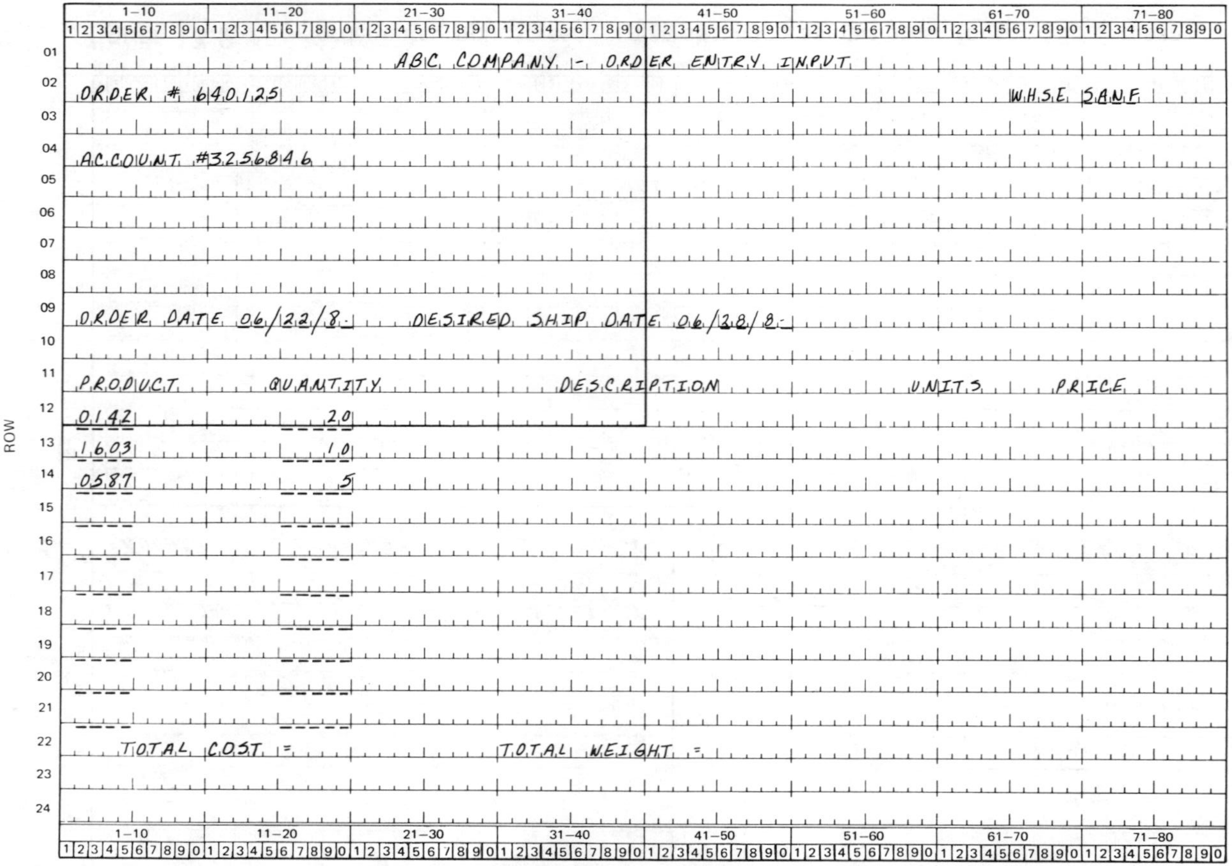

figure 13.3 Order-entry input: the initial input for a sample order for the ABC Company.

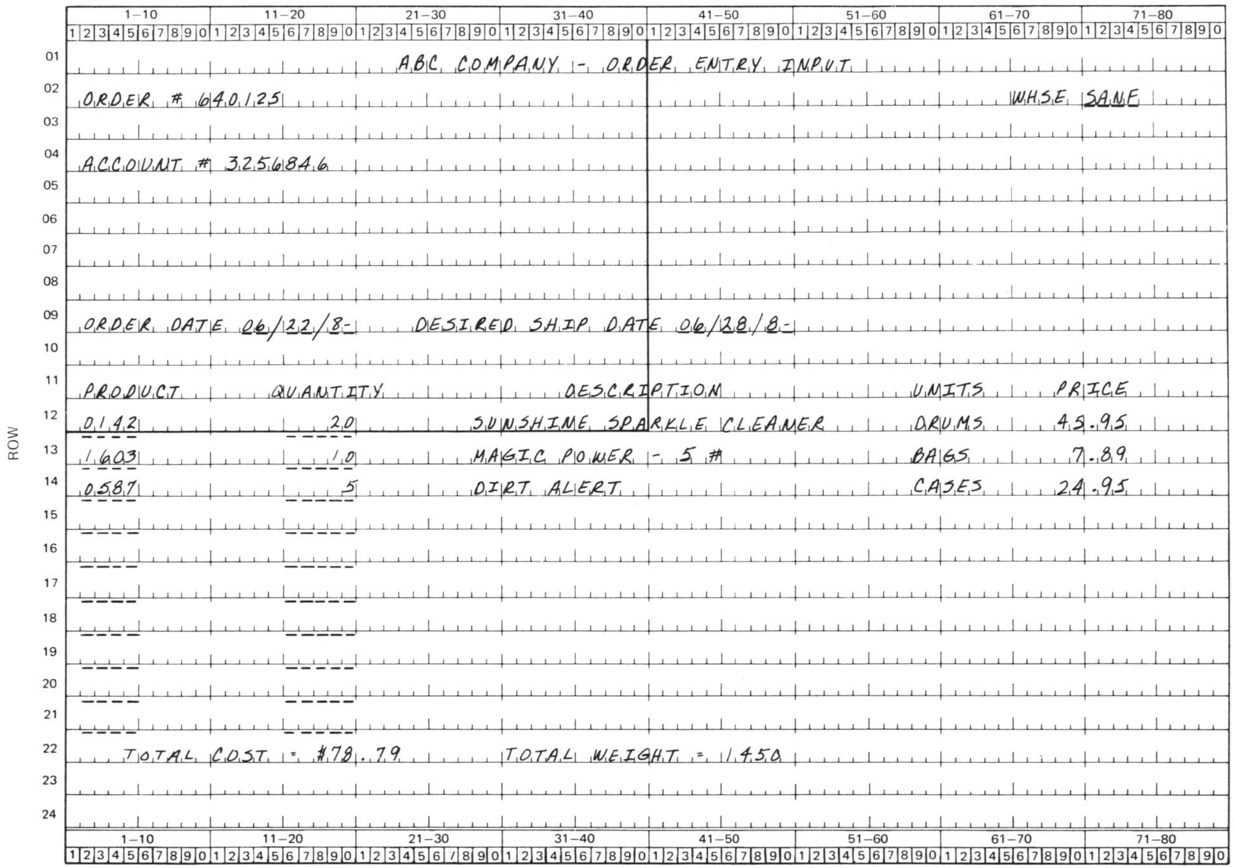

```
ORDER # 040125
ACCOUNT # 3256846                                    WHSE SAME
                        ABC COMPANY - ORDER ENTRY INPUT

ORDER DATE 06/22/8-    DESIRED SHIP DATE 06/28/8-

PRODUCT      QUANTITY       DESCRIPTION              UNITS    PRICE
0142         20             SUNSHINE SPARKLE CLEANER DRUMS    45.95
1603         10             MAGIC POWER - 5 #        BAGS     7.89
0587         5              DIRT ALERT               CASES    24.95



TOTAL COST = $78.79        TOTAL WEIGHT = 1450
```

figure 13.4 Order-entry input: the input which follows that of figure 13.3 for a sample order for the ABC Company.

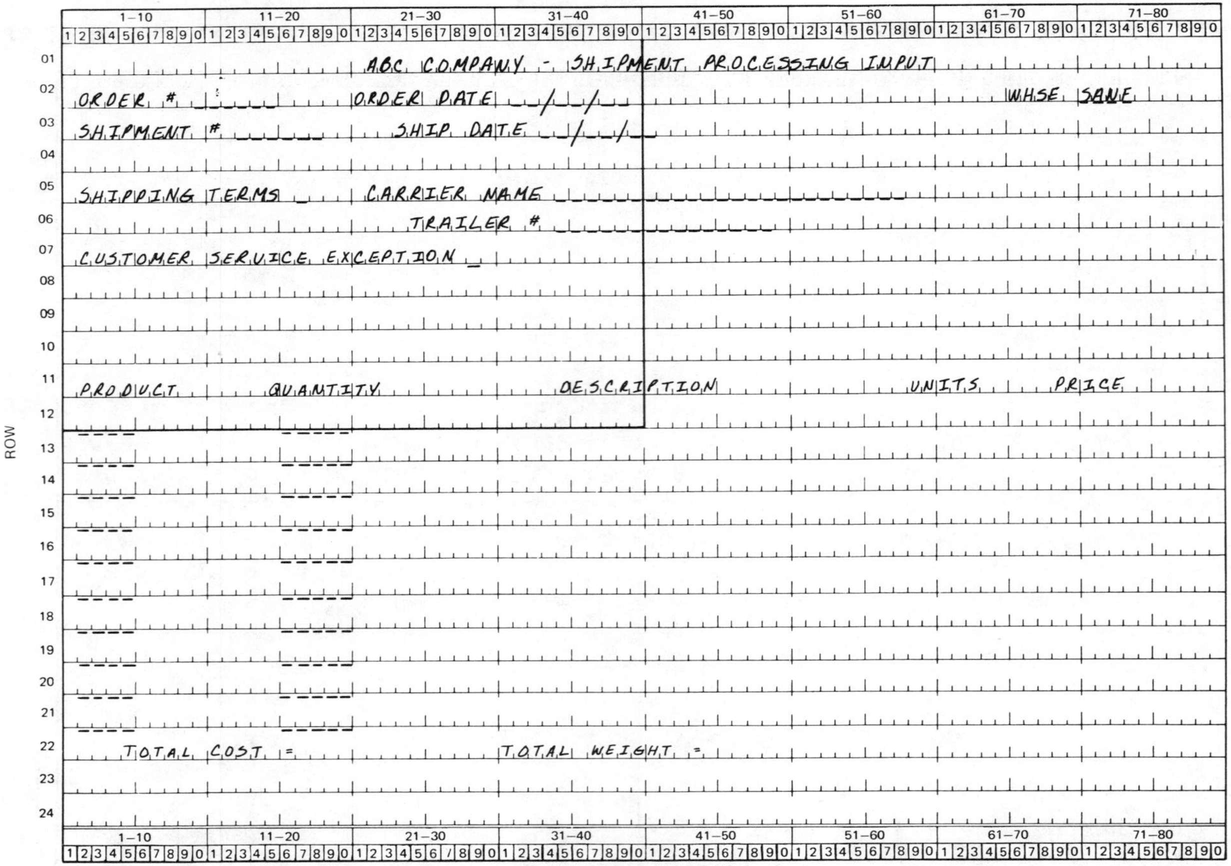

```
         1-10      11-20      21-30      31-40      41-50      51-60      61-70      71-80
01                     ABC COMPANY - SHIPMENT PROCESSING INPUT
02  ORDER #            ORDER DATE _/_/_
03  SHIPMENT #         SHIP DATE _/_/_
04
05  SHIPPING TERMS     CARRIER NAME                                              WHSE BANE
06                     TRAILER #
07  CUSTOMER SERVICE EXCEPTION _
08
09
10
11  PRODUCT    QUANTITY          DESCRIPTION              UNITS         PRICE
12  _____    _____             _____            _____         _____
13  _____    _____             _____            _____         _____
14  _____    _____             _____            _____         _____
15  _____    _____             _____            _____         _____
16  _____    _____             _____            _____         _____
17  _____    _____             _____            _____         _____
18  _____    _____             _____            _____         _____
19  _____    _____             _____            _____         _____
20  _____    _____             _____            _____         _____
21
22  TOTAL COST =                 TOTAL WEIGHT =
23
24
```

figure 13.5 The shipment processing input screen format, showing a sample shipment for the ABC Company.

```
ORDER # 040125    ABC COMPANY - SHIPMENT PROCESSING INPUT
SHIPMENT # 435604    ORDER DATE 06/22/8-             WHSE SAME
                     SHIP DATE 06/28/8-
SHIPPING TERMS C    CARRIER NAME XPRESS TRUCKING
                    TRAILER # LNX740329
CUSTOMER SERVICE EXCEPTION


PRODUCT       QUANTITY         DESCRIPTION              UNITS    PRICE
0142             20
1603             10
0587              5




         TOTAL COST =              TOTAL WEIGHT =
```

figure 13.6 Shipment processing: the initial input for a sample shipment for the ABC Company.

Row	Content
02	ORDER # 010125 ABC COMPANY - SHIPMENT PROCESSING INPUT
03	SHIPMENT # 435604 ORDER DATE 06/212/8- W.S.E SAME
04	SHIP DATE 06/28/8-
05	SHIPPING TERMS C CARRIER NAME XPRESS TRUCKING
06	TRAILER # LNX740329
07	CUSTOMER SERVICE EXCEPTION
11	PRODUCT QUANTITY DESCRIPTION UNITS PRICE
12	0142 20 SUNSHINE SPARKLE CLEANER DRUMS 45.95
13	1603 10 MAGIC POWER 5# BAGS 7.89
14	0587 5 DIRT ALERT CASES 24.95
22	TOTAL COST = $78.19 TOTAL WEIGHT = 1450

figure 13.7 Shipment processing: the input which follows that of figure 13.6 for a sample shipment for the ABC Company.

Input record lengths The order entry and shipment processing input record lengths can be determined by adding up the number of bytes for each field on the files. Including an additional three bytes at the start of each record as the record identification code, the result is 122 bytes for the order-entry input record and 162 bytes for the shipment processing input record. With the current average of thirty orders and shipments per day, approximately 8,520 bytes (30 records × 122 bytes + 30 records × 162 bytes) must be used to store the information for order and shipment processing by the intelligent CRT terminal. The product code file requires 30,000 bytes (300 × 100 bytes) of storage. Thus, the total required storage is about 38,520 bytes versus the available 96,000 bytes on the terminal.

Input codes Certain key fields of information have been assigned codes to represent specific values. The advantage of the coded representation is that it reduces the number of characters which must be entered into the system. This, in turn, lowers the possibility of input errors as well as reduces the amount of storage required to hold input information. Three fields to which codes have been assigned are the warehouse identification, the shipping terms indicator, and the customer service exception indicator.

In addition to the manufacturing plants and attached warehouses in Los Angeles, Minneapolis, and Philadelphia, the ABC Company has five field warehouses located in Atlanta, Cincinnati, Denver, New York, and San Francisco. The four character field **warehouse identification** codes assigned are:

- ATLA   Atlanta
- CINC   Cincinnati
- DENV   Denver
- NEWY   New York
- SANF   San Francisco

The ABC Company handles shipping charges in two ways: (1) freight prepaid by the company and added to the customer's invoice, and (2) freight collect whereby the customer pays freight charges. The two codes used to indicate **shipping terms** are

- P   freight prepaid
- C   freight collect

The **customer service exception** code is used to indicate the reason why the customer service standard level of a five-day total order cycle time was not met. The codes include

- Q   product held due to a failure to meet quality standards
- P   product held due to a failure of packaging to meet quality standards
- A   product allotment exceeded
- O   product out of stock
- C   carrier delay

Overview—Design of Order-Entry Input—ABC Company

The key input of the new order-entry system for the ABC Company has been specified as follows:
1. The **order-entry input** involves a direct telephone communication (via a WATS network) between the salesperson and a trained order-entry clerk. It employs data entry via an intelligent CRT terminal where predefined data input formats are

employed to assist the operator. In this manner, there is accurate and complete input of all information needed for order entry.

2. The **shipment processing input** is initiated by the yellow copy of the shipment processing form, used by the stock loader when pulling ordered items. It employs an intelligent CRT terminal (like order entry) whereby predefined data input formats are displayed to assist the terminal operator. These predefined formats assure accurate and complete input of all information needed for shipment processing.

These inputs are compatible with the order entry and shipment processing data files and outputs. In the next section, these inputs will serve as the starting point for the new order-entry methods and procedures of the ABC Company.

SUMMARY: Design of Order-Entry Input—ABC Company

Consider the Human Factor in Designing Order-Entry Input Centers on the users of the order-entry system—order-entry clerks, salespeople, field warehouse accounting clerks, and shipping clerks—who will be involved in its daily operation.

Incorporate Underlying Principles of Order-Entry Input Relates to including those essential design principles that make the new order-entry system easier to use.

Specify Input Requirements—Relate Output Through Files and Procedures to Input Focuses on specifying the input requirements for the new-order entry system. Inputs essential to this system are order-entry input and shipment processing input. These inputs are capable of producing the desired order-entry output by employing specified procedures and various system data files.

Determine Specific System Input Refers to specifying in detail the contents of the order-entry system input. System input analysis forms are prepared and include detailed information on each element.

Determine the Method of Order Entry Relates to determining the use of distributed processing for new order entry. The equipment to be installed at each field warehouse is an intelligent terminal with a CRT screen for displaying entered data.

Design of Order-Entry Input Centers on the source document formats of inputs for the new order-entry system. Also, input record lengths and input codes are included in this final input design step.

DESIGN OF ORDER-ENTRY PROCEDURES—ABC COMPANY

Now that the inputs, data base, and outputs have been designed, this part of the chapter centers on devising those methods and procedures necessary to complete the design. The design process for relating input to output and data files is as follows:

1. Consider the human factor in designing order-entry procedures.
2. Incorporate underlying principles of order-entry procedures.
3. Specify order-entry methods and procedures.

4. Determine the steps in order-entry procedures.
5. Devise order-entry back-up procedures.

As with the first part of the chapter, these steps are summarized by an overview section.

Consider the Human Factor in Designing Order-Entry Procedures

In the design of order-entry methods and procedures, as with the other design areas, the MIS task force agreed to place the utmost priority on the human factor. Additionally, it is recognized that the order-entry operator/terminal interface is critical to the success of the system. The system will be developed in such a way that this interface will be simple and easy for the order-entry operator to use—even at the detriment of some system efficiency.

Incorporate Underlying Principles of Order-Entry Procedures

In addition to the human aspect discussed above, the MIS task force adopted the system principles set forth in chapter 12 for system methods and procedures. Below, each of these principles is related to the new system.

Principle of modularity of methods and procedures The MIS task force will modularize the order-entry system by breaking down each function into its component tasks. The component tasks will then be carefully studied and logically grouped together in order to reduce the complexity of the system and facilitate potential future changes.

Principle of automated methods and procedures In designing the new order-entry system, the MIS task force is prepared to change or recombine methods and procedures that cut across established organizational boundaries whenever appropriate. They will recommend necessary changes in the organizational structure to insure a more efficient or more economical operation.

Principle of economy of methods and procedures The new order-entry system will be designed with methods and procedures that are relatively low in operating cost. An important factor in meeting the objective of an economical system is the capturing of the order entry and shipment processing input very close to the originating source of information.

Principle of simplicity of methods and procedures Simplicity will be achieved by following the KISS (Keep It Simple, Stupid!) guideline in the design of all methods and procedures. Needless steps and duplication of effort will be eliminated in the new system.

Principle of flexibility of methods and procedures The methods and procedures of the new system will be adaptable to changes in the existing business environment. Expansion or contraction of business operations—such as, the number of orders and shipments per day and the number of plants or warehouses—should have little impact on the system.

Principle of speed of methods and procedures One of the most important objectives of the new order-entry system is to reduce the average order cycle time from twelve to five days. The new methods and procedures must be capable of meeting this goal; otherwise, the feasibility study must be revised at this point to recognize the inability to meet this requirement.

Principle of reliable methods and procedures Another important objective of the new system is to reduce the percent of orders incorrectly billed from 7 to 1 percent. As noted in the first part of the chapter, predefined order-entry formats for the intelligent CRT terminals will be extremely valuable in reaching this objective.

Principle of provision for corrections To meet the accuracy goal of the new system, it is clear that provision must be made for not only reliable and accurate detection of errors, but also for simple and accurate correction of errors.

Specify Order-Entry Methods and Procedures

The specification of methods and procedures requirements relates to the order processing flow from input to final output. Within this design framework, the main thrust of order-entry methods and procedures is to produce output that is meaningful to the users. To assist the MIS task force in achieving this objective, it is helpful to consult the order-entry system output analysis forms (tables 10.1 through 10.5), the system data file analysis forms (tables 10.7 through 10.10), and the system input analysis forms (tables 13.1 and 13.2) previously discussed. Because these forms state the important factors to be included in the final design, the MIS task force is in a better position to specify efficient and economical methods and procedures that relate data-entry input elements to data files and final outputs.

As a means of relating input to output and data files, an overview of the new order-entry methods and procedures is provided in figure 13.8. Sales order information is sent to the field warehouse—now, primarily by telephone, but also by mail, teletype, or hand. This information is captured by a trained order-entry clerk. Next, a credit check is performed; if it fails, the order is flagged and suspended from further processing until a decision can be made by marketing management. Warehouse shipment information about shipments to customers or other field warehouses is entered by a trained shipment processing clerk. All data are transmitted to the central computer in St. Louis.

At corporate headquarters, computer processing of the new order is undertaken. The customer name and address and the product code files are utilized in order processing. New orders are placed on the open order file. Shipment information triggers the purge of an order from the open order file and the creation of a shipment record for that order on the shipment file.

Based upon the completion of order-entry processing procedures by the intelligent terminals at the five field warehouses and computer processing at St. Louis, a multicopy order and shipment processing form is created for each new order. The customer invoices and the shipments by carrier report are prepared from the shipment information. Once a month, a customer service report is prepared. As a part of daily processing, daily shipment and open order data are passed to the finished product inventory system and daily shipment data are transferred to the accounts receivable system to provide them with current information.

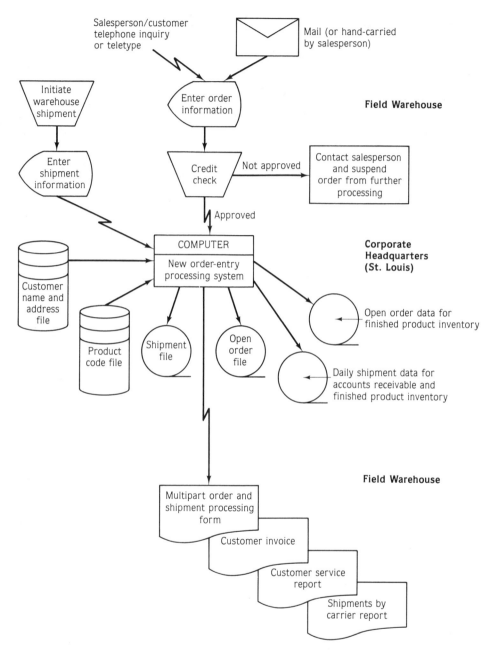

figure 13.8 Sales order processing and warehouse shipment data flow for the new order-entry system of the ABC Company.

377

Determine the Steps in Order-Entry Procedures

Once order-entry methods and procedures have been specified, the next item in the design process is to determine their detailed steps. The new order-entry system methods and procedures are illustrated in figure 13.9. Because the new methods and procedures have been designed based upon the foregoing principles, they are much simpler, more flexible and accurate, and quicker than the old ones. Furthermore, they will assist in reducing the order cycle time to five days and improving the order and shipment processing accuracy.

On the **first** day, the customer or salesperson uses the WATS telephone line to call in an order to the nearest field warehouse. The person phoning in the order speaks directly to a trained order-entry clerk who enters the complete information required for order processing. Any data-entry errors or lack of complete information are detected by either the order-entry clerk or the intelligent CRT terminal and can be corrected before the telephone conversation is ended. Throughout the day, the day's order-entry input is saved and each night is transmitted to central headquarters for processing by the new order-entry system, as noted in figure 13.8.

On the **second** day, multicopy order and shipment processing forms are printed at the field warehouse from data processed the previous evening and transmitted from the central headquarters in St. Louis. The original copy of this form is mailed to the customer as confirmation that the order has been received and is being processed. In turn, the red copy is kept by the field warehouse accounting clerk to answer questions from the salesperson or customer regarding the order. The remaining three copies of this form are sent to the shipping section. The shipping clerk determines the number of trucks needed for the next day's shipments, contacts the carriers, and arranges for a pick-up time. Finally, this individual notifies the field warehouse manager of the next day's loading schedule.

On the **third** day, the blue copy of the shipment processing form is filed by the shipping clerk and two copies (yellow copy and bill of lading copy) are given to the warehouse stock loader. As the items are picked to fill the order, the stock loader notes any items which cannot be shipped as ordered on the yellow copy. This copy is returned to the shipping clerk. The bill of lading copy is also annotated for corrections and is given to the carrier as the bill of lading.

The prior procedures have been concerned with pulling the order and shipping the product; whereas, the remaining ones are concerned with completing the informational shipment processing procedures. Thus, the shipping clerk pulls the blue copy of the order and shipment processing form and matches it to the yellow copy. The shipping clerk makes any corrections noted by the stock loader, assigns the customer service exception code (if necessary), and refiles the blue copy. The shipping clerk then uses the yellow copy to enter shipment processing information via the intelligent CRT terminal. The yellow copy is then forwarded to the field warehouse accounting clerk. At this time, the accounting clerk pulls the red copy and destroys it as the yellow copy now serves as the official company record as to what was shipped. The day's shipment input is saved and transmitted to St. Louis for processing on the same day that materials were shipped.

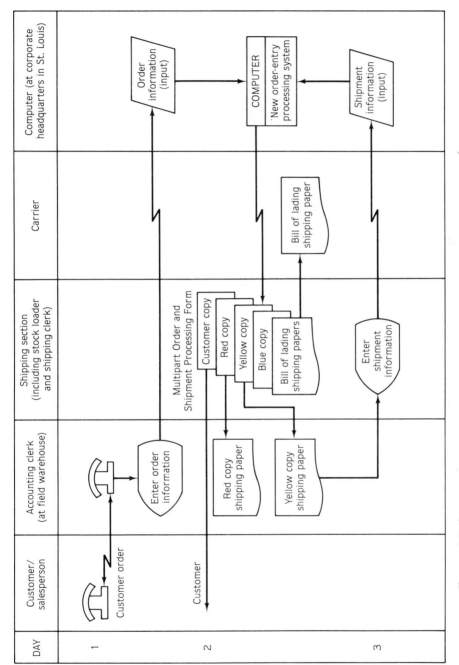

figure 13.9 A document flowchart of the new order-entry system of the ABC Company.

table 13.3 Order-entry and shipment processing procedures system analysis form for the ABC Company.

SYSTEM PROCEDURES	
Page 1 of 2 Date: May 1, 198_ Preparer: G. W. Reynolds	Type of procedure: ORDER-ENTRY AND SHIPMENT PROCESSING

Locations affected: Five (all) field warehouses

SPECIFIC INSTRUCTIONS:

First day:

Salesperson   Call an order to the nearest field warehouse by using the company's WATS telephone network.

Order-entry clerk   Speak directly with the company salesperson in order to enter information via an intelligent CRT terminal for a new order. Also, mail, teletype, and manually prepared orders are entered by the order-entry clerk. Any data-entry errors or missing information are detected by the order-entry clerk on the intelligent CRT terminal. All of the day's order-entry input is transmitted to the central computer system in St. Louis for processing in the evening.

Second day:

Field warehouse accounting clerk   Use the five copy order and shipment processing forms received from the central computer system in St. Louis (processed the previous evening). The **original** copy is mailed to the customer as a confirmation that the order has been received and is being processed. The **red** (second) copy is kept by the field warehouse accounting clerk to answer questions from the salesperson or customer regarding the order.

The preceding day-to-day operations are set forth in the order-entry system and shipment processing procedures analysis form, illustrated in table 13.3. This form is compatible with the input, data file, and output analysis forms referenced in this chapter and in chapter 10.

Devise Order-Entry Back-Up Procedures

Although the MIS task force has carefully evaluated the reliability of the new order-entry system, they felt it necessary to devise back-up procedures in case of possible system problems. Procedures were developed to cover the possibility of a computer failure at St. Louis. In the event the main computer would be inoperable, an agreement has been made with a time-sharing service bureau to obtain computer time to handle the ABC Company's data processing operations.

In the event of problems with the intelligent CRT terminals at any field warehouse, the sales order-entry clerk will instruct the salespeople to telephone their orders to another field warehouse. The order-entry clerk at the other field warehouse will input the order data with the appropriate field warehouse code specified. The order-entry system will

table 13.3 Continued.

SYSTEM PROCEDURES	
Page 2 of 2 Date: May 1, 198_ Preparer: G. W. Reynolds	Type of procedure: ORDER-ENTRY AND SHIPMENT PROCESSING

Shipping clerk Receive the remaining three copies of the order and shipment processing forms; determine the number of trucks needed for loading the next day's shipments.
Third day:
Shipping clerk File the **blue** (fourth) copy of the order and shipment processing form.
Warehouse stock loader Pick the items to fill orders per the **yellow** (third) copy and note any items which cannot be shipped as ordered.
Shipping clerk Match the yellow copy to the blue copy, make any corrections noted by the stock loader, assign the customer service exception (if necessary), and refile the blue copy. The **bill of lading** (fifth) copy should also be annotated with corrections and given to the carrier.
The yellow copy is then used to enter shipment processing information via the intelligent CRT terminal. In turn, this copy is forwarded to the field warehouse accounting clerk. All data regarding the day's shipments are transmitted to St. Louis in the evening for processing.
Field warehouse accounting clerk Pull and destroy the red copy.

process the data and use the field warehouse code to route the order and shipment processing forms to the correct warehouse for subsequent handling according to the standard procedures.

Overview—Design of Order-Entry Procedures—ABC Company

Within this overview of order-entry methods and procedures, emphasis is placed on relating the entire data processing flow from input through data files to output. This flow for the ABC Company's methods and procedures can be summarized as follows:

1. **Order-entry input** Starts the order processing flow for the ABC Company. Orders are entered via an intelligent CRT terminal where predefined data input formats are employed to assist order-entry clerks. For the new order-entry system, computer processing at central headquarters in St. Louis consists of **customer name and address file** and **product code file** as input, along with order-entry information for the current day from the five field warehouses. The evening output of this processing run are the **multipart order and shipment processing forms** that are forwarded to the appropriate warehouse early the next morning for shipment of orders to the customer.

2. **Shipment processing input** Starts the billing and related processing flow for the ABC Company. The yellow copy of the multipart order and shipment processing form is used at the field warehouse for entering shipment information via the intelligent CRT terminal. As with order-entry input, data are forwarded each evening to the central headquarters for processing by the new order-entry system. Again, the customer name and address and product code files are used. Output of this evening processing includes not only forms mentioned above, but also **customer invoices**, the **shipments by carrier report**, and the **customer service report** (monthly) on which shipments which fail to meet the new customer service standard will be highlighted. Also, output includes the **open order file** and the **shipment file**, along with **daily shipment data** and **open order data** for the finished product inventory system and **daily shipment data** for the accounts receivable system.

The new order-entry system at central headquarters in St. Louis makes daily use of the inputs, data files, and outputs that have been designed in this chapter and in chapter 10. This entire data processing flow is illustrated in figure 13.8. Thus, all design elements in the new order-entry system for the ABC Company, except for system controls, are now complete.

SUMMARY: Design of Order-Entry Procedures—ABC Company

Consider the Human Factor in Designing Order-Entry Procedures Refers not only to assisting order-entry personnel in the performance of their jobs, but also to considering the human over the technical elements in order-entry procedures.

Incorporate Underlying Principles of Order-Entry Procedures Centers on incorporating essential design principles that make order-entry procedures easy and economical for customers and company personnel to use.

Specify Order-Entry Methods and Procedures Refers to setting forth methods and procedures that link order-entry inputs with data files and outputs. To assist in this task, order-entry system input, data file, and output analysis forms are reviewed.

Determine the Steps in Order-Entry Procedures Focuses on the determination of specific order-entry methods and procedures. Detailed procedures require three days for internal processing. Table 13.3 provides an overview of day-by-day procedures from order entry to shipment processing.

Devise Order-Entry Back-Up Procedures Centers on back-up computer procedures at central headquarters in St. Louis and at each field warehouse in the event of system breakdown.

CHAPTER SUMMARY:

The order-entry and shipment processing input for the new system was developed according to the design process discussed in chapter 11. Although the information required for this input is basically the same as under the old system, the method of obtaining the input is based on the use of distributed processing, CRT displays, and interactive computing. As a result, there will be more accurate and timely order and shipment information.

In a similar manner, the order-entry and shipment processing procedures for the new system were developed following the design process presented in chapter 12. Although these procedures represent a major change, careful communication of these plans and thorough training of the order-entry clerks will help insure their success. Overall, the order-entry and shipment processing procedures link system inputs, data files, and outputs in an efficient and economical manner so that the new system is a major improvement over the existing one.

Questions

1. a. What was the primary design principle followed in designing both the order-entry system inputs and procedures?
   b. Why is this principle so important?
2. What are the basic factors that led the MIS task force to recommend the use of distributed processing for input to the new order-entry system?
3. What steps were taken by the MIS task force to verify the decision to use distributed processing before making a final recommendation to top management?
4. Describe the basic equipment to be installed at each field warehouse to support the new method of data entry.
5. What is the purpose of entering the customer service exception on the shipment processing input?
6. What provisions have been made in processing orders to handle credit checking?
7. a. What is the internal order processing time for the new order-entry system?
   b. What are the major changes that have reduced the total order processing time?
8. What are the major benefits of the new order-entry and shipment processing system over the present one?

Self-Study Exercise

True-False:

1. _____ The Principle of Acceptability of Inputs refers mainly to minimizing subsequent manual tasks, as in the new order-entry system.
2. _____ The Principle of Simplicity of Input assures that clarity will be kept foremost in mind in the order-entry system.
3. _____ Specifying input requirements for the new order-entry system disregards outputs, whether they be reports or data files.
4. _____ The two basic inputs for the new system are order-entry input and shipment processing input.
5. _____ The only elements of the order-entry system input analysis form are informational elements and number of bytes.
6. _____ PFK, in an intelligent order-entry CRT terminal, stands for program function key.

7. _____ The most important aspect of the design of order-entry input is the source document format.
8. _____ The product code file includes product description, units, and prices.
9. _____ A logical starting point for specifying order-entry methods and procedures is to start with order-entry system analysis forms for inputs, data files, and outputs.
10. _____ Order-entry procedures of the second day include the salesperson's using the company's WATS telephone network.

Fill-In:

1. The guiding design principle for devising new order-entry system input is the Principle of the _____ _____ of Data Input.
2. Required inputs for the new order-entry system are order-entry input and _____ _____ input.
3. Remote _____ CRT terminals provide the capability to continue order-entry processing even if the central computer is down.
4. The information processing requirements of each _____ _____ are insufficient to warrant the installation of a computer to handle the new order-entry system.
5. _____ data input formats are stored by the intelligent CRT terminal to assist the data-entry clerks in their work.
6. The intelligent CRT terminals of the new order-entry system calculate the _____ cost and shipping weight of each order.
7. The _____ code file requires 30 thousand bytes of storage.
8. The MIS task force utilizes the Principle of _____ of Methods and Procedures to break down each function to be performed into its components.
9. The _____ of order-entry methods and procedures requirements relate to the informational flow from the input stage to final output.
10. The function of the system _____ analysis form is to present systematic methods and procedures that are to be performed by company personnel.

chapter 13

case study 1

finished product inventory

Responsibility for finished product inventory will be centralized under a single group reporting to the vice-president of physical distribution and located at company headquarters in St. Louis. This group will use the reports set forth in chapter 10's case study—today's plant and warehouse inventory status report, projected plant and warehouse inventory status report, recommended warehouse stock replenishment orders, and recommended plant manufacturing schedules—to help implement a comprehensive policy for the management of finished product inventory.

The shipment processing clerk (or shipping clerk) at each field warehouse provides the primary input to maintain control over inventory. The shipping clerk must enter the following types of information:

1. The quantities and product codes of finished products for each customer shipment must be entered into the system. This input is based on the information from the yellow copy of the order and shipment processing form used by the loader who picked the items for the order. (The format of the screen used to enter these data was illustrated in figure 13.5.)
2. The quantities and product codes of plant–warehouse and interwarehouse shipments to replenish the field warehouse inventory must be entered.
3. The quantities and product codes of shipments outward from the field warehouse to other field warehouses or returns to plants must be inputed. These shipments reduce the particular field warehouse inventory just as customer shipments do.
4. The quantities and product codes of customer return shipments must also be entered. Along with the basic inventory information, management would like to capture information on the reason for each customer return. Some of the more common reasons might include: product received damaged, product not included on original customer order (wrong product shipped), and customer refused to pay collect charges.

Assignment: Design of Finished Product Inventory System Input and Procedures

1. Devise appropriate CRT terminal screen input formats (figure a is a sample) to handle input data that must be entered by the shipping clerk at the field warehouse. Modify the screen

format used to input customer shipment data (figure 13.5) to include the capability to process input for shipments outward from the field warehouse (which are deductions from inventory). Similarly, design another screen input format to process input for all shipments into the field warehouse (which are additions to inventory). In both cases, start with the system input analysis forms (table a is a sample) to define the basic input requirements.

2. Occasionally, warehouse–warehouse or warehouse–plant shipments must be made to alleviate the temporary imbalance of inventory. Prepare a document flowchart (similar to figure 13.9) and the appropriate system procedures analysis forms (table b is a sample) for these types of shipments. Orders for these types of shipments are placed by the central physical distribution group based on their analysis of the various finished product inventory reports.

case study 1-table a System input analysis form for the ABC Company.

SYSTEM INPUT				
Date: Preparer:		Type of input:		
Informational elements	Type of characters	Number of characters	Source	Comments
Comments:				

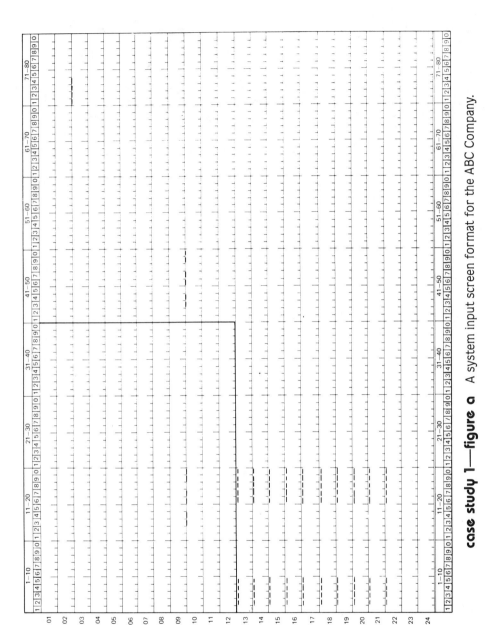

case study 1—figure a A system input screen format for the ABC Company.

case study 1-table b System procedures analysis form for the ABC Company.

SYSTEM PROCEDURES	
Date: Preparer:	Type of procedure:

chapter 13

case study

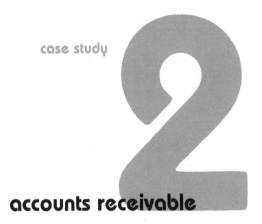

accounts receivable

Responsibility for the accounts receivable function will remain centralized in the Accounts Payable and Accounts Receivable Section at St. Louis, under the vice-president of accounting and finance. This group will prepare the monthly customer statements, the accounts receivable aging schedule report, and uncollected balances report. A major activity of this group, of course, is the maintenance of accurate customer account records.

The primary input data required by the group to maintain customer accounts are the daily shipment and the customer remittance data. Shipment data are passed from the order-entry system to the accounts receivable system. Customers are instructed to direct their payments, as well as questions about their accounts, to the central group in St. Louis. The accounting clerks will use an intelligent interactive terminal to apply cash received to specific open invoices on the customer's account.

The intelligent terminal will have the capability to retrieve key information from the accounts receivable master file in order to assist in the posting of cash receipts to specific open invoices. In this way, the accounting clerks will be assisted in their work, just as the order-entry clerks and shipment processing clerks are at the five field warehouses.

Assignment: Design of Accounts Receivable System Input and Procedures

1. Devise the CRT terminal screen input formats (case study 1, figure a is a sample) to handle the input required to post customer cash payments to specific open invoices. This process will begin when the accounting clerk enters the customer account number, depresses a program function key, and receives a CRT screen of information about that account. The information displayed must include the date and amount of all open invoices and must allow space for entry of customer cash remittance data. It will be important to enter the date of cash application so that the discount earned or forfeited can be determined. Start with the system input analysis forms (case study 1, table a is a sample) to define the basic input requirements.
2. Occasionally, products will be returned from the customer to the plant or warehouse. At this time, it is necessary to credit the customer account. Prepare a procedures analysis form (case study 1, table b is a sample) to show how these procedures might be handled. Assume that the shipping clerk at the field warehouse enters information about the returned shipment via a

CRT to the Finished Product Inventory System to adjust the inventory records. The Finished Product Inventory System then sends a computerized record to the accounts receivable system to adjust the customer's account.

chapter

design of system controls

OBJECTIVES:

- To examine the relationship of information value to its cost as it applies to designing a new system.
- To discuss the essentials of internal control and the relationship of internal control to system control.
- To specify system control design principles to be an integral part of any management information system.
- To delineate the specific types of system controls that must be designed and incorporated into a new system.

IN THIS CHAPTER:

INTRODUCTION TO DESIGN OF SYSTEM CONTROLS
    Information Value Versus Cost
    Cost of Accuracy
INTERNAL CONTROL
    Internal Check
    Internal Accounting Control
    Internal Administrative Control
INCORPORATE SYSTEM DESIGN PRINCIPLES—CONTROLS
DESIGN OF SYSTEM CONTROLS
    Input Controls
    Computer-Programmed Controls
    Data Base Controls
    Output Controls
    Interactive Controls
    Security Controls
QUESTIONNAIRE FOR EVALUATING SYSTEM CONTROLS
CHAPTER SUMMARY
QUESTIONS
SELF-STUDY EXERCISE

The design of a new system—an interactive management information system, in particular—is not complete until adequate provision has been made for system controls. This need is more pronounced with newer DP equipment since fewer people will be involved in the new operation and, therefore, the organization will rely heavily on machines to process data. Management must be assured that the new methods and procedures devised will control the required data in an accurate manner. Basically, this job is the responsibility of the systems analyst who must consider internal control when finalizing the new system design. Failure of the systems analyst to integrate internal control within the framework of the new system is an open invitation to future problems detrimental to its successful operation.

Initially, in this chapter, the value and relative cost of information are examined, since this relationship determines the control points and types of controls that must be designed into the system. Internal control is discussed from both an overview and an accounting standpoint. Within this broad framework of system controls, important design principles are set forth. Lastly, the design of system controls—input, computer-programmed, data base, output, interactive, and security—are discussed in detail. This final section focuses on the type of controls that should be designed and where they should be located for most effective usage. In essence, these system controls provide an underlying structure for finalizing the design of the order-entry system of the ABC Company.

INTRODUCTION TO DESIGN OF SYSTEM CONTROLS

No system design job is complete without adequate provision for internal control. The systems analyst should make certain that the final design allows no one person full responsibility over an entire operation; this should be apparent in such areas as cash and payroll because one person with complete responsibility can defraud an organization. Control points must be built into the system. Checks at control points insure that what has been processed agrees with predetermined totals. Controls of this type insure accuracy during processing, resulting in reliable output.

Design of a new system must include the following controls:
1. input
2. computer-programmed

3. data base
4. output
5. interactive
6. security

The successful incorporation of these controls into the final systems design will reduce the need for human auditing procedures. A sufficient number of built-in safeguards is required to eliminate fraud and inaccuracies.

Individuals who have a special interest in the design of system controls are the internal and external auditors, as well as management, stockholders, creditors, and the government; they all want assurance that processed data are reliable when transmitted from one source to another. Control points for checking the accuracy of processed information and the distribution of control over one specific area are vital for effective internal control.

Information Value Versus Cost

The difficulties in achieving **accuracy**—defined as freedom from error or mistake—tend to increase with the complexity of the environment in which an organization must operate. Newer methods and procedures must be invented to handle demands originating from adding new products, distribution channels, and the like, resulting in additional controls over inputs, files, methods, procedures, and outputs. The appropriate level of accuracy and its related costs must be determined for these new operating conditions. Errors and mistakes can be caused by the human element, incorrect instructions to personnel and machines, a malfunction of the equipment, or a combination of these at any stage from data origination through reporting. Hence, the systems analyst must ascertain the accuracy desired and devise the new system accordingly.

The question of how much accuracy is actually required to operate a reliable DP system must be answered. An examination of an organization's functional areas will reveal a need for 100 percent accuracy in some and less in others. Activities such as payroll, sales commissions, accounts receivable, and accounts payable require the highest degree of accuracy. On the other hand, inaccuracies for low-value inventory items—cartons, supplies, and similar items—can be tolerated since the cost of accurately controlling these items would exceed their value. Based on these two extremes, the systems analyst needs a general guide to relate the value of the data to the corresponding cost of obtaining it. The most desirable degree of accuracy is reached when the value exceeds the cost of acquiring the data by the greatest margin. However, from a practical viewpoint, maximum accuracy is desired only to the point at which the value derived from the information is still greater than its corresponding cost (figure 14.1).

In view of this value–cost relationship, the best methods for achieving a high degree of accuracy must: (1) eliminate most errors as they occur, and (2) detect the remaining ones, thereby providing a means for making the appropriate correction. The systems analyst must develop control points to accomplish the degree of accuracy required. The receipt of payments, for example, means establishing a control point for banking an organization's daily receipts; these same deposits must be balanced at a later time when posted to the customer accounts. Even though the control totals agree as the data move

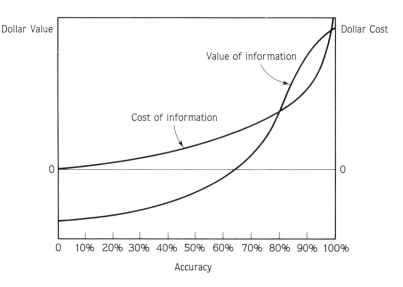

figure 14.1 The relationship of value, cost, and accuracy to information. From a practical viewpoint, **optimum** accuracy is obtained up to the point where the value of information is still greater than its corresponding cost.

through the successive stages of the system (which means most errors were eliminated when they first occurred), there is still no assurance that the correct accounts were posted. The systems analyst must make use of control routines, such as the self-checking digit, within the accounts receivable program so that all accounts are properly credited. Only in this manner will accuracy be assured in the new system.

Cost of Accuracy

The degree of accuracy, as indicated in figure 14.1, depends upon the cost of obtaining the information and the value of having a certain percentage of accuracy. Inaccurate information is of little value to management; in fact, inaccurate information can lead to profit-losing decisions. As noted in the illustration, the value of information rises fast as the degree of accuracy increases, particularly at the higher levels. More accurate information facilitates better decisions; however, decision making is not sensitive to slight improvements in near-perfect accuracy.

Costs increase at a faster rate as the degree of accuracy increases because more operations and control points must be established for nearly continuous checking of data. Generally, costs increase very rapidly as output accuracy is pushed beyond 95 percent. Data-entry accuracy, for example, may be increased from 95 to 98 percent by having the operators of data-entry equipment reduce their rate of output; to increase accuracy even further requires extensive verification. In essence, the cost of eliminating the last possible mistake is often prohibitive.

In the preceding section, the accounts receivable example indicated the higher costs associated with 100 percent accuracy. Even in this area, it is possible to design an effective system of less than 100-percent accuracy: instead of spending money on procedures that require the detail to equal the control figures, some organizations employ accounts receivable adjusters to handle customer complaints on errors not corrected as the data are processed. This cost is considerably less than those programs and procedures which yield 100-percent accuracy. The important point here or for any other activity is that increased accuracy adds to the cost of originating and processing the data. It is the job of the systems analyst to devise a system that meets the accuracy and cost requirements set forth in the objectives of the system project.

INTERNAL CONTROL

Once the systems analyst has resolved the problem of what level of accuracy the new system will require, the necessary internal control can be incorporated into the system design work. Even though the systems analyst is not an accountant, he or she must be cognizant of the essential elements of internal control. **Internal control,** as defined by the American Institute of Certified Public Accountants, comprises the plan of organization and all of the coordinate methods and measures within a business to: (1) safeguard its assets, (2) check the accuracy and reliability (defined as suitable or fit to be relied on) of accounting data, and (3) promote operational efficiency and encourage adherence to prescribed managerial policies. This three-part division of internal control is referred to, respectively, as **internal check, internal accounting control,** and **internal administrative control.** Each of these parts will be discussed in subsequent sections of the chapter.

A comprehensive system of internal control extends beyond the finance and accounting functions. It includes promoting operational efficiency in the DP system and control over the operating programs and procedures. In fact, all information developed by the system is covered by the concept of internal control.

Ideally, internal control is integrated as an essential part of the methods and procedures in processing an organization's activities. The self-correcting mechanism of the system allows DP personnel to take appropriate action when errors, mistakes, or equipment malfunctions arise. In such a system, the amount of internal and external auditing is kept to a minimum. The adequacy of internal control, then, has a direct bearing on the amount of auditing required and serves as a measure of the system's effectiveness in following prescribed procedures.

Internal Check

Internal check safeguards an organization's assets against defalcation in the form of accounting procedures, physical controls, or other means that allow for the segregation of functional responsibilities. No matter what approach is used, the separation of data processing functions permits no one individual to have complete control over one area or activity. However, the use of computers tends to reduce the amount of activities that can be separated since many are under the control of the equipment itself and fewer workers are involved. Nevertheless, internal control procedures can be devised that will safeguard an organization's assets.

Many organizations utilizing DP equipment have experienced large defalcations in the areas of payroll and inventory caused by the computer programmer who operates the program that he or she has written. If one has complete control, a loophole can be built into the system for carrying out fraudulent schemes. An example is a computer programmer who runs the weekly payroll program; this individual can create payroll checks for fictitious employees (and remove the fraudulent checks before turning them over to a designated company employee), can pay extra overtime and wages to oneself, and can deduct a few cents of income tax from every employee in the plant for payment to oneself. Another way the computer programmer/operator can steal without being caught is by changing a few instructions in the computer inventory program; the machine can be made to report high inventory losses as normal merchandise breakage, enabling accomplices in the plant to take large amounts of goods from the warehouse.

The problem of safeguarding an organization's assets has become more acute with interactive systems since data, such as inventories and supplies, are constantly changing. It is difficult enough for the auditors to reconstruct specific data for an audit trail, not to mention what goes on inside the machine that never appears on a computer printout. In view of these existing conditions, the systems analyst must devise DP activities from several points of view, simultaneously. These are set forth in figure 14.2.

Establish the DP department as a separate unit so that it does not have direct control over an organization's assets (power to disburse funds, control over inventory, and so forth). This reduces attempts on the part of computer programmers and machine operators to engage in fraudulent activity since several people must be involved.

Require that systems analysts, programmers, equipment operators, and record librarians be distinct and separate groups.

Provide for the transfer of computer programmers and machine operators frequently to different programs and machines, respectively. The rationale is that the individual is less likely to undertake changes in the program or the data being processed.

Have a periodic surprise audit of the computer room where details are checked out thoroughly for irregularities. The fact that work can be checked at any time is a deterrent to fraudulent activity.

Set up checking procedures to verify that data received by the user are accurate. This requirement can be accomplished by comparing batch totals as the data are processed.

Use any other economically feasible means to assign personnel to check on the activities of each other. The more persons involved in one activity, the lower the possibility of collusion.

Require fidelity bonds for DP personnel in positions of trust.

---

figure 14.2 Recommended procedures for effective internal control to safeguard an organization's assets.

Several organizations have built into their systems a means of printing audit information. A computer executive audit routine is utilized that spots apparent irregularities in operating procedures which immediately trigger a printout of the questionable transaction for examination by the auditor. The output of the computer's console typewriter is locked and under the control of the internal auditor. Another method involves feeding test data into a computer and checking to determine whether anything interferes with the routine processing. The net effect of either approach is internal checking on the computer system.

Internal Accounting Control

The systems analyst must devise methods and procedures that provide for internal accounting control. Accounting controls that check the accuracy and reliability of data processed by the system must be an integral part of it. At certain control points in the system, batch totals, subtotals, hash totals, grand totals, record counts, count of prenumbered documents, or a combination of these are used to check the accuracy of data as they move from one stage to another. Within the system, accounting control can be performed using manual methods, mechanical equipment, or computers. The approach for checking the accuracy and reliability of data is contingent upon the system designed.

The amount of processing performed at one time differs greatly for manual and computer systems. Manual systems have a great need for more control points since processing ability is limited. The output at one stage serves as input for the next stage. Since processing is performed for one small stage at a time, intermediate totals become an integral part of internal accounting control. However, the same cannot be said for computer systems, since reliance is placed on hardware and software and not on manual processing. More processing is performed simultaneously, resulting in fewer stages for intermediate totals. Many times, intermediate results are not available in a readable format since they may be contained on a magnetic disk, drum, or tape file. However, printouts may be incorporated into the new system design for internal control purposes.

Essential characteristics of internal accounting control are control total techniques, control by comparison, and control by authorizations and approvals. Each is explained below.

Control total techniques   Control total techniques assure that data have been transmitted accurately. Although the systems analyst provides batching techniques and similar methods at various control points for noncomputer activities, computers require a different approach since they provide a means for obtaining unprecedented accuracy. Although computer totals may no longer be needed to assure processing accuracy, they are still required to check accurate transmission of data from one processing run to another both within the computer room and to and from the computer equipment.

Control by comparison   Control by comparison takes a variety of forms. Comparisons can be made manually between vendor's invoices and receiving department reports to verify the receipt of materials; cash receipts can be compared with accounts receivables to determine the accuracy of incoming receipts. Likewise, the computer can make logical decisions for a greater than, equal to, or less than condition which allows it to check the data as they are processed.

Control by authorizations and approvals   Control by authorizations and approvals can be performed manually or electronically. The most promising area is the computer with its logical comparative ability. Since the computer has been programmed for a set of predetermined criteria, it serves as a basis for acceptance or rejection; in effect, the computer is capable of making routine decisions. These include reordering raw materials, checking the customer's credit status, approving vendors' invoices, and printing checks. By no means do these activities exhaust the computer's potential for control by authorization and approval.

Internal Administrative Control

Administrative controls comprise the plan of organization and all methods and procedures concerned with operational efficiency and adherence to managerial policies. They are distinguishable from the other two areas of internal control discussed previously in that they originate in and are an essential part of all operating departments, not just finance or accounting. A popular method for reviewing the adequacy of internal administrative control with respect to the DP system is a management audit, which locates problem areas for the purpose of identifying those areas needing improvement through further, detailed study. The management audit, then, is a tool for determining managerial weaknesses, and later forms the basis for making suggestions and recommendations for improving an organization's operations.*

As in the case of internal check and internal accounting control, the basis of the investigation for a management audit is a questionnaire. The questions are constructed so that a YES answer indicates a favorable point and a NO answer is unfavorable. Most questionnaires provide ample room to explain a negative answer since this represents a weakness that must be investigated. Many times, the question asked will not be applicable; in such cases, the column for NA is checked. The answers provide a framework for a comprehensive evaluation of an organization.

Although one of the basic objectives of any DP system is to assist management in making decisions, the function of its equipment is to provide analytical information on a timely basis. The need for correctly organized information available for the right personnel is essential. The objective of the management audit for internal administrative control, then, is (1) to determine that an organization is realizing optimum utilization of its computer facility, and (2) to insure that well-designed functional applications which adhere to managerial policies are meeting management information and control needs.

---

*For a comprehensive study of management audits, see Robert J. Thierauf, Robert C. Klekamp, and Daniel W. Geeding, **Management Principles and Practices, A Contingency and Questionnaire Approach** (Santa Barbara, Calif.: Wiley/Hamilton, 1977), chapters 2, 8, 12, 16, and 20.

SUMMARY: Internal Control

Internal Check Concerned with safeguarding an organization's assets against defalcation by separating DP functions so that no one individual has complete control over one area or activity.

Internal Accounting Control Concerned with checking the accuracy and reliability of data processed by the system. This is accomplished at certain control points in the system by employing techniques that check on accuracy as data move from one stage to another.

> **Internal Administrative Control** Concerned with promoting operational efficiency and adherence to managerial policies through the use of a management audit questionnaire to locate problem areas requiring improvement through further, detailed study.

INCORPORATE SYSTEM DESIGN PRINCIPLES—CONTROLS

Foremost among system design control principles is the **Principle of the Acceptability of Controls,** stating that system controls should be acceptable to organization personnel who are to use them. The nonacceptance of system controls may result in out-of-control conditions leading to a failure to accomplish specific organization objectives. Hence, the acceptability of system controls involves the human element. As indicated many times before, organization personnel can make or break any system project.

To overcome the problems associated with the human element, it is highly recommended that systems analysts develop system controls jointly with personnel who will be using them. In this manner, all problems with new system controls will have been corrected before implementation. Similarly, if organization personnel can see their ideas integrated in the new system, they will be more receptive to it, especially during the critical change-over phase. Overall, the first design principle is critical to the success of any new system.

An important question in system design is **where** standards, such as production tolerances and flexible budgets, are to be measured, that is, controlled. This involves a second design principle—the **Principle of Establishing Strategic Control Points**—which has a number of basic characteristics. First, a central point is established for **key operations** or events. For example, in a manufacturing process, quality control points may be located in the receiving department, in certain manufacturing departments critical to the product's quality, and in those departments that process the product just before it undergoes highly expensive operations.

The **timing** for comparing actual results against standards at strategic control points should permit the manufacturing process to be stopped or altered before serious damage is done to the product or the mistake is compounded. In short, control points should be set up which enable management to retain the degree of control necessary for efficiency in manufacturing.

Control points should also be **comprehensive** and **economical.** To be comprehensive, control points must include every major operation that can be measured. Using the preceding manufacturing example, all products are subjected to the same major control points throughout their manufacture. A factor closely related to comprehensiveness is economy. If all manufactured goods were subjected to measurement after each production step, there would be little economy in manufacturing operations. Hence, only the critical points—those with an important impact on final product quality—are employed.

Finally, the selection of various strategic control points should be **balanced.** There is a tendency to overcontrol the quantitative factors and undercontrol the qualitative factors: marketing, manufacturing, and finance are generally controlled very closely, whereas personnel development and leadership are not. This condition can lead to a state of imbalance in which line executives are given precise standards of performance, but staff executives are not held to specific standards.

A third important design principle relating to system controls is the **Principle of Control Accountability.** Inasmuch as various types of strategic control points are established, the managers at those operating levels can be held accountable for the process being controlled. The fact that an individual is accountable for results enhances the chances of more effective control over on-going operations. It should be noted that control accountability utilizes both quantitative and qualitative measurements.

A fourth principle is the **Principle of Management by Exception** by which actual performance is measured against established standards, relating "what is happening" to "what should be happening." If standards are appropriately established and if the supervision is adequate for determining what subordinates are doing, comparison of actual against expected performance is fairly straightforward. This comparison often takes the form of reports that highlight exceptional items, hence the name of this principle. A typical example is a monthly production department cost report that compares actual against budgeted amounts. All variances (favorable and unfavorable) of 5 percent or more are starred on the report and reviewed by the respective supervisor and the plant manager. In this manner, all exception items are brought to the attention of the appropriate management level for possible corrective action.

With the present techniques of time and motion study, for example, hourly standards for mass-produced items can be developed for reliable measurement and comparison. If items are custom-made, however, comparing performance results may be a formidable task. Also, there are many activities for which it is extremely difficult to develop sound standards of performance measurement; in less technical kinds of work, both standards and meaningful comparison may be difficult to develop. For example, detailed standards cannot be easily developed for the job of an industrial relations director, nor can results be measured in a precise way. If the industrial relations department is within its budget, has shown evidence of sound management, and is rated high on general standards of behavior, this overall comparison may be adequate under the circumstances. The important point is that, as jobs move away from the assembly line or any other type of routine, the design of system controls using "management by exception" becomes more difficult.

Other principles that must be considered by systems analysts focus on a wide range of items. A fifth design principle, the **Principle of the Flexibility of Controls,** recognizes the importance of built-in flexibility that, if environmental changes occur (caused either by external or internal conditions), system controls can be adapted as the system, itself, is adapted. In a somewhat similar manner, the sixth design principle, the **Principle of the Suitability of Controls,** provides that controls reflect the job they are designed to perform—that is, controls should serve the needs of organization personnel to be more effective. From this view, information from the system will be valuable in accomplishing specific tasks. The seventh design principle, the **Principle of the Reliability of Controls,** means that controls should be reliable in order to assist management in accomplishing its assigned tasks. This helps management develop confidence in the system for controlling daily operations and meeting desired organization objectives.

DESIGN OF SYSTEM CONTROLS

An essential part of the design phase is the integration of controls within the new system. As noted earlier, systems design encompasses the following controls: (1) input, (2) computer-programmed, (3) data base, (4) output, (5) interactive, and (6) security. Each of these important system controls are discussed at some length below.

The relationship of system controls to a computer is illustrated in figure 14.3. Many of these controls center around the accurate and efficient processing of data. Even though input may be 100 percent accurate, the accuracy of output is contingent upon the proper manipulation and handling of data within the various computer programs. For effective control in this area, programmers must have full knowledge of what these controls are and how they should be incorporated within a program. Otherwise, substantial program reworking may be necessary after the program is considered operational.

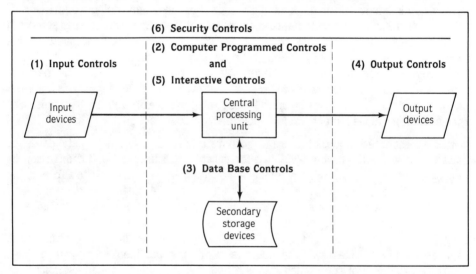

figure 14.3 The relationship of system controls to a computer, in both batch and interactive processing modes.

SUMMARY: Incorporate System Design Principles—Controls

1. **Principle of the Acceptability of Controls** Controls should be acceptable to organization personnel; otherwise, they will not accomplish the job for which they were designed.
2. **Principle of Establishing Strategic Control Points** The purpose of establishing strategic control points during the systems design phase is to monitor the work activities being performed and report any deviations from established standards. If there are important discrepancies at a control point, the work process should be corrected before additional effort is expended. This approach allows for economy and efficiency of on-going operations.
3. **Principle of Control Accountability** The location of strategic control points is an important way of holding the manager accountable for the process being

controlled. Control accountability is applicable to quantitative and qualitative measurements.

4. **Principle of Management by Exception**   An essential task of system controls is to detect potential and actual deviations from predetermined plans on a timely basis so that corrective action can be taken. This principle is concerned with analyzing only significant deviations, thereby ignoring those items that are "on target."
5. **Principle of the Flexibility of Controls**   Controls should be flexible in order to meet external and internal changing conditions.
6. **Principle of the Suitability of Controls**   Controls should reflect the job they are designed to perform; that is, they should suit the needs of organizational personnel. In this manner, information from the system will be valuable in accomplishing assigned tasks.
7. **Principle of the Reliability of Controls**   Controls should be reliable and consistent so that they can accomplish the tasks for which they were designed.

Input Controls

Control of input is defined as the procedural controls necessary to handle data prior to computer processing. Input data must be handled very carefully since they are the most probable source of errors in the entire DP system. If errors are created anywhere between the origination point and input into the computer equipment, they will be carried forward throughout the entire system. In order to keep errors to a minimum, the following procedural controls over input data are available for a computer installation:
    verification methods
    input control totals
    external labels

Verification methods   Some degree of input verification is absolutely necessary for consistently reliable results. The extent and methods employed are best determined by comparing the level of undetected errors against the cost of verification. Only in this manner can verification devices assure the accuracy and validity of input data. Several methods of input verification include key verification, visual verification, self-checking numbers, and tabular listings.

Key verification is widely used for verifying input data. For punched cards, a card verifier is employed whereby new cards are prepared when errors are found. For the newer approaches to data entry, such as key-to-disk and key-to-tape, input data are stored on a disk in locations appropriate to the keystation of original entry. Once the recorded data have been verified by rekeying the same data and making the appropriate corrections, verified data can be transferred from the disk onto magnetic tape or disk for entry into a computer system.

Visual verification is the least preferred verifying method because of the tendency to hasten the job at the expense of accuracy. In essence, the eye is not as accurate as equipment. Also, newer data-entry devices transfer source data to magnetic tape or disk which cannot be verified visually.

The **self-checking digit** is a special type of verification for reference numbers, such as account numbers and inventory numbers. This method employs information in the form of an extra digit that is mathematically designed to provide detection of transposition or substitution of digits within a number. Card punches and paper tape machines are currently available that will perform this checking of critical data.*

Tabular listings refer to a method of comparing printed output totals with original batch totals. This method can be slow since those output batches that disagree with the original batches must be corrected for a new printout. However, a large number of accounts receivable payment data, for example, could be sorted by batches and then printed on some off-line equipment. An error could be located quickly since input figures have been grouped on the same basis as the printout.

Input control totals Control totals which aid in determining the accuracy of input data are generally obtained from adding machine tapes or from totals established in originating departments. Careful system design can provide a high level of accuracy with very little additional effort and expense. Control totals—batch totals, hash totals, and record count totals—are usually taken on batches or groups of source documents and is the most common method for insuring input accuracy.

Batch controls refer to source documents which are accumulated into batches, constituting physical groups for processing data. The number of items within a batch should be limited in order to efficiently reconcile discrepancies; generally, batches of approximately one hundred transactions are a convenient size. The transactions within the batch should be as homogeneous as possible to insure compatibility of control totals. If volume necessitates the use of many batches, it may be advisable to have the batch number on all input data in order to permit reconciliation, especially after the continuity of the batch has been lost. Batch control totals should be attached to related input data when forwarded for processing. Consideration should also be given to computer verification of batch controls through programming to prove the detail against the control totals.

Hash totals may be used for control purposes where batch totals (dollars or units) cannot be calculated. They are items which are not normally added together—such as account numbers, employee numbers, and unit prices—and represent totals of some data field common to all documents in the batch.

To illustrate the use of hash totals, consider the weekly payroll for a manufacturing firm. The hashing of employee social security numbers—that is, the adding of all numbers for the past pay period—would provide a starting point for this period's payroll processing. The addition of new employees and dropping of terminated employees for the current period through hashing employee social security numbers would be merged with the hash totals of the prior period. This new total should agree with the current period's hash totals of the payroll. By utilizing this input control technique, there can be assurance that all employees have been paid. However, it should be noted that there is no assurance that all values for weekly hours worked and rates of pay are correct.

---

*For more information on the self-checking digit, see Robert J. Thierauf, **Data Processing for Business and Management** (New York: John Wiley & Sons, Inc., 1973), pp. 108–110.

Record count totals can be found in several forms, depending on the system in use. Prenumbered forms are a type of record count because of the computer's ability to control them. Serial numbers of documents, such as vouchers and requisitions which constitute input, might be introduced along with account codes, quantities, and similar data for storage within the computer. At designated intervals, serial numbers of those documents which have not been processed, but should have been, can be determined by the computer program for immediate follow up. This input technique assures that all data are processed through the computer system.

Minimum control for any system should include some type of input count, such as a transaction or a card count, because an independent count assures that all input data have been processed off line and are ready for computer processing. Record counts can also be used to check the accuracy of the data processed on line. Likewise, record counts which are carried at the end of each on-line file can be compared before the computer program is terminated. This insures that file data are transferred from one source to another without loss of records.

External labels External labels, another form of input controls, are visible ones attached to a magnetic tape or magnetic disk. Certain identifying information can be written on the tape or disk label. Information which may be a part of the label includes: name of run, type of information, density, reel number, number of reels in the file, frequency of use, date created, drive number, earliest date of possible reuse, record count, and the name of the individual responsible for the magnetic tape or disk. Thus, the purpose of external labels is to assure that the correct input file is being processed by the computer operator, and data are not being destroyed prematurely.

Computer-Programmed Controls

A high degree of programmed controls is always advisable on the computer since undetected errors can have serious and far-reaching consequences. Programmed control steps should be included as a part of the machine's internally stored instructions. The extent of controls depends upon the increased costs in programming and machine time versus the resulting increased accuracy. In addition to this consideration, the extent of computer-programmed controls includes the programming ability of the DP section, the requirements of the particular run, and the capabilities of the equipment. Since most computer applications are slowed down by the speeds of input/output units, the central processor has generally more than sufficient time to perform the necessary control checks. Therefore, the number of programmed controls is dependent upon the amount of program memory available. If sufficient space is available for program instructions, the programmer should include as many controls as is necessary. They can be classified as follows:

    validation checks and tests
    computer control totals
    internal labels
    error routines

Validation checks and tests Validation comprises a series of checks and tests that can be applied to verify input, file, and calculated data during the computer processing run. It includes sequence checks, character mode tests, self-checking digit tests, limit checks, code validity tests, blank transmission tests, alteration tests, and checkpoints. The inclusion of all or part of these processing controls is contingent upon the program itself.

Sequence checks are incorporated into a computer program to check data ascending or descending sequence while it is processed. Data might be identified by customer number, inventory number, or another acceptable basis for processing. This programmed approach highlights data out of sequence, occurrence of duplicate numbers, and gaps in the sequence of processing.

Character mode tests are extremely valuable before manipulating data within the computer program. Fields which must contain numeric, alpha, zero, special characters, or some combination of these can be checked internally against a transaction code look-up table in order to detect erroneous data before executing detailed computer instructions. These tests are a must for programs requiring 100-percent accuracy. The addition of numeric and alphabetic data, needless to say, results in incorrect totals.

Self-checking digit tests should be verified by the computer, using the same formula as for input checking; approximately 95-percent reliability can be expected. This technique is used extensively for code numbers that are recorded initially by hand and later converted into a machine acceptable language. As with most validation tests, provision should be made for reporting incorrect errors when located.

Limit checks, sometimes called **reasonableness tests** or **tolerance checks,** can be incorporated as part of a computer program. These are magnitude tests designed to determine whether or not processed data fall within predetermined limits. The machine can be instructed how to handle processed data outside these established value limits. Limits, for example, can be set for maximum and minimum levels of inventory, largest dollar amount for a payroll period, and highest amount of credit extension to a customer. Limit checks also include checking the capacity of equipment accumulators or registers when performing arithmetic calculations.

Because limit checks are widely used in an interactive mode, the following example will help to highlight their importance. Order processing clerks, using CRT terminals, enter items ordered from customer order forms. Before deductions are made for items ordered, the computer program checks to make certain that the inventory number is legitimate. Next, the computer program checks the quantity ordered: if the quantity ordered is 5000, and the normal range is from 1 to 500, a message is flashed on the CRT screen; or, if the quantity is 5 units, and the normal range is from 50 to 500, the CRT unit would ask the operator to check the quantity entered for correctness. Overall, limit checks can be integrated into any program to test for the reasonableness of entered data.

Code validity tests involve the testing of significant code numbers to establish their validity. The computer program will reject invalid codes by comparing the code on the input data with a list of valid codes stored on line. For example, an invalid employee number that is read by the high-speed reader cannot be located within the payroll master file.

Blank transmission tests are used to monitor data fields at transfer points for blank or zero positions. They might also be used to detect the loss of data and prevent the premature destruction of existing records in file storage.

Alteration tests indicate the failure to update a file properly. This can be determined by comparing the contents of the file before and after each posting.

Checkpoints permit the computer to restart processing from the last checkpoint, rather than from the beginning of the run, in the event of an error, irregularity, or interruption in the program. This can result in a considerable savings of time and money for extremely long processing runs. The best time to establish check points is during the flowcharting and programming phases. A typical check point procedure requires that processed data be dumped periodically into a temporary storage area which is available for immediate processing if a restart is necessitated.

Computer control totals Computer control totals are a continuation of the input control techniques described previously—batch controls, hash totals, and record count totals. Errors uncovered by any one of these control techniques during computer processing may be treated as the program requires: the entire group of data may be rejected immediately, or it may be allowed to pass through for future processing. In the latter case, it is customary to store the error conditions and report it externally as a printout. Additional computer totaling techniques which provide procedural control during processing are balancing controls, cross-footing balance checks, zero balancing, and proof figures.

Balancing controls, normally included in many accounting computer programs, are comparable to those used in controlling manual processing procedures. Beginning balances plus and minus the transactions produce a new ending balance for each record. The total of new balances for all records should agree with the manual control totals. Inventory, accounts receivable, accounts payable, and payroll are functions that rely on this computer control technique.

Cross-footing balance checks, long used by accountants, are employed to check the accuracy of individual footings. This is accomplished by adding all figures first in a horizontal direction, then in a vertical direction. When the computer has completed these footings, the sum of the horizontal totals should equal the sum of the vertical totals.

Zero balancing is a method to insure that a multiplication has been correctly calculated. The value "$X \times Y = Z$" may be checked by multiplying Y by X, then subtracting Z to prove that the result is zero. Another method is to add a series of figures in two different directions (as in cross-footing balance checks), then subtract one from the other for a resulting zero.

Proof figures can be used to check a series of multiplications. An arbitrary figure, larger than a multiplier, is selected. Each multiplicand is multiplied once by its true multiplier, then by the difference between the true multiplier and the proof figure. Upon completion of a series of multiplications, the total of the products resulting from both multiplications is compared with the product of the total of the multiplicands and the proof figure. They should be equal.

Internal labels The internal label of a magnetic disk or tape is an extension of the external data label. Certain identifying information, under the control of the computer program in the form of a lead record, can be written on the disk or tape. Before actual computer processing begins, the program reads the lead record to ascertain that the correct magnetic disk or magnetic tape has been mounted properly. If the computer stops and indicates an incorrect input disk or tape, one of two possible errors have occurred: either the wrong input device was used or the external label does not correspond to the internal label.

Error routines There are several ways of handling error routines. The most common is to treat the error routine as an integral part of the internally stored program. If errors are detected as a result of the validation checks, processing tests, and computer control totals, the program can instruct the computer to store the data, punch a data card, print the data, or to initiate some other method for bringing the error condition to light for external handling and correction. Under no circumstances would the processing run be halted; this is in contrast to other error routines which halt the computer for major problems like: the magnetic tape file is out of sequence, input cards are out of sequence, magnetic bits of data are being lost, or an incorrect magnetic disk is being addressed.

Data Base Controls

Just as computer-programmed controls are essential for an effective system, the same can be said for data base controls. Without adequate controls in this area, important data items found in on-line and off-line files can be lost or destroyed. If loss or destruction should occur, certain recovery procedures must be undertaken to recreate these essential data items. Through the use of the foregoing programmed controls, the systems analyst can help ensure, to a large degree, that file data are not destroyed during regular processing. However, there is still need for additional controls over those factors relating to file processing and storage. These include:

    supervisory protection programs
    file record controls
    physical safeguard controls

Supervisory protection programs Supervisory protection programs, like IBM's "exclusive control" program, solve the problem of concurrent updating and the resulting loss of data. It permits only one data transaction to update an on-line file at a time. Basically, this software package requires that each data set request permission of the supervisory program for updating a specific item. If the file is available, the proper machine instructions are executed. At no time during this updating process will the supervisory program allow other transactions access to this particular data file. Other manufacturers have comparable routines.

File record controls File record control relates to master computer files that must be retained for future processing. They, along with the program tapes, should be under the control of the DP librarian because of the cost incurred by their construction. Computer systems should adhere to the "grandparent-parent-child" concept of file maintenance.

Assuming daily updating, this approach requires the preservation of three days of successive output files along with the corresponding data transactions. There should also be a plan to store this data away from the data processing center in a fireproof vault.

The following example will illustrate the "grandparent-parent-child" file maintenance concept. A company updates its inventory master files once a day; each evening all transactions for the current day (recorded on magnetic tape) are processed against the master file (also on magnetic tape). Based upon these processing procedures, the new updated inventory master file represents the "child" file, and the previous day's inventory master file represents the "parent" file. The file prior to these is the "grandparent" file which has been stored in another location in case there is need to reconstruct the inventory master file. If there are any difficulties with the "parent" file during the daily updating process, the "grandparent" file can be processed with the appropriate day's transactions to produce a duplicate "parent" file which, in turn, can be used to produce the "child" or current file. Thus, not only must the "grandparent" file be available for computer processing, but also the appropriate transactions must be in machine-processable form in order to produce the "parent" file.

With the pervasive use of direct access or magnetic disk files, the basic method for file reconstruction is to periodically dump the file onto magnetic tape. The frequency of file dumps depends upon the amount of activity. For example, if there is a great amount of daily activity for the inventory data base, it should be dumped daily so that file reconstruction is simplified inasmuch as yesterday's transactions can be processed with today's to produce the current inventory data base. However, a word of warning is appropriate—the greater the frequency of dumping, the higher the processing cost due to the amount of computer time required for the dump.

Physical safeguard controls Physical safeguard controls are necessary to ensure that data files are not dumped, lost, or misused while awaiting processing. This entails storage in a fireproof location with controlled environmental factors, such as temperature and humidity. For vital files, including computer programs, off-premises storage should be provided for further safeguards. As indicated above, physical safeguards should incorporate the "grandparent-parent-child" concept of file maintenance.

For magnetic tape files, **tape rings** should be used. They are small plastic rings which fit inside the reel holding the tape. When absent, they prevent the contents of the file from being erased; when present, old information can be automatically erased and new information written when the internally stored program executes magnetic tape write instructions. It should be noted that some computers are designed on the opposite basis; that is, the lack of a ring allows data to be written while the insertion of a ring indicates the data will remain unchanged.

Output Controls

The role of the computer as a control center is further accentuated when output controls are present. Whereas input controls insure that all data are processed, output controls assure that results are reliable. The latter promotes operational efficiency within the

computer area over programs, processed data files, and machine operations. Output controls also insure that no unauthorized alterations have been made to data under computer control. They can be classified as follows:
  output control totals
  control by exception
  control over operator intervention

Output control totals The major categories for output control totals, which were listed above under input and computer programmed controls, are essential for control of output data. The most basic of all output controls is the comparison of **batch control totals** with figures that preceded the computer processing stage. If the batch totals agree, the output data must represent input data that have been processed accurately. In the absence of batch control totals for certain programs, the comparison of **output hash totals** or **record count totals** to predetermined figures will substantiate the validity of the data. No matter what output approach is used for control totals, the job of checking their accuracy belongs to the internal auditor or a designated employee.

Periodically, a computer processing audit of selected individual items should be made by the auditor. Specific transactions are traceable from the originating department, through the computer and its related files, and finally to output control totals. Such tests assure that data are being processed accurately and in accordance with prescribed methods and procedures.

Control by exception The individuals responsible for output control would also be responsible for investigating exceptions. These exceptions could include errors, excessive inventories, sales to customers in excess of their credit limits, higher prices for raw material purchases, and deviations from established sales prices. If computer programs are properly designed and programmed for controlling routine items and exceptions, time can be spent on the exceptional data for more effective decision making. Most advanced computer systems rely heavily on this output control technique.

Control over operator intervention Control over operator intervention is a problem common to all computer faciltiies. Generally, programmed controls do not prevent the console operator from interrupting the data being processed and manually introducing information via the computer console. Even if the internally stored program does possess a routine for printing all information introduced, the operator still is able to suppress the printout on the console typewriter. However, if it is discovered that the typewriter has been turned off, the operator would immediately be suspect. In those systems where changes can be effected without a hard copy record, the need for supervising the computer's operation increases. However, research has indicated that unauthorized console intervention is minimized by rotating the computer operators and, in particular, by exercising output controls over all computer printouts. The need for effective internal check and internal accounting control, then, is apparent.

Interactive Controls

The foregoing computer system controls are generally applicable to all systems, whether batch or interactive processing. However, additional controls are necessary for a two-way flow of information in an interactive environment. In fact, these systems create many problems not found in batch processing systems. How can confidential data be made accessible to only authorized personnel? What happens to data in the system when the computer is down for repairs? How can accuracy be assured with real-time processing? Generally, controls will be dictated by such factors as the requirements of the system, the equipment itself, and security specifications. The areas set forth below are not all-inclusive, but are representative of the control requirements found in a typical interactive system. They include:

on-line processing controls
data protection controls
diagnostic controls

On-line processing controls   On-line processing controls are necessary since messages from and to the input/output terminal devices can be lost or garbled. It is possible that the terminal will go out of order during transmission or receipt of data. To guard against working with incorrect data under these conditions, the system should provide program routines for checking messages. They include message identification, message transmission control, and message parity check.

Message identification is used to identify each message received by the computer. Message number, terminal identification, date, and message code is the usual information sent which permits directing the data to the correct program for processing. If a message is received with an incorrect identification, it should be routed to an error routine for corrective action or rejection from the system. In general, rejection necessitates retransmission of the entire data.

Message transmission control requires that all messages transmitted are actually received. One method is to assign a number to each message and periodically have the computer check for missing numbers and out-of-sequence messages; unaccounted-for message numbers are printed for investigation. Another method is the confirmation of all messages received whether through the computer or the input/output terminal.

A message parity check is one which verifies the accuracy of the message sent. Since it originates at the sending terminal, a check digit is added to the end of the message, representing the number of bits in the message. In a similar manner, the receiving terminal compiles a check digit on the number of bits in the message received. If both check digits are equal, a correctly received signal is sent to the terminal or computer. If there is a difference, most systems will ask for a retransmission of the data.

Data protection controls   Data protection controls provide answers to many interactive problems. A typical problem is: what assurance is there that unauthorized personnel are prevented from using the system? The software for data protection controls includes lockwords and authority lists.

Lockwords and **authority lists** are means of preventing unauthorized access to an interactive system. Lockwords, or passwords, are several characters of a data file that must be matched by the sender before access is granted to the file. Since the password may become common knowledge after a period of time, it must be changed periodically. Another type of security is an authority list. In this case, the lockword identifies the sender. When reference is made to the authority list stored in the computer system, it indicates what type of data the sender is permitted to receive. Whichever method of data security is employed, a control routine should be established within the system to count the number of attempts to send a message. If the unsuccessful attempts have exceeded a certain number, perhaps 3 or 4, it is possible that some unauthorized person is tampering with a terminal device. It should be noted that these methods do not exhaust the list of data security controls, but are the more popular ones.

Diagnostic controls One of the difficulties with any interactive system is a malfunction of the equipment or a programming error occurring during the system's operations. It is best to keep the system operating if the trouble can be circumvented. This can be accomplished through the use of diagnostic programs to detect and isolate error conditions for proper corrective action. Once the problem has been determined, the supervisory program should make the necessary adjustment: it can restart the program in question, reexecute the faulty instruction, switch control to an error routine, initiate a switch-over to another system, or shut down part or all of the system. The first three are used to overcome software problems, and the rest are necessary to control hardware malfunctions.

Diagnostic programs in an interactive system are dependent upon the system design and the equipment. They are absolutely necessary when a terminal breaks down. While one diagnostic program checks the communication network and establishes that there is a problem in the network, another checks each line until the down terminal(s) is (are) located. Having determined the problem, terminal control is returned to the supervisory program which can close down the line for repairs and route all output messages to an adjoining terminal.

When a system must halt because of a breakdown of a major piece of equipment, emergency procedures are needed to handle the work until the malfunctioning equipment has been repaired. Restart procedures must also be devised, which include the use of a checkpoint record. This record, usually a magnetic disk or tape file, is a complete log of all messages and pertinent data processed up to a certain time. When a restart is necessitated, the checkpoint record restores the system to the time this record was written. Every terminal is advised about the last message number that was properly processed. All subsequent messages must be resent in order for the system to operate on a current basis.

Security Controls

Computer system controls are essential for a successful computer system. The same can be said for security controls. Basically, these controls cover contingencies not adequately encompassed by other controls. For security controls to be effective, they must be standardized for daily operations. When not followed precisely, there will be poor system control, resulting in operational difficulties and inconsistent corrective action. Most computer installations have the following security controls:

program control
equipment log
preventive maintenance

Program control Program control is a necessity after each program has become operational. Computer programs which are generally stored on magnetic tape are duplicated and at least one copy is retained in a safe storage area as a back-up tape for emergency conditions. Of equal importance is the possibility of deliberate or accidental alteration as the program is being run. All programs which are properly controlled are altered only with great difficulty.

Control procedures include making someone responsible for its safekeeping (usually a DP librarian), issuing it to only authorized personnel, and having the individual sign a log book. A computer-run book for each program should also be controlled, as well as instructions written for the console operator that are necessary to meet any contingency during the processing of the program. An organization's auditors should check the various programs on a surprise basis by using test decks of input data whose outcome have been predetermined. This approach should detect changes that have been made without proper authorization.

Equipment log A daily equipment log should be prepared so that computer operating time can be monitored. Some manufacturers have installed devices on the computer for recording time usage. This permits the auditor to check one source against another for falsification of computer time which may have been used for unauthorized program changes. Also, the auditor should check for any time discrepancies between the equipment log (when the run was completed) and the librarian record (when the program tape was returned); these discrepencies may indicate that extra computer time was taken to alter the program for future processing.

In conjunction with the equipment log, a copy of the console typewriter printout should be delivered daily to the internal auditor. If at all possible, the auditor should check and initial the first typewriter console printout and pick it up at the end of the day. One continuous printout should be available; if cuts appear in the printout, it may indicate unauthorized alterations were made during the day. Also, any unusual items found on the printout should be fully explained by the responsible console operator.

Preventive maintenance Preventive maintenance, although not a part of the hardware per se, is extremely important in minimizing equipment downtime. It consists of devoting a certain amount of time periodically, perhaps every other day, to routine hardware tests. Test problems run by the manufacturer's field maintenance staff are fed into the computer to check the machine's circuitry. Those system components which are beginning to show signs of weakness are replaced. This reduces the amount of lost production time and increases the reliability of the equipment during normal processing.

QUESTIONNAIRE FOR EVALUATING SYSTEM CONTROLS

A questionnaire for evaluating system controls in computer (batch and interactive) systems provides a basic framework for determining how effective controls are within the computer system. Extenuating circumstances usually exist in every system, creating the need for questions and investigations not covered in the normal questionnaire. The individual reviewing the system must exercise his or her judgment in this regard.

Whether or not additional questions are appended, the questionnaire is so designed that at least a minimum degree of control is maintained in a system.

SUMMARY: Design of System Controls

Input Controls Designed basically to control data before entry into a computer system.
1. Verification methods: key verification, visual verification, self-checking digit, and tabular listings
2. Input control totals: batch controls, hash totals, and record count totals
3. External labels

Computer-Programmed Controls Designed to control computer processing as well as uncover previously undetected errors.
1. Validation checks and tests: sequence checks, character mode tests, self-checking digit tests, limit checks, code validity tests, blank transmission tests, alteration tests, and checkpoints
2. Computer control totals: batch controls, hash totals, record count totals, balancing controls, cross-footing balance checks, zero balancing, and proof figures
3. Internal labels
4. Error routines

Data Base Controls Designed to ensure that data file items are not lost or destroyed.
1. Supervisory protection programs
2. File record controls
3. Physical safeguard controls

Output Controls Designed to ensure the reliability of output from a computer system.
1. Output control totals: batch controls, hash totals, and record count totals
2. Control by exception
3. Control over operator intervention

Interactive Controls Designed to facilitate a two-way directional flow of information.
1. On-line processing controls: message identification, message transmission control, and message parity check
2. Data protection controls: lockwords and authority lists
3. Diagnostic controls

Security Controls Designed to cover contingencies that are not adequately encompassed by the other controls.
1. Program control
2. Equipment log
3. Preventive maintenance

All questions in the questionnaire must be answered by checking the appropriate YES, NO, or Not Applicable. A YES answer indicates good control and a NO, poor control. (The same procedure is followed in a Management Audit Questionnaire). The questionnaire follows basically the same sequence as found in the preceding sections.*

CHAPTER SUMMARY:

Once the system analysts have established the level of accuracy, they must devise a system compatible with it. This means implementing adequate verification and input control procedures to obtain the degree of accuracy needed for data processing inputs; the same must be accomplished for computer processing, the data base, and output stages. The entire system design must embody the requirements for accuracy. The internal and external auditors can be of great assistance to system analysts in this important area. It is more economical to incorporate these concepts into the design than after the system is installed; otherwise, the procedures for safeguarding an organization's assets, providing reliable financial records, promoting operational efficiency, and maintaining adherence to managerial policies will be ineffective for meeting desired organization objectives.

Questions

1. How does the systems analyst determine the percentage accuracy to incorporate into a new system? Explain.
2. What are the principal causes of inaccuracies in a typical DP system? Explain.
3. a. Define internal control and explain its essential components.
   b. Who is interested in internal control and why?
4. Is there such a thing as too much internal control? Explain.
5. a. Distinguish between input controls and output controls.
   b. Distinguish between batch processing controls and interactive processing controls.
6. Of all the system controls given in the chapter, which group is the most important and why?
7. What effect do computer system controls have on the internal auditor and the external auditor?
8. Describe the checks and tests that can be built into a computer program.
9. How important is control over operator's intervention in a computer system?

---

*For a comprehensive questionnaire to evaluate system controls, see Robert J. Thierauf, **Data Processing for Business and Management** (New York: John Wiley & Sons, Inc., 1973), pp. 593–597.

Self-Study Exercise

True-False:

1. _____ Accuracy is increased to the point where the cost of the information is greater than its corresponding value.
2. _____ Another name for internal control is internal admistrative control.
3. _____ Internal check is concerned with safeguarding an organization's assets.
4. _____ Control total techniques are an essential part of internal accounting control.
5. _____ There are few, if any, design principles that are applicable to system controls.
6. _____ Output controls are designed to control computer processing as well as uncover previously undetected errors.
7. _____ Control by exception is normally an important part of output controls.
8. _____ Interactive controls are an essential part of batch processing controls.
9. _____ Log of computer time is a part of security controls.
10. _____ There is no need for a computer programmer to have a comprehensive knowledge of system controls.

Fill-In:

1. The need for 100-percent accuracy will be found in certain areas of _____, such as payroll and accounts payable.
2. The reason for costs growing at a faster rate as accuracy increases is that more _____ _____ must be set up for continuous checking of data.
3. Internal _____ _____ checks the accuracy and reliability of data processed by the system.
4. A _____ _____ questionnaire is a tool determining managerial weaknesses when evaluating internal administrative control.
5. The Principle of the _____ of Controls refers to the fact that controls should suit the needs of management.
6. The Principle of the _____ of Controls refers to the fact that controls should be consistent in accomplishing the tasks for which they were designed.
7. The most widely used form of verification methods is some type of _____ verification.
8. _____ controls refer to accumulating data into feasible groups for processing.
9. _____ control steps should be included as a part of the computer's internally stored instructions.
10. _____ _____ controls are necessary to keep from losing or destroying on-line and off-line file data.

Problems:

The American Company is audited by a national CPA firm. When the company was using an integrated information system (batch processing), the outside auditors criticized its handling of inventory and accounts receivable. Now that the company has implemented an interactive information system, the vice-president of information systems is wondering what problems may be incurred. The problems below center on interactive controls for inventory and accounts receivable.

1. Under the old system, the problems encountered by The American Company in the inventory area were considered substantial by the outside auditors. In the past, many of the physical inventory counts differed substantially from the perpetual figures. With an interactive information system, most anyone can alter (add or subtract) the inventory data since numerous on-line CRT terminals are available for use; even if entry were restricted to certain personnel, the vice-president is wondering about routine errors that would never be caught, since they are handled directly by the computer and not by company personnel.

   In view of these difficulties, the vice-president of information systems has requested that you supply answers to the inventory problem and, in particular, the type of controls necessary to keep inventory errors to a minimum.

2. The problems encountered by The American Company in the accounts receivable area were also considered to be material, based upon a review by the outside auditors under the old system. In the past, many charges, payments, and adjustments were posted to the wrong accounts, although the detail of the accounts agreed in total. With an interactive information system, more company personnel will be working with the on-line accounts receivable file. The vice-president of information systems is wondering whether the number of errors will increase under the changed conditions.

   The vice-president of information systems has requested that you supply answers to this problem. Specifically, she wants controls that will minimize accounts receivable problems and incorrect postings to the accounts receivable file.

chapter

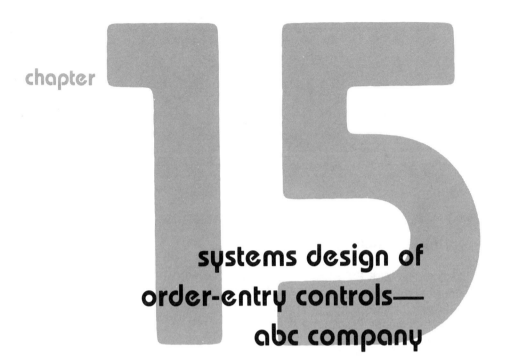

systems design of order-entry controls— abc company

OBJECTIVES:

- To complete the design of the new order-entry system for the ABC Company.
- To apply the underlying design principles of system controls to the new order-entry system.
- To define the order-entry controls that comprise effective internal control.
- To design order-entry controls for the new order-entry system.

IN THIS CHAPTER:

DESIGN OF ORDER-ENTRY CONTROLS—ABC COMPANY
   Incorporate Underlying Principles of Order-Entry Controls
   Design Input Controls
   Design Computer-Programmed Controls
   Design Data Base Controls
   Design Output Controls
   Design Interactive Controls
   Design Security Controls
   Overview—Design of Order-Entry Controls—ABC Company
CHAPTER SUMMARY
QUESTIONS
SELF-STUDY EXERCISE
CASE STUDY 1—FINISHED PRODUCT INVENTORY
CASE STUDY 2—ACCOUNTS RECEIVABLE

The previous chapter presented the basic design principles and concepts associated with the development of system controls. This chapter will further develop the design of the ABC Company's order-entry system by focusing on its control aspects. Initially, the underlying principles of system controls, as they apply to the new system, will be presented. Next, practical applications of these principles will be illustrated. Finally, the controls for the ABC Company's order-entry system will be specified; these include the input, computer programmed, data base, output, interactive, and security controls.

Just as in the prior chapters on the ABC Company, additional information will be presented about the finished product inventory and accounts receivable systems. This background will prepare students to complete the design of their system controls. These last assignments complete the text's case studies.

DESIGN OF ORDER-ENTRY CONTROLS—ABC COMPANY

In order for any new system, like the order-entry system of the ABC Company, to be complete, it is necessary to incorporate system controls. As explained in the previous chapter, this means provision for internal control, that is, strategic control points where data are compared for accuracy from one point to another (internal accounting control) and distribution of control so that no one person has complete control over processing (internal check). In essence, there is need to incorporate system controls that relate input through methods and procedures to the data base and output for the new order-entry system.

The outline below for designing order-entry controls for the ABC Company is comparable to that in the prior chapter. However, a beginning section has been added to highlight the important principles underlying the new system controls. The design sections to be presented, then, are:
1. Incorporate underlying principles of order-entry controls.
2. Design input controls.
3. Design computer-programmed controls.
4. Design data base controls.
5. Design output controls.

6. Design interactive controls.
7. Design security controls.

As with earlier chapters, an overview section summarizes these design sections.

Incorporate Underlying Principles of Order-Entry Controls

For the new system, one of the basic objectives is to reduce the amount of orders incorrectly billed from 7 to 1 percent. Since the initial source of information for billing is the order-entry system, it is essential that the system be designed with adequate controls to insure a high degree of accuracy. The controls must further insure that the system operates in the manner intended by management. In order to design the system controls to meet these goals, the MIS task force agreed to follow the principles set forth in the prior chapter.

Principle of the acceptability of controls The edits, checks, and balances of the new order-entry system must be carefully designed in such a way that they are useful to the order-entry clerk and others who use the system. The system controls contribute to increasing the accuracy and integrity of the system without over-complicating an individual's ability to use it. Order-entry personnel must have an opportunity to review the system controls as well as provide their own ideas during the design process.

Principle of establishing strategic control points Key control points for monitoring order-entry input and subsequent processing must be established. These control points must be comprehensive, covering every major operation that can be measured. Otherwise, data values at one point in the order-entry system will not agree with the same values found in other related systems.

Principle of control accountability The MIS task force will design the system controls so that, at each strategic control point, there is an individual responsible for insuring that the controls are used correctly. This will help to insure that the controls operate in the manner intended. No "short-cuts" around the system are taken.

Principle of management by exception During the design, implementation, and early operational stages of the system, operational standards will be set. Actual system performance can be measured against these standards to determine if the system is operating as expected. For example, a typical standard might specify invoicing cost, determined by dividing total system operating cost by the number of invoices produced. Significant violation of the standards is a signal to management to undertake appropriate corrective action.

Principle of the flexibility of controls The new order-entry system controls must be flexible enough to allow a high degree of accuracy and system integrity, even if some of the basic system operational parameters change. Parameter changes could be an increase in the number of customers or the volume of invoices, the creation of new products, or a changeover in the order-entry personnel. In effect, the new system must be adaptable to changing conditions.

Principle of the suitability of controls The controls must serve the needs of the company by providing effective control to meet the accuracy goal of a one-percent error rate in billing and to insure system integrity. From this view, accurate and complete information will be valuable in accomplishing order-entry tasks.

Principle of the reliability of controls System controls must be highly dependable and remain capable of insuring the desired degree of accuracy and system integrity. Over a period of time, the controls must prove worthy of management's confidence. In this manner, they will accomplish the order-entry tasks for which they were designed.

Design Input Controls

Control of input is essential to the successful operation of the new order-entry system. Figure 15.1 depicts the primary components of the order-entry system associated with

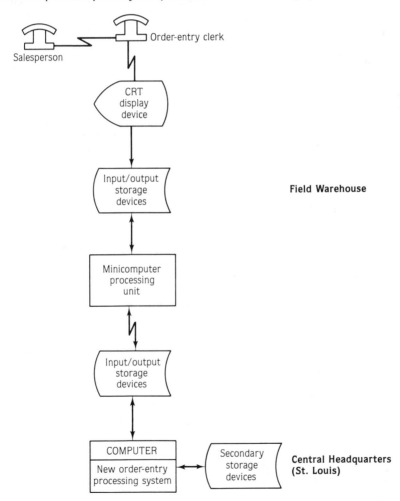

figure 15.1 The primary components of the new order-entry system which are associated with the processing of input for the ABC Company.

the processing of input. For effective control of system input, the most important step is the receipt of the initial order information from the salesperson. The order-entry clerk must communicate clearly with the salesperson in order to obtain accurate information required to process an order. While the salesperson is still on the telephone, the order-entry clerk will enter information about the order via the CRT display device. The clerk will visually inspect the entered data and verify with the salesperson certain key fields, such as customer account number, desired shipment date, and total units to be shipped.

The capabilities of the intelligent CRT terminals to edit input data will assist the order-entry clerk in detecting input errors, such as entering numeric data into a field defined to contain alphabetic data only (or vice versa), invalid customer account numbers, invalid product codes, and invalid dates. Comparable input edit controls can be applied to the shipment processing input. The shipping clerk is the key individual in insuring that accurate and complete information is captured for each shipment. Any questions about exactly what was loaded on any shipment must be referred to the loader responsible for that shipment. The intelligent CRT terminal can perform many edits on the basic shipment input to help guarantee its accuracy and completeness.

From this, it is obvious that the new system minimizes the number of people responsible for correct entry of order data. Also, it provides direct communication between these two people—the salesperson and the order-entry clerk—and uses intelligent CRT terminals to capture the data.

Design Computer-Programmed Controls

A number of computer programmed controls can be designed for the new order-entry system. As demonstrated below, they will focus on order-entry input and shipment processing input. Also illustrated are examples of incorrect entry items for both types of input.

Order-entry input edits The edits to be performed for each information element on each order are summarized in table 15.1 (See also table 13.1.) All these edits can be programmed and performed by the intelligent CRT terminal. For the order-entry system, each field found to be in error will be indicated by an asterisk to the right of the field.

The order numbers must be assigned in ascending sequence. The intelligent CRT terminal can store the last order number used and automatically display the next order number in the sequence. To avoid duplication of order numbers, Atlanta order numbers will be assigned consecutively beginning with 000001 and ending with 150000; Cincinnati orders will range from 150001 to 300000; Denver order numbers, from 300001 to 450000; New York, from 450001 to 600000; and San Francisco order numbers will be over 600000.

The order date will be checked by the intelligent CRT terminal using computer-programmed controls, as follows:

table 15.1 Computer-programmed control edits for the order-entry input of the ABC Company.

Informational element	Type of field	Type of edit
Order number	Numeric	Order number assigned by intelligent CRT terminal, that is, consecutive numbers in ascending sequence.
Order date	Numeric	Must be a valid date.
Field warehouse code	Alphabetic	Field warehouse code must be valid; current codes are **ATLA** (Atlanta), **CINC** (Cincinnati), **DENV** (Denver), **NEWY** (New York), and **SANF** (San Francisco).
Customer account number	Numeric	Last digit is a check digit.
Desired shipment date	Numeric	Must be a valid date; must be later than the order date.
Product code	Numeric	Must be a valid code present in the product code file.
Quantity	Numeric	Must be greater than zero and less than 5000.

1. month—between 01 and 12
2. day—less than 32 for months 01, 03, 05, 07, 08, 10, and 12
   —less than 31 for months 04, 06, 09, and 11
   —less than 30 for month 02 in leap years only
   —less than 29 for month 02, not in leap year
3. year—equal to or greater than the current year; last two digits only

The field warehouse code will be assigned to the intelligent CRT terminal based on its location. Thus, ATLA (Atlanta), CINC (Cincinnati), DENV (Denver), NEWY (New York), and SANF (San Francisco) will be automatically displayed as part of the order input screen format. The order-entry clerk may override the displayed code due to equipment problems at another location. However, if the warehouse code is overriden, computer program controls will only accept ATLA, CINC, DENV, NEWY, or SANF.

The seven-digit customer account number will consist of six digits plus one check digit to insure the account number is valid. The check digit will be constructed using modulus eleven with prime number weighting logic. The most significant digit is multiplied by seventeen; the second most significant digit is multiplied by thirteen; the third, by seven; the fourth, by five; the fifth, by three; and the least significant digit is multiplied by one. The products are then added, and the sum is subtracted from the next higher multiple of eleven. The remainder is the check digit. For example, if the first six digits are 3 2 5 6 8 4, then the check digit is determined as follows:

$$\begin{array}{cccccc}
3 & 2 & 5 & 6 & 8 & 4 \\
\times 17 & \times 13 & \times 7 & \times 5 & \times 3 & \times 1 \\
\hline
51 & 26 & 35 & 30 & 24 & 4 \\
51 & +26 & +35 & +30 & +24 & +4 & = 170
\end{array}$$

The next multiple of 11 higher than 170 is 176 (= 11 × 16): 176 − 170 = 6

Hence, the check digit is 6, and the customer account number is 3256846.

This logic will insure that transposition and other common input errors will be detected. The order-entry clerk must enter the entire seven-digit number. The intelligent CRT terminal will then calculate the check digit based on the first six digits. If the calculated check digit agrees with the seventh digit entered by the order-entry clerk, it is virtually certain that the customer account number was entered correctly.

The desired shipment date will be checked in a manner similar to the order date. Additionally, the desired shipment date must be a later date than the order date.

The product code entered by the order-entry clerk will be searched in the product code file by the intelligent CRT terminal. If not found, the product code will be flagged with an asterisk as invalid.

The last item for editing, as noted in table 15.1, is the product quantity ordered. It must be in the range of 1 to 5000 units or it will be flagged as invalid.

Incorrect order-entry input A sample of incorrect order-entry input is shown in figure 15.2. The asterisks to the right of the various fields indicate that those particular data items failed to pass one or more of the computer-programmed controls discussed above. Incorrect order-entry input includes the following:

  customer account number—the check digit is incorrect.
  desired shipment date—shipment date is earlier than order date.
  product code 0143—not found in product code file.
  quantity 5001—greater than 5000.

It is important to note that all fields in error were detected and flagged as invalid. The edit process did not stop when the first invalid field was found, since processing inefficiencies for orders with multiple input errors would result. Also, the total cost and total weight of the order cannot be determined due to input errors in the product code and quantity fields. An error message, then, is displayed informing the order-entry clerk that the input is invalid and must be corrected before the order can pass the edits of the intelligent CRT terminal and be stored for transmission to St. Louis that evening.

Shipment processing input edits Accurate and complete shipment information, like order-entry information, is extremely important since this input provides the basis for preparing the customer invoice. Table 15.2 lists the edits to be performed for each informational element on a shipment. (See also table 13.2.) The existence of the open order is checked on the open order file; all the rest of these edits can be programmed on the intelligent CRT terminal.

When shipment data are transmitted to St. Louis for further processing, one of the first checks performed by the central computer is a verification that there is an order on the open order file for this shipment information: the open order file is searched for an order with this order number, order date, and field warehouse code. If it is found, processing continues; otherwise, the shipment information is rejected as invalid.

Carrier name and carrier trailer number cannot be edited since both alphabetic and numeric data could appear. The only check made on these fields at shipment processing time is to insure that they are not left blank.

If there has been no change in the products and quantities ordered, the product code and quantity information does not have to be entered. The order-entry system will accept the

```
01  ORDER # 340126                           ABC COMPANY = ORDER ENTRY INPUT
02
03                                                                                   WHSE = ANE
04  ACCOUNT # 3256847 *
05
06
07
08
09  ORDER DATE 09/30/3=    DESIRED SHIP DATE 10/01/8=*
10
11  PRODUCT     QUANTITY            DESCRIPTION                 UNITS       PRICE
12   0143 *        20
13   1603         5001 *       MAGIC POWER = 5#               BAGS         7.89
14   0587          5          DIRT ALERT                      CASES       24.95
15
...
22    TOTAL COST = ***              TOTAL WEIGHT = ***
23
24  ORDER CANNOT BE SUBMITTED DUE TO INVALID INPUT
```

figure 15.2 The results of incorrect order-entry input for the ABC Company.

427

table 15.2 Computer-programmed control edits for the shipment processing input of the ABC Company.

Informational element	Type of field	Type of edit
Order number	Numeric	Order number must be found on the open order file.
Order date	Numeric	Must be a valid date.
Field warehouse code	Alphabetic	Field warehouse code must be valid; current codes are **ATLA** (Atlanta), **CINC** (Cincinnati), **DENV** (Denver), **NEWY** (New York), and **SANF** (San Francisco).
Shipment number	Numeric	Shipment numbers are assigned consecutive numbers in ascending sequence.
Shipment date	Numeric	Must be a valid date: must be later than the order date.
Shipping terms indicator	Alphabetic	Only **P** and **C** are acceptable.
Carrier name	Unedited	Carrier name field must not be left blank.
Trailer number	Unedited	Trailer number field must not be left blank.
Product code	Numeric	Must be a valid product code present in product code file; leave blank if order shipped is exactly as originally ordered.
Quantity	Numeric	Must be greater than zero and less than 5000; leave blank if order shipped is exactly as originally ordered.
Customer exception code	Alphabetic	Only **Q, P, A, O,** and **C** are acceptable.

shipment input as the order shipped, that is, as originally ordered. If one product code or one quantity is different than originally specified, all product codes and quantities must be entered exactly as shipped. The other data fields for shipment processing are entered in a manner very similar to the order-entry input.

Incorrect shipment processing input An example of invalid shipment processing input is illustrated in figure 15.3. Specifically, the incorrect shipment fields are as follows:
   shipment date—shipment date is invalid, because no month has 32 days.
   shipping terms indicator—must be P for prepaid or C for collect.
   carrier name—this field was left blank.
   product code 0143—not found in product code file.
   quantity A3—not all numeric.

It should be noted that all fields in error were detected and flagged in one processing step to avoid a separate processing step to detect each error. Total cost and total weight were not calculated because of the input errors in quantity and product code. Also, an error message is displayed to alert the shipping clerk that the input is invalid.

Design Data Base Controls

Data base controls are required to insure that files are not lost or destroyed. The "grandparent-parent-child" approach of saving three days of key system files—open order, shipment, customer name and address, and product code files—will be used. Additionally, each field warehouse will keep three days of order and shipment input

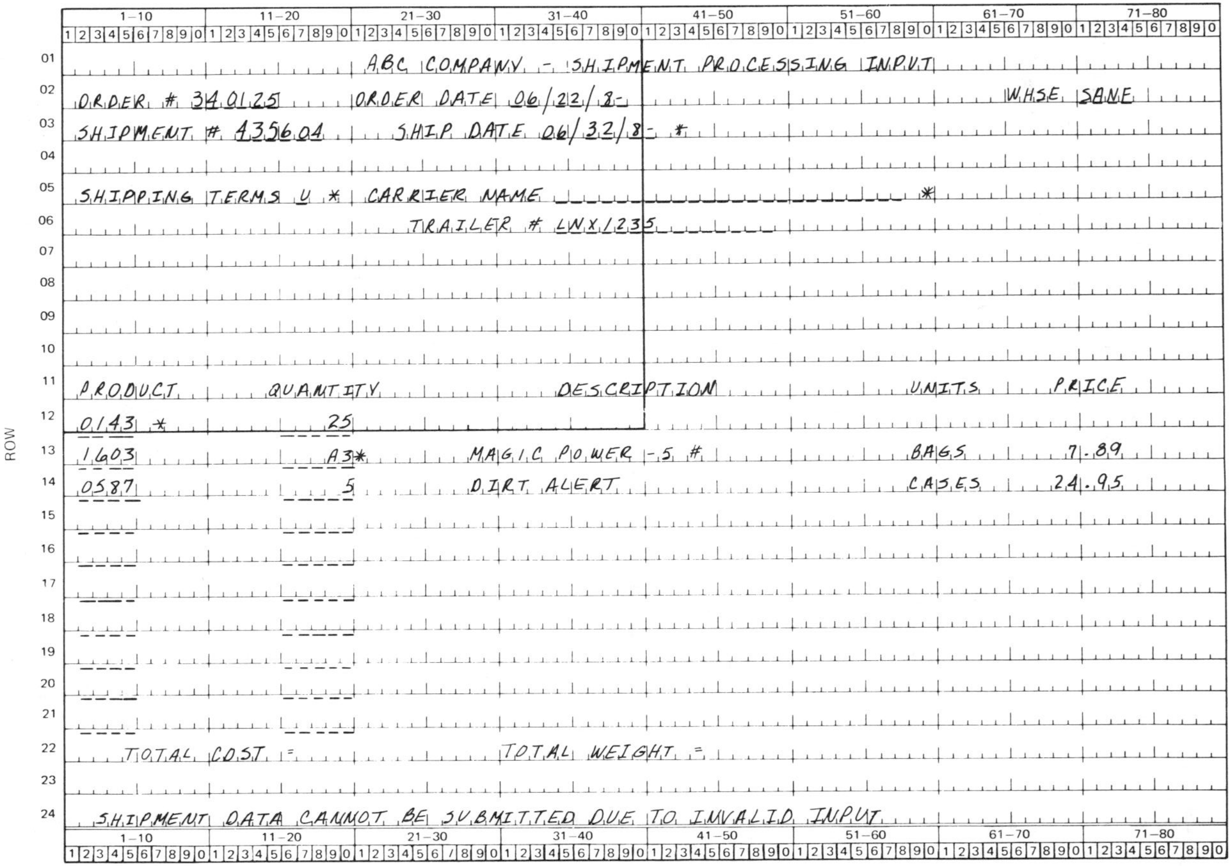

```
ROW
01  ORDER # 34.0125    ABC COMPANY - SHIPMENT PROCESSING INPUT
02                     ORDER DATE 06/22/8-                       WHSE SOME
03  SHIPMENT # 435604  SHIP DATE 06/32/8- *
04
05  SHIPPING TERMS U * CARRIER NAME _____*
06                     TRAILER # LNX1235
07
08
09
10
11  PRODUCT    QUANTITY      DESCRIPTION              UNITS     PRICE
12  0143 *     25
13  1603       A3*           MAGIC POWER -5 #         BAGS      7.89
14  0587       5             DIRT ALERT               CASES     24.95
15
16
17
18
19
20
21
22  TOTAL COST =                  TOTAL WEIGHT =
23
24  SHIPMENT DATA CANNOT BE SUBMITTED DUE TO INVALID INPUT
```

figure 15.3 The results of incorrect shipment processing input for the ABC Company.

available for reprocessing of transactions against a previous generation of the master file.

Design Output Controls

Control of system output is directed toward insuring that the processing results are accurate, complete, and reliable. Output controls must be designed for each connection between major components of the order-entry system. In figure 15.4, primary components of the new system associated with the handling of output are illustrated.

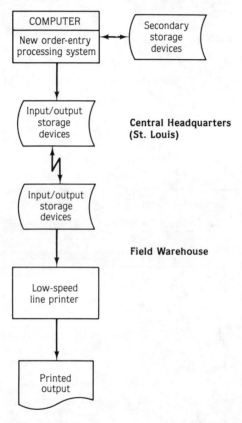

figure 15.4 The primary components of the order-entry system which are associated with the handling of output for the ABC Company.

Controls are required to insure that output transmission from St. Louis to each of the field warehouses is complete. Output record count of the number of order and shipment processing papers transmitted from St. Louis to each field warehouse is used as an output control mechanism. An output control total will be transmitted from St. Louis to each field warehouse. The printer operator at each field warehouse can verify the control total against the number of printed order and shipment processing forms to insure that none are lost. Causes of lost data are transmission problems and equipment problems at St. Louis or the field warehouse. If the control total does not check with the printed output, then the field warehouse operator can contact St. Louis and request retransmission of the day's output.

Some errors in order and shipment input not caught by the intelligent CRT terminal will be detected at St. Louis. These exceptions will be highlighted on an error report that is transmitted to and printed at the appropriate field warehouse. Responsible personnel at the field warehouse can then follow up on these exceptions.

Design Interactive Controls

The interactive controls for the new order-entry system are designed to meet the primary objective of insuring that system data are accessed only by authorized personnel. Each field warehouse will have two terminals; one for order-entry input and the other for shipment processing input. The terminals will be locked and inoperable without a key device. Once the terminal is turned on, the user must enter a password to display the proper input formats. There will be a unique password for each field warehouse and also a different password for the two inputs. The passwords will be changed every three months.

Design Security Controls

Security controls are required to guard the system and its data from unauthorized access or from use in a way not originally intended. The open order, shipment, customer name and address, and product code files will incorporate password protection: no one will be able to access any of these files unless they specify the unique password in the control deck used to request processing of the file. Furthermore, the second generation of each file (that is, the second most current version) will be transported daily to a fire-proof vault located across town from the main computer location in St. Louis.

Source listings of all programs will be kept in a locked room maintained by a DP librarian, responsible for controlling unauthorized access and insuring that all listings are kept current.

Overview—Design of Order-Entry Controls—ABC Company

The main thrust of the order-entry system controls for the ABC Company has been the editing capabilities of the intelligent CRT terminals. In particular, the focus has been on **computer-programmed controls** of two types.

1. The **order-entry edits** detect incorrect order input entered by the order-entry clerk. Informational elements edited include order number, order date, field warehouse code, customer account number, desired shipment date, product code, and quantity. An asterisk to the right of each indicates the data field failed to pass one or more of the above computer programmed controls.
2. The **shipment processing edits** detect inaccurate shipment input entered by the shipping clerk. Informational elements subject to the editing process are order

number, order date, field warehouse code, shipment number, shipment date, shipping terms indicator, carrier name, trailer number, product code, quantity, and customer exception code. As with order-entry inputs, an asterisk indicates that the data field is invalid.

These computer-programmed controls are compatible with the other system controls enumerated in this chapter. Further, all of the system controls in the chapter tie in with the inputs, methods, procedures, data files, and outputs of prior chapters. Thus, the design of system controls has been completed for the new order-entry word system of the ABC Company.

SUMMARY: Design of Order-Entry Controls—ABC Company

Incorporate Underlying Principles of Order-Entry Controls Focuses on incorporating essential design principles that make order-entry system controls effective for day-to-day operations.

Design Input Controls Allows the order-entry CRT operator to detect input errors before they enter the system. Similarly, input controls insure that invalid shipment processing data are detected before entry into the system.

Design Computer-Programmed Controls Relates to performing certain edits of input data before being accepted by the minicomputer for storage at each field warehouse. The types of edits are found in tables 15.1 and 15.2 for order-entry input and shipment processing input, respectively.

Design Data Base Controls Refers to utilizing the "grandparent-parent-child" approach of saving three days of open order, shipment, customer name and address, and product code files.

Design Output Controls Centers on requiring that output transmission both to and from central headquarters in St. Louis utilize record count totals to insure that data are not lost in transmission.

Design Interactive Controls Refers to the use of locks and passwords to allow access to the two terminals at each field warehouse.

Design Security Controls Allows only authorized users to reference the open order, shipment, customer name and address, and product code files by employing passwords. Also, security will be maintained over order-entry programs.

CHAPTER SUMMARY:

For the ABC Company, basic principles have been applied to designing controls for the new order-entry system. Fundamentally, the system controls must achieve a high degree of accuracy, insure that the system operates in the manner intended, and safeguard the system and its data. The use of an intelligent CRT terminal with computer programmed controls that can edit data (including range checks, check digits, record count totals, and table searches) will greatly improve the order and shipment data accuracy. Multigeneration data sets, use of password-protected data sets, cycle control of data sets to an off-site location, key-lock terminals requiring passwords for access, and controlling access to program source listings will help to safeguard the system and its data. The inclusion of the system controls will insure that the new order-entry system of the ABC Company has been designed as originally intended in the feasibility study.

Questions

1. What are the basic objectives to be met by the controls of the new order-entry system of the ABC Company?
2. How is duplication of numbers avoided in assigning order numbers at the five field warehouses?
3. a. How is the check digit for the customer account number determined?
   b. How is this check digit used?
4. What controls are used to insure that the output transmission from St. Louis to each field warehouse is complete?
5. How are multigeneration data sets and off-site cycling of data sets combined to protect the system data?
6. Relate the importance of order-entry system controls to the design of order-entry inputs, methods, procedures, data files, and outputs for the ABC Company.

Self-Study Exercise

True-False:

1. _____ The Principle of Establishing Strategic Control Points relates to monitoring order-entry data at key control points for the ABC Company.
2. _____ The Principle of the Suitability of Controls refers to maintaining order-entry accuracy and system integrity for the ABC Company.
3. _____ The type of editing for the order date relates to testing for a valid date.
4. _____ The type of editing for quantity refers to determining if the quantity is greater than zero and less than 5000.
5. _____ The editing of shipment processing data is not undertaken in the new order-entry system.
6. _____ The first digit of the customer account number is the check digit.
7. _____ When shipment processing data are transmitted to St. Louis, an initial check is made to verify that there is an order on the open order file.
8. _____ Data base controls are not necessary in the new order-entry system.
9. _____ Output controls are identical to the input controls for the new order-entry system.
10. _____ Security controls refer only to the four data files of the new order-entry system.

Systems Design of Order-Entry Controls—ABC Company

Fill-In:

1. The Principle of _____ _____ in the new order-entry system of the ABC Company refers to making someone responsible at each strategic control point.
2. The Principle of _____ _____ _____ in the new order-entry system of the ABC Company refers to measuring actual system performance against established performance standards.
3. The last digit of the customer account number is a _____ digit.
4. The _____ of the product code refers to determining if it is present in the product code file.
5. Many of the edits for order-entry input are the same as for the _____ _____ input.
6. The only acceptable codes for the _____ _____ indicator are P or C.
7. The "grandparent-parent-child" approach is found in the design of the _____ _____ controls for the new order-entry system.
8. Output _____ _____ totals in the new order-entry system are employed as a part of output controls.
9. The primary objective of _____ _____ is to insure that system data are accessed only by authorized company personnel.
10. Security controls include keeping all _____ _____ of order-entry computer programs in a locked room maintained by a DP librarian.

chapter 15   case study

finished product inventory

In the new finished product inventory system, responsibility for all finished goods inventory will be centralized under a single group reporting to the vice-president of physical distribution. The group will have specific responsibility for a periodic review of current additions and deletions to the finished product inventory. The purpose of such a review is to check on the validity of these inventory changes, especially deletions since the ABC Company's products can be sold easily on the outside.

For additional information on the finished product inventory system, reference can be made to case study 1 of chapters 10 and 13.

Assignment: Design of Finished Product Inventory System Controls

1. List the basic reasons for product return shipments from customers and develop a code for each reason.
2. Develop a table (similar to table 15.2) for both input screen formats—additions and deletions—of the finished product inventory system.

case study

chapter 15

accounts receivable

As with the prior system, the accounts receivable section will be responsible for applying each remittance to specific billed items on the appropriate customer's account so that the discount earned or forfeited can be determined. Based upon this procedure, the internal auditor will be responsible for the accuracy of these postings. Thus, this individual will check on the validity of these postings on a weekly basis in order to maintain good customer relations.

For additional information on the accounts receivable system, reference can be made to case study 2 of chapters 10 and 13.

Assignment: Design of Accounts Receivable System Controls

1. Develop a table (similar to table 15.2) for the accounts receivable cash remittance screen developed in case study 2 of chapter 13.
2. Identify additional specific controls that should be applied to the processing of customer cash remittances.

part four

beyond
systems analysis and design

chapter

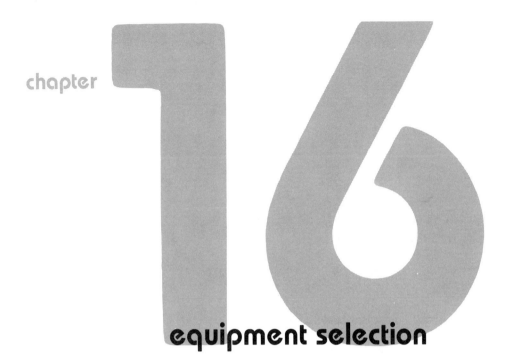

equipment selection

OBJECTIVES:

- To set forth the proper approach for the equipment selection process.
- To enumerate the essential elements for submitting bid invitations to equipment manufacturers.
- To examine the important items for evaluating proposals submitted by equipment manufacturers.
- To explore the evaluation methods for selecting the equipment manufacturer(s).

IN THIS CHAPTER:

   Equipment Specifications
EQUIPMENT SELECTION PROCESS
   Approaches to Equipment Selection
   Determine Equipment Manufacturers
   Submit Bid Invitations to Manufacturers
   Evaluate Manufacturers' Proposals
   Select Equipment Manufacturer(s)
CHAPTER SUMMARY
QUESTIONS
SELF-STUDY EXERCISE

Once the first three steps of the feasibility study—introductory investigation, systems analysis, and systems design—are completed, the MIS task force is ready to undertake the last major step: equipment selection. This usually takes 10 percent of the total time for a system project. The final selection of the most suitable equipment may require some system modifications. In effect, the original exploratory survey report may have to be modified as a result of the findings of this step; these modifications may be major, but generally they are not. The rationale for including equipment selection within the framework of the feasibility study is that the acceptance or nonacceptance of a system project is not completely established until the order(s) is (are) placed with the respective equipment manufacturer(s).

Initially, this chapter presents the important equipment specifications. After exploring the basic approaches to equipment selection, a recommended approach is set forth. The remainder of the chapter centers on the equipment selection process: determining equipment manufacturers, submitting bid invitations to manufacturers, evaluating manufacturers' proposals, and selecting equipment manufacturer(s). Included in this process is the signing of the equipment contract by a top-level executive.

Equipment Specifications

Up to this point, the MIS feasibility study effort has been directed toward the analysis of the present system and the detailed design of the new system. Requirements for the system have already been determined; by way of review, they are: new policies consistent with organization objectives, output needs, a data base structure (on-line and off-line files), new methods and procedures, planned inputs, a common data representation, internal control (including computer system controls), and new system performance specifications. These requirements formed the foundation for designing the new system on a modular subsystem basis that considers the human element.

The emphasis at this point in the feasibility study is to specify the equipment to be used in the newly designed system. Specifically, the systems analysts will state in general terms these hardware requirements, which include:

1. a communications network—refers to the type of communication channels that will be used to send data or request information from input/output terminals to the central computer.

2. terminal and control devices—relates to the wide range of terminal equipment that accept keyed or punched data as input to the computer and/or produce printed, displayed, punched, or audio messages as output from the computer.
3. auxiliary storage devices—centers on storing important data on line in auxiliary or secondary storage and less important data in other low-cost storage media.
4. central processing unit(s)—refers to the major source of computing capabilities whereby on-line input/output devices can communicate with the computer for solving a wide range of problems.
5. other CPU peripherals—includes card and OCR readers, printers, tape drives, disk units that are an integral part of a central processing unit and are necessary to complete a typical computer system.

As will be demonstrated below, these specifications are an integral part of the bid invitations to the various equipment manufacturers.*

EQUIPMENT SELECTION PROCESS

Equipment selection should be undertaken by the MIS task force upon the completion of the system design and equipment specification phases. It is accomplished in four phases.
1. Determine equipment manufacturers.
2. Submit bid invitations to manufacturers.
3. Evaluate manufacturers' proposals.
4. Select equipment manufacturer(s).

Approaches to Equipment Selection

Although there are two basic approaches to equipment selection, the **recommended** one is to submit the information compiled to date on the feasibility study, with each area of the system outlined, to each manufacturer. General information on the organization, its future processing plans, and list of new system specifications also should be included. The particulars of these specifications will be covered in subsequent sections.

An **alternative** approach is to request that equipment manufacturers bring in their own system personnel to study the present system and devise a new one tailored to their own equipment. This approach should be taken only when the organization has neither the experienced personnel nor the time to conduct their own feasibility study. The operations will be timed and cost savings will be calculated on this basis. Generally, different approaches by each equipment manufacturer are the result, thereby making a final evaluation virtually impossible when placed on a common basis. Most manufacturers will direct their proposals to highlight the specific features of their own equipment over competition.

---

*For a more detailed explanation of these specifications, see Robert J. Thierauf, **Systems Analysis and Design of Real-Time Management Information Systems** (Englewood Cliffs, N. J.: Prentice-Hall, Inc., 1975), chapter 5.

The problem of time is another important consideration since each manufacturer must conduct a lengthy system review. For example, five manufacturers are involved and each spends a month to review the present system. This means about six months of lost time and continual disruption to current operations. After department heads, supervisors, and operating personnel have been through the same set of questions five times, their attitude toward any type of new system, needless to say, is negative, and morale has reached an all-time low. Despite all the manufacturers' efforts, one month is still not ample time to learn a system in sufficient detail, especially to anticipate exceptions and problem areas. Because of the manufacturers' inabilities to gather all the pertinent facts in the time allotted, their recommendations can be poor, and many times, impractical systems are advocated. Thus, this second approach should generally be discarded.

Determine Equipment Manufacturers

Most organizations that are undertaking a system project have specific computer equipment under consideration based on the exploratory survey report. Since they have computer and related peripheral equipment salespersons calling on them at various times, there has been previous contact with most of the manufacturers.

Equipment representatives should be contacted and invited to an orientation meeting on the proposed system. During the course of the meeting, they should be instructed about the applications to be covered, general problems that will be encountered, approximate volumes (present and future), and other pertinent information. All manufacturers should indicate in writing whether they wish to receive a bid invitation. The reason for this approach should be obvious. There is no need to prepare a packet of specifications, flowcharts, decision tables, and comparable material for a manufacturer who has no interest in bidding.

Submit Bid Invitations to Manufacturers

Now that letters of intent to bid are on file from equipment manufacturers, an organization then submits bid invitations to interested equipment suppliers. When sending bid invitations, the preferred approach is to mail the same set of data to all competing manufacturers. This informs the manufacturers what requirements they must meet, keeps the number of questions to a minimum, and provides a fair basis for comparison of all responses. The manufacturers may need additional information and assistance from the prospective customer as they progress with the preparation of their proposals. Generally, one person from the MIS task force will perform this consultative function for each manufacturer.

Through this approach, the manufacturers should have ample information to familiarize themselves with an organization and its peculiar DP problems. The recommendations made in their proposals should show clearly how the equipment will meet the customer's needs. If the specifications lack clear definition from the beginning, the bid invitation will come back as proposals with standard approaches; in effect, all of the preliminary work will have been wasted. The equipment manufacturers cannot prepare proposals tailored specifically to a particular customer if the data contained in the bid invitation are deficient. It is of utmost importance that data submitted to manufacturers be as complete and self-explanatory as possible, especially in view of the complexities inherent in an interactive management information system.

Contents of bid invitation   Much of the material needed for the bid invitation can be taken directly from the data contained in the exploratory survey report and developed during the systems design and equipment specification phases. The contents of the bid invitation include

> general information about the company.
> future data processing plans.
> a list of new system specifications.
> new system flowcharts and decision tables.
> a list of equipment specifications.
> data to be forwarded by each manufacturer.

The detail for each major topic is shown in figure 16.1. Sections I and II should be brief so that attention can be focused on the remaining parts of the bid invitation. Data necessary for a thorough study are contained in Sections III, IV, and V; these form the basis for the manufacturer's proposal. Section III is composed of five essential parts: planned inputs, methods and procedures for handling data, data base (on- and off-line files) to be maintained, output needs, and other requirements and considerations of the new system. If proper documentation was undertaken for systems design, minimal time will be necessary to complete this section since much of the material can be used in its present form.

New system flowcharts and decision tables are part of Section IV. System flowcharts illustrate the methods and procedures of each functional area under study and also define the interrelationships among the areas. Decision tables will delineate the programming effort envisioned and help the manufacturer determine the hardware needed under the existing conditions. Lastly, this section of the bid invitation should contain a general flowchart depicting overall aspects of the new system to provide the equipment manufacturer with an overview of the system and its subsystems.

I. **General Information about the Company**
   A. Description of the company and its activities.
   B. Overview of present DP equipment and applications.
   C. Unusual DP exceptions and problems.
   D. Other important general information.
II. **Future Data Processing Plans**
   A. Listing of areas encompassed by the new system.
   B. Target date for installation of new system.
   C. Deadline date for submitting proposals.
   D. Equipment decision date by the company.
   E. Criteria to be employed in analyzing and comparing manufacturers' proposals.
III. **List of New System Specifications**
   A. Planned inputs:
      1. Where data originate within the system.
      2. Name and content of input data (such as documents and forms).
      3. Hourly rates of input data.
      4. Volume of inputs, including high and low points.
   B. Methods and procedures for handling data:
      1. Transmission of local and distant data from I/O terminal devices.
      2. Types of transactions handled.

---

figure 16.1   Contents of a bid invitation to an equipment manufacturer. (Robert J. Systems, © 1975. Reprinted by permission of Prentice-Hall, Inc., Englewood Cliffs,

3. Computations and logical decisions required.
4. New data generation within the system.
5. Control points to test accuracy of data and eliminate processing of fraudulent data.
C. Data base (on- and off-line files) to be maintained:
1. Where data are to be stored—on line in the data base and off line.
2. Names and contents of files to be maintained.
3. Methods and procedures for updating files.
4. Size of files to be maintained.
D. Output needs:
1. Names and content of output (such as reports and summaries).
2. Timely distribution of output data.
3. Hourly rates of output data.
4. Volume of outputs, including high and low points.
E. Other requirements and considerations:
1. Changes in policies to conform with new system.
2. Compatibility of common data processing language.
3. Limitations of the human factor.
4. Special internal control considerations, in particular, attention to interactive controls.
5. System performance—ability to handle the company's future growth, including peak loads.
6. Lease or cost of equipment not to exceed a stated figure.
7. Additional special requirements and considerations.

IV. **New System Flowcharts and Decision Tables**
A. Brief description of the system approach for each functional area under study.
B. System flowcharts and/or decision tables for each area.
C. System flowcharts that show the interrelationships of various areas for the new system.
D. A flowchart that depicts an overview of the new system.

V. **List of Equipment Specifications**
A. A communications network to handle the new system.
B. Various types of input and output terminal devices.
C. Auxiliary storage devices, such as magnetic disk, magnetic drum, and mass storage.
D. Specific requirements of the central processing unit, such as multiprocessing and multi-programming capabilities.
E. Other CPU peripherals, such as tape drives, disk units, card readers, and printers.

VI. **Data to be Forwarded by Each Manufacturer**
A. Processing time for each area on the equipment.
B. Proposed computer hardware:
1. Capabilities and technical features of basic and peripheral equipment; in particular, the data communications network, terminal devices, auxiliary storage devices, and the central processing unit.
2. Expansion capabilities (modularity).
3. Purchase price and monthly rental figures on a one-, two-, and three-shift basis for basic and peripheral equipment.
4. Alternative purchase and lease option plan (third-party leasing).
5. Estimated delivery and installation date.
6. Number of disk packs, magnetic tapes, etc. required and their costs.
7. Equipment cancellation terms.
C. Site preparation and installation requirements:
1. Amount of space needed.
2. Electrical power, air conditioning, and humidity control requirements.
3. Flooring requirements and enclosure of equipment.
D. Extent of manufacturer's assistance:
1. Cost of manufacturer's personnel to assist in the installation and for how long.
2. The availability and location of programming classes.
3. The possibility of on-site training classes.
4. Availability of higher-level programming languages, programming aids, and program libraries.
5. Nearest testing facilities and on what shifts.
6. Amount of equipment time for compiling and testing programs without charge.
E. Maintenance service to be provided.
F. Equipment support for emergency processing.
G. Other pertinent information.

Thierauf, **Systems Analysis and Design of Real-Time Management Information**
New Jersey.)

Section V contains a general listing of equipment specifications. Competing manufacturers must have a basic understanding of the data communications network, I/O terminal devices, auxiliary storage devices, the central processing unit, and other CPU peripherals contemplated. This section not only details presently owned equipment compatible with the new system, but also facilitates the comparison of bids from competing manufacturers.

Data to be included in each manufacturer's proposal are listed in Section VI. Specifying in advance what the proposals should contain ensures that comparable information for a final evaluation will be forthcoming.

Conferences with manufacturers Even though bid invitations specify numerous details of the new system, legitimate questions will be raised by the various equipment manufacturers. Many of the questions center on those areas which may need modification to take advantage of the equipment's special features. The result may be cost-saving benefits. Conferences between the manufacturer and the potential customer, then, can prove beneficial to both parties. However, caution is necessary since salespeople may use these conferences to sell the company and the final proposal. The final evaluation of the manufacturers' proposals should be objective, not subjective.

Evaluate Manufacturers' Proposals

The manufacturers should be given a reasonable amount of time to prepare their proposals. In most cases, approximately two to three months is adequate. When the proposals are completed, several copies are mailed to the customer for review, followed by an oral presentation by the manufacturer's representative. At this meeting, the salesperson will stress the important points of the proposal and answer questions. After this procedure has been followed for all competing manufacturers, the MIS task force is prepared to evaluate the information contained in the various proposals.

There are many criteria for evaluating a manufacturer's proposal. The important ones are listed below.

1. **Extent of automation proposed** This refers to the amount of newer DP equipment proposed for the new system. This criterion gives the study group an overview of what is being advocated by each equipment manufacturer.
2. **Throughput performance** This specifies the amount of data that can be processed by the computer system within a specific time period. In view of the difficulties in running computer programs—such as, malfunctions of the computer's components, paper jams, error stops, I/O units out of cards, among others—most experienced DP managers add 50 percent to the manufacturer's time estimates for processing computer programs. To check the validity of time estimates after the 50 percent adjustment, the **benchmark** test can be employed, which means selecting a representative job to be performed by the new system. The results of the benchmark test reveal how well the equipment meets the specific application in terms of total processing time, which is throughput performance.
3. **Type of equipment** The make, model number, and quantity of basic and peripheral equipment along with their capabilities, operating characteristics, and technical features are specified. Data on internal memory, operating speeds,

storage capacity, and hardware controls are a part of the manufacturer's proposals on computer and related equipment.

4. **Rental, purchase, or other basis** Various means of acquiring DP equipment—currently including rental, outright purchase, option to buy, and third-party leasing (lease-back arrangements)—are examined. The decision to purchase or lease can be based upon a number of factors, such as availability of capital funds, obsolescence of equipment, and the proposed usage.

5. **Delivery of equipment** A definite delivery date along with ample time to check out the equipment on the user's premises are specified. Delivery dates range from several weeks to two (or more) years. The delivery date is contingent upon the manufacturer, the type of hardware, and (of course) the order date.

6. **Installation requirements** This refers to the dimensions, weight, and necessary power and wiring requirements of each piece of equipment. Any false-flooring, under-flooring, or overhead wiring necessary to connect the computer units together is also explained. Most computers require air conditioning and humidity control.

7. **Manufacturer's assistance** This specifies the amount and type of assistance that can be expected from the equipment manufacturer. Assistance includes: programmers, systems analysts, and engineers to implement the new system; training schools for the client's managers, programmers, and operators; software packages to simplify programming; and equipment for program testing prior to installation. It should be noted that services for DP education and systems engineering are billed separately to the user.

8. **Availability and quality of software** This refers to the software which will support the hardware. Software includes: programming languages for business and scientific applications; program packages for reading and punching cards, reading and writing magnetic tapes, sort and merge routines, and others; compilers to assist in writing the final programs; executive routines to aid the computer operator during program debugging and to handle successive programs during production runs; monitor or dump routines for tracing the execution of program instructions during program testing; and other predetermined programs, such as linear programming and random number generators, for handling specialized problems. The availability of reliable software cuts the user's time and expense since the DP staff can work at a reasonably high level of programming efficiency.

9. **Maintenance contracts** The service necessary to maintain the equipment in good operating condition is usually provided free under a rental contract, but not when the equipment is purchased. For small equipment installations, the maintenance function will be performed from the manufacturer's sales office. For larger systems, the manufacturer may assign maintenance personnel to one or more systems.

10. **Other considerations** The manufacturer should also discuss the system's ability to deal with exceptions and unusual items, the nearness of other customers with compatible hardware to help in case of emergencies, the amount of computer overtime included free, and similar items.

Compliance with terms of bid invitations One last important question must be answered: How well has each manufacturer complied with the terms of the bid invitation? This involves completeness, clarity, and responsiveness. Does each proposal cover all points set forth in the bid invitation? Is the proposal clear in every respect? Are all estimates of time and cost for peak, medium, and low workloads accurate? Does the proposal reflect a proper understanding of the bid? Inadequate coverage of any of these points indicates a weakness that may be indicative of problems in the future.

Select Equipment Manufacturer(s)

Selection of equipment manufacturer(s) is a difficult job for the MIS study team. The selection process is much easier if the equipment proposed is identical for all practical purposes; in such cases, the choice is normally based on cost. However, this approach is rarely possible since most manufacturers have unique equipment features. For this reason, each manufacturer will propose a slightly different approach to meet the customer's needs. The MIS task force will need to employ evaluation techniques for comparing different approaches.

Methods of evaluation One method of evaluation utilizes a decision table as shown in table 16.1. A decision table for a final evaluation not only defines the important criteria in compact notation, but also permits an objective evaluation since the values have been determined before receipt of the manufacturers' proposals. In the illustration, the highest possible score is one hundred points for each of the five competing manufacturers. Ten points are deducted for each NO answer of a major criterion, and five points are subtracted for each NO answer of a minor criterion. The major criteria represent factors that have long-range effects on an organization in terms of profits and return on investment. Values for another organization might be different from those at this company. For the study currently being evaluated, this is a realistic approach to making a final decision.

Another evaluation method assigns different weighting factors to each criterion. Each manufacturer is given a score for each weighting factor; in most cases, the score is lower than the absolute value of the weighting factor. The values are totaled for all criteria, yielding the total points for each manufacturer. As with decision tables, the competitor with the highest score is selected.

Yet another method evaluates equipment superiority by performance per dollar. The task force must be careful to include all aspects of the machine's performance: various hardware speeds, reliability of the equipment, efficient software, and similar considerations. Otherwise, the scoring method will put too much emphasis on the machine's characteristics and not enough on the supporting parts, which may be just as important in the final analysis.

There are many factors to consider in equipment selection. In certain cases, some criteria are more important than others. Many of them are closely related to the hardware or software, whereas others encompass the operating environment of the new system; still other criteria revolve around the manufacturer or people. No matter what factors are

table 16.1 Criteria used in a feasibility study to select an equipment manufacturer. (Robert J. Thierauf, **Systems Analysis and Design of Real-Time Management Information Systems,** © 1975. Reprinted by permission of Prentice-Hall, Inc., Englewood Cliffs, New Jersey.)

DECISION TABLE

Table Name: CRITERIA TO SELECT EQUIPMENT MANUFACTURER
Date: February 25, 198_
Preparer: Robert J. Thierauf

	Rule number													
	1	2	3	4	5	6	7	8	9	10	11	12	13
Condition													
High degree of automation proposed	Y	Y	Y	N	Y								
✓ Low-cost throughput performance	Y	Y	Y	Y	N								
✓ Expandability of equipment	Y	Y	Y	N	Y								
✓ Reasonable monthly rental cost	N	Y	Y	Y	N								
Availability of equipment when needed	Y	Y	N	Y	Y								
Capable of meeting installation requirements	Y	Y	Y	Y	Y								
Available programming assistance	N	N	N	Y	N								
Good quality training offered	Y	Y	Y	Y	N								
✓ Dependable and efficient software	Y	Y	Y	N	Y								
Available equipment for compiling programs	N	Y	Y	Y	N								
Adequate equipment maintenance	Y	N	Y	Y	Y								
✓ Equipment backup in local area	N	Y	N	Y	Y								
✓ Availability of operating personnel	Y	Y	Y	Y	Y								
Compliance with terms of bid invitation	Y	Y	Y	N	Y								
Action													
Subtract 5 points for each no (N) answer	X	X	X	X	X								
Subtract an additional 5 points for each no (N) answer checked	X	—	X	X	X								

Other Information:
Competitor's total points: **1**—70, **2**—90, **3**—80, **4**—70, **5**—65

deemed critical in the evaluation process, the method must be objective for selection of one manufacturer (or more) over the others. The MIS study team will be better able to present its final recommendation to the executive steering committee if it was determined on a logical and methodical basis. In the final analysis, the purpose of spending so much time, effort, and expense on the feasibility study is to obtain the best DP equipment.

Signing of equipment contract(s) The signing of the equipment contract by a top-level executive, who has been the guiding force for both groups—the executive steering committee and the MIS task force—brings the MIS feasibility study to a formal close. The introductory investigation, systems analysis, systems design, and equipment selection, being the major steps of the feasibility study, represent approximately one-third of the total time spent on a system project. The period just ahead will require not only more time than the feasibility study, but also more involvement of the organization's resources, its operations and personnel. The problem of how to coordinate and control the activities during this interim period is a challenging task for even the most seasoned data processing manager and associates.

SUMMARY: Equipment Selection Process

Determine Equipment Manufacturers The process of determining which equipment manufacturers should receive bid invitations.

Submit Bid Invitations to Manufacturers The submission of bid invitations to selected equipment manufacturers should include these areas:
1. general information about the company: overview of the company and DP operations.
2. future data processing plans: areas to be implemented, completion dates, and the like.
3. list of new system specifications: planned inputs, methods and procedures for handling data, data base (on-line and off-line files) to be maintained, output needs, and other important items.
4. new system flowcharts and decision tables: applicable to each functional area to be implemented and an overview of the proposed system.
5. list of equipment specifications: a communications network, terminal and control devices, auxiliary storage devices, a central processing unit(s), and other CPU peripherals.
6. data to be forwarded by each manufacturer: computer hardware and processing times, site preparation requirements, manufacturer's assistance, maintenance, and other important information.

Evaluate Manufacturers' Proposals The criteria that can be utilized for evaluation purposes include:
1. extent of automation proposed
2. throughput performance
3. type of equipment
4. rental, purchase, or other basis
5. delivery of equipment
6. installation requirements
7. manufacturer's assistance
8. availability and quality of software
9. maintenance contracts
10. other considerations

Select Equipment Manufacturer(s) The selection process is relatively easy if the equipment proposed is identical. If the equipment proposals are somewhat dissimilar, the selection process can be difficult. Whether the proposals are somewhat alike or totally different, the dilemma can be resolved by utilizing a decision table or the weighting method. The manufacturer(s) scoring the highest number of points is (are) selected.

CHAPTER SUMMARY:

The final step of the feasibility study is the selection of equipment—in particular, a communications network, terminal devices, auxiliary storage devices, a central processing unit, and other CPU peripherals. Bid invitations describing the important aspects of the new system provide the basis for manufacturers' proposals. Proposals are then compared for selected criteria which have been assigned specific numerical values. All things being equal, the manufacturer who has the highest score is awarded the contract. The ultimate responsibility for determining equipment, however, is not the manufacturer's, but the user's.

Questions

1. Why is equipment selection the final phase of the feasibility study? Explain.
2. Of what importance are equipment specifications to the MIS task force?
3. Are there any problems associated with having various computer manufacturers draw up the specifications for an organization and then submit bids on this basis?
4. Of the various contents of a bid invitation, what is the most important for assisting the MIS task force in evaluating final equipment proposals?
5. What are the most important criteria for evaluating manufacturers' proposals?
6. a. What methods can be utilized to evaluate manufacturers' proposals?
   b. Is one method preferred over others?

Self-Study Exercise

True-False:

1. _____ The equipment selection process is the final phase of the feasibility study.
2. _____ Generally, the equipment selection process takes about 10 percent of the total system project time.
3. _____ There is generally no preferred approach to the selection of new DP equipment.
4. _____ The focus of the bid invitation to equipment manufacturers is on future data processing plans.
5. _____ Generally, planned input and output needs are not contained in a bid invitation to equipment manufacturers.
6. _____ An essential part of a bid invitation is a brief description of the system approach for each functional area under study.
7. _____ Data to be forwarded by manufacturers never specify the extent of their assistance throughout the system project.
8. _____ Throughput performance is an important criterion for evaluating a manufacturers' proposal.

452 Equipment Selection

9. _____ Compliance with the terms of bid invitations by equipment manufacturers is not very important from the company's viewpoint.
10. _____ Signing the equipment contract brings the feasibility study to a formal close.

Fill-In:

1. The next major step after systems design in a feasibility study is _____.
2. A list of new _____ _____ contained in a bid invitation is taken directly from the systems analysis and design phases.
3. The first item of a bid invitation to equipment manufacturers is _____ _____ about the company.
4. An essential part of a bid invitation to equipment manufacturers is the new _____ _____ and decision tables.
5. The last part of a bid invitation to equipment manufacturers contains the _____ to be forwarded by each manufacturer.
6. _____ _____ relates to the amount of data that can be processed by a computer system within a specific time period.
7. A logical starting point for evaluation of manufacturers' proposals is the extent of _____ proposed.
8. Often, equipment manufacturers are dropped from the evaluation process because of long _____ times.
9. One method of evaluating manufacturers' proposals is utilization of a _____ _____ which defines the important criteria in compact notation and permits an objective evaluation.
10. The introductory investigation, systems analysis, systems design, and equipment selection represent approximately _____ _____ of the total time expended on a system project.

Problem:

Before The American Company converted to the present interactive information system, it undertook an exhaustive study of current equipment available. After determining those equipment manufacturers who were interested in bidding, invitations were submitted to these vendors. Several methods were considered to evaluate the manufacturer's proposals. After a lengthy discussion, the executive committee and the feasibility study group decided to utilize a decision table to select the equipment manufacturer. The completion of this activity brought the feasibility study to a formal close for The Amercian Company.

The feasibility study group of The American Company has developed criteria for selecting the appropriate equipment manufacturer before mailing bid invitations. Each **major** criterion will be valued at 10 points and each **minor** criterion at 5 points. These criteria were reviewed and approved by the executive committee. Both committees have agreed upon the number of equipment manufacturers (four) to receive bid invitations. Based upon the four manufacturers' proposals for an interactive computer system, the feasibility study group has determined the following evaluation. An **X** mark below indicates that the competing firm has received a value of 10 or 5 points for the specific criterion listed.

Using the foregoing criteria, prepare a decision table that determines which equipment manufacturer should be selected to receive the computer equipment contract.

	Equipment Manufacturer			
Major criteria:	1	2	3	4
High degree of automation proposed	X	X	X	
Low-cost throughput performance	X		X	X
Low monthly rental cost	X		X	X
Programming assistance availability		X	X	X
Dependable and efficient software	X	X	X	X
Backup equipment availability		X	X	X
Good reputation for meeting commitments	X	X	X	X
Compliance with terms of bid invitation	X	X	X	
Minor criteria:				
Availability of equipment		X		X
Acceptable quality of training	X	X	X	X
Adequate equipment maintenance	X	X	X	X
Free programming assistance				X

DECISION TABLE

Table name: CRITERIA TO SELECT EQUIPMENT MANUFACTURER
Chart no.: FS-SEM-1
Prepared by:
Page 1 of 1
Date:

Rule number: 1 2 3 4 5 6 7 8 9 10 11 12

Condition

Action

Other Information:

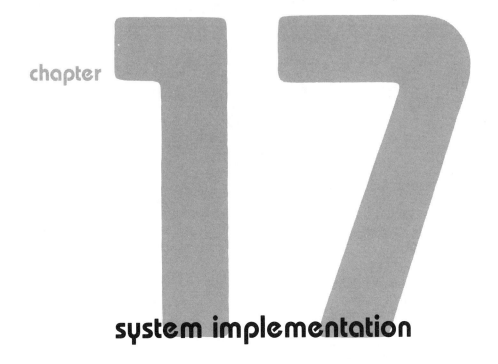

chapter 17
system implementation

OBJECTIVES:

- To enumerate the major steps involved in the implementation of a new system.
- To identify important human factors that help to build a strong implementation team.
- To discuss the activities necessary to implement a system.
- To emphasize the continuing need to review the operational aspects of a system.

IN THIS CHAPTER:

SYSTEM IMPLEMENTATION
   Preparatory Work for System Implementation
   Program Development
   Equipment Acquisition and Installation
   System Level Testing
   System Installation and Conversion
OPERATIONAL REVIEW OF SYSTEM
   Examination of New System Approaches
   Evaluation of New Equipment
   Benefit and Cost Analysis of System
CHAPTER SUMMARY
QUESTIONS
SELF-STUDY EXERCISE

Even though a substantial amount of time and cost was expended on the feasibility study (introductory investigation, systems analysis, systems design, and equipment selection), considerably more organizational resources will be spent on system implementation. In reality, long hours for DP and non-DP personnel are just beginning and extreme patience is required. The best laid feasibility study plans can be upset by those personnel who fail to cooperate in executing the plans devised.

In this chapter, the detailed steps involved in system implementation are enumerated, thereby providing a logical framework for getting a system off the ground. The employment of parallel operational procedures for checking the new system against the old system is discussed, followed by the conversion of the various subsystems and their related modules to daily operations. A periodic review for possible improvements to the new system concludes the chapter.

SYSTEM IMPLEMENTATION

The task of system implementation for a comprehensive system, such as an interactive management information system, is a major undertaking since it generally involves the entire organization structure. This results in a great need for implementation planning. A logical starting point for this type of planning involves knowledge of the following areas: personnel needs, programming, equipment selected, physical requirements, and conversion activities. An understanding of these areas and the relationships among them establishes the specific tasks to be undertaken and delineates any anticipated problems and exceptions. This background permits the detailed planning of each task in the implementation in order to arrive at specific deadlines. The scheduling method should follow the natural flow of the work to be undertaken. The usual questions of **who, what, where, when, how,** and **why** must be answered in developing the schedule. Implementation planning should include provision for reviewing both completed and uncompleted tasks so that the entire system project will be under control at all times.

The major steps for system implementation are:
    preparatory work for system implementation
    program development

equipment acquisition and installation
system level testing
system installation and conversion

Preparatory Work for System Implementation

A considerable amount of preparatory work is required before implementating a new system. Any temptation to avoid these activities and "get on with it" must be resisted. The important preparatory activities discussed below are
1. implementation planning
2. team-building activities
3. development of standard subroutines

Implementation planning   Initially, a preliminary implementation plan is prepared to schedule and manage the many different activities that must be completed for a successful system installation. The preliminary plan serves as a basis for the initial scheduling and assignment of resources to important implementation activities. The preliminary plan will be updated throughout the implementation phase in order to reflect the current status. A complete implementation plan will include the following items:
1. **Select qualified personnel**   Personnel qualified for system implementation and normal operations are usually selected by the manager of the MIS group. Since implementation requires a team effort, it is imperative that a compatible group be brought together.
2. **System test plan**   The purpose of the test is to verify that the system will produce the expected results. The system test will evalute the system so that "standard responses" to all input will be produced. As many "exception conditions" as possible should be tested. The system test plan should be a formal, written document that is reviewed and approved by the implementation team and by system users. It should specify the input required and who will provide it; the output expected, how it will be evaluated, and by whom; the number and purpose of each system test variation; and the date each system test will be conducted.
3. **Personnel training plan**   The plan should include everyone associated with the implementation, use, operation, or maintenance of the new system. The personnel training plan must specify the individuals who will need training, what type of training program and materials are appropriate for each, who will develop the training program and materials, and when the training will be presented and by whom.
4. **Equipment installation plan**   This should include site preparation, equipment installation, and equipment check-out. The equipment installation plan must specify who will complete these activities for each piece of new equipment and the date by which they will be completed.
5. **System conversion plan**   This plan specifies the activities which must take place to convert from the old system to the new system to provide an orderly way to convert from the old procedures, files, and system to the new ones. A good system conversion plan will assure everyone that the transition will be smooth and efficient.
6. **Overall implementation plan**   The schedule of the implementation work should be sufficiently detailed that each important activity can be controlled. This involves determining appropriate start and completion dates for each activity and sequenc-

ing the activities in an efficient manner. A PERT network chart is typically prepared for activities requiring careful timing or significant effort.

Team-building activities The implementation of a complex system requires a coordinated team effort. The members of the implementation team must work closely together, often under pressure, to meet the desired objectives. It is important for the implementation project manager to recognize this need for close cooperation and schedule "team-building" activities to include both the implementation team and the system users. Important team-building activities include the following:

1. **Provide system overview** Everyone associated with the new system should be given an opportunity to see the "big picture" and know where they fit in. Each person should have some basic project background, know the primary objectives of the system, be able to relate to the system benefits, and know the projected installation date. Also, each should know what his or her job is during the system implementation, conversion, and operation phases. Members of the implementation team and key system users will need more detailed understanding of the system.

2. **Expectation sharing** Both the implementation team and the system users should candidly share their expectations about the system implementation phases. This is easier said than done. However, it can be accomplished, thereby providing a strong foundation for a successful project.

3. **Set system implementation standards** Standards to be followed throughout the implementation phase should be collectively developed in the following areas: collection and summarization of project status information from the implementation team, program coding, program and system documentation, and testing. There are two key points to setting standards: (1) The standards must be developed by a team effort to ensure their acceptance; and (2) The implementation stages must be monitored to ensure conformance to the standards.

4. **Complete implementation planning** The responsibility for each part of the implementation plan is assigned to a different member of the implementation team so that the project manager has more time to coordinate the total team effort. Also, the individual team members have an opportunity for their own growth and development. They are able to plan in much more detail than the project manager. Their participation in the planning process elicits a higher level of commitment to the plan.

Referring to the second activity above, the project manager should share his or her expectations with each individual on the team. For example, the project manager may discuss the assignment each team member is likely to get prior to making the final decision. In turn, the individual can describe the type of assignment which would contribute most toward his or her personal development or which he or she is best prepared to undertake. In this manner, both project manager and team members can work together.

Additionally, the project manager and the system users should share expectations about the need to change system specifications and how such changes will be handled. If the implementation requires more than a few months, the need to change the system specifications is highly probable. Typical reasons for changes include: identification of a way of doing things better than the original plan, the environment in which the system will operate has changed, or an oversight was discovered during the design phase which must be corrected. Hopefully, a change in design specifications will affect only some portion of the system not yet implemented. The project manager and system user should decide together that a change will be necessary and how best to effect it. In one approach, the implementation team installs the system as planned in the original specifications; upon completion of the installation, the project manager and user evaluate potential changes and implement them on a priority basis. In another approach, proposed changes are evaluated and implemented as they occur.

Development of standard subroutines   The implementation team should identify basic types of processing to be performed in many computer programs. Modular subroutines can be written, tested, debugged, and stored for use by other team members. This can lead to a great savings in the implementation effort by allowing team members complete access to modular subroutines, thereby avoiding "recoding." Examples of processing that can be handled by standard program modules, or subroutines, are: simple error handling—print image of input in error, flag the field, and print an error message; and print report title and heading information—title of report, date, time, and page number.

Program Development

Program development is the process of taking a set of program specifications (input, processing requirements, and output) and converting it into a tested and documented computer program. The activities associated with program development include:

1. finalize program design
2. develop program code and debug
3. test program
4. review program code for adherence to standards
5. document program

Finalize program design   The first step in the development of a computer program is to review its specifications for clarity and completeness. Programmers should review the program's input, processing, and output with the implementation project manager and/or system users. The review should be sufficiently detailed so that the programmer, the project manager, and the system user agree on what the computer program is to accomplish. The project manager should comment on any special programming techniques which might be useful; similarly, the system user can caution the programmer about unusual processing conditions which might arise. Both the project manager and system user should offer suggestions on how to test the program. Overall, this first step can make the remaining steps easier.

Develop program code and debug   A modular approach similar to the modular design method used throughout the text should be employed to make the job of programming and debugging easier. **Modular programming** refers to the technique in which the logical parts of a problem are divided into a series of individual routines so that each routine may be programmed independently. Access to the individual routines is controlled by a single routine, commonly known as the **"mainline program"** in a batch processing mode. The mainline program governs the flow of data to the proper processing routines, as depicted in figure 17.1.

The success of modular programming hinges largely on the adherence to this sequence—modular analysis precedes modular design which in turn, precedes modular programming. The design of a system must be approached from the onset with

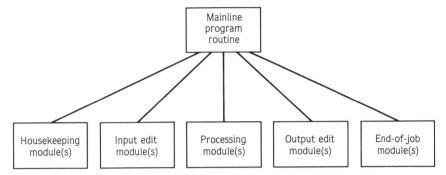

figure 17.1 An overview of modular programming in a batch processing mode. (Robert J. Thierauf, **Systems Analysis and Design of Real-Time Management Information Systems,** © 1975. Reprinted by permission of Prentice-Hall, Inc., Englewood Cliffs, New Jersey.)

modularity in mind since the probability of successfully breaking a monolithic system into modules **after** the design is finished is very small.

In a modular programming approach, a program consists of several modules, each of which is generally limited to 50 to 200 machine language instructions, or 1 to 6 pages of coding. A **program module** is defined as a closed subroutine to which control is passed by a **calling program** (mainline program routine) and which, when it completes its processing, returns control to the calling program. Decisions may be made in a module that will cause a change in the flow of the system, but the module will not actually execute the branching; it will communicate the decision to the calling program to execute the branching. The creation of a modular design is relatively easy because a computer problem consists of functional DP modules. The actual processing in the modules is a function of the relationship between data elements and results from the analysis. The only functions actually programmed are the data processing functions at the bottom ends of the branches, because they are the building blocks on which the higher-level functions are made.

The placement of often-used functions within an overlay structure is an important element of the modular structural design. For example, consider the three-level structure below. One module from each level will be in memory at any given moment,

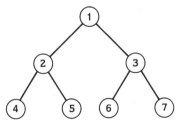

overlaying a previously called module. Hence, the execution of module 5 requires that modules 2 and 1 be in memory. If module 5 were a generalized function which must also be used by module 6, the situation would not be feasible since one is not derived from the other; whereas if module 1 is the generalized module, it is available to all other modules. Execution of the modular system requires at least one module that is not on the analysis

462 System Implementation

tree above, the mainline program routine mentioned earlier. Most of the fundamental system flow logic is in this module which controls the execution of the functional modules.

To illustrate the modular programming concept (batch processing mode), figure 17.2 depicts a typical weekly payroll program broken down into eleven modules. These individual modules are separate tasks varying in size and complexity. For example, the highest-level module, module 1—(Process weekly employee payroll file), represents the solution to the weekly payroll program. Each module of the next level (modules 2 through 6) and succeeding levels (modules 7 and 8, and modules 9 through 11) solve one of the tasks to be performed in order to process the program. Within this sample programming project, four programmers handle several modules within the same program. Each programmer is responsible for coding and testing his or her own modules, and the lead programmer supervises and is responsible for final program testing (figure 17.3).

So far we have described only one approach to modular programming—that in which every transfer of control is made in a mainline program. However, for an interactive (real-time) environment, a more important approach exists. In real time, for the sake of speed and efficiency, each module controls the program flow by ordering a RTCP (real-time control program) to call in another module. The RTCP is not a mainline program because it functions as a housekeeper and does not control the logical system flow. This is, then, a type of modular programming in which the flow continues on a smooth path through a chain of modules, rather than flowing in and out of the mainline control program.

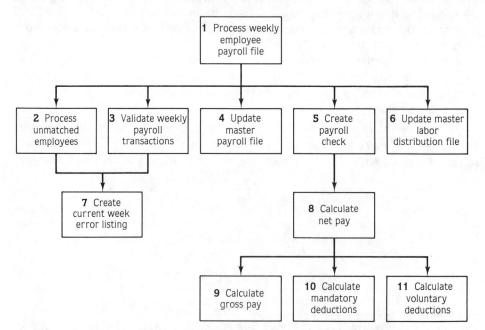

figure 17.2 The major programming modules for a typical weekly payroll program. (Robert J. Thierauf, **Systems Analysis and Design of Real-Time Management Information Systems**, © 1975. Reprinted by permission of Prentice-Hall, Inc., Englewood Cliffs, New Jersey.)

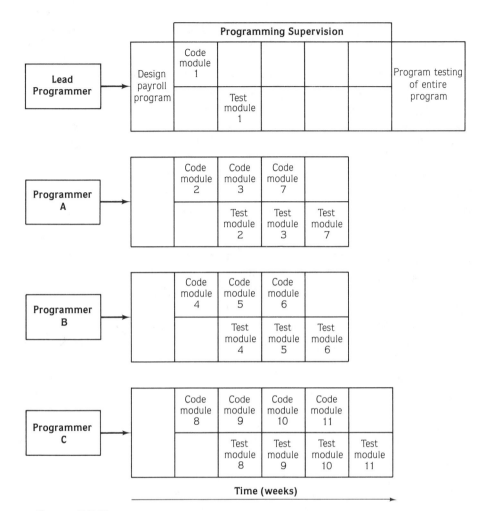

figure 17.3 The organization of modular programming for a typical weekly payroll program, as depicted in figure 17.2. (Robert J. Thierauf, **Systems Analysis and Design of Real-Time Management Information Systems**, © 1975. Reprinted by permission of Prentice-Hall, Inc., Englewood Cliffs, New Jersey.)

The benefit of a modular design is that it accommodates changes occurring after the system design phase (coding, testing, learning, operating, and maintaining the system). Given that both fundamental system logic and specific procedures invariably change throughout the system, modularity enhances the identification of these changes and therefore the control of the system evaluation. The modular design approach helps the programming staff to build an interactive operating environment, including the necessary batch functions, that is efficient and capable of being maintained.

The process of **debugging** a program refers to correcting programming language syntax and diagnostic errors so that the program "compiles cleanly." A **clean compile** means that the program can be successfully converted from the source code written by the programmer into machine language instructions. Once the programmer achieves a clean compile, the program is ready for testing.

Debugging can be a tedious task. It consists of four steps: inputing the source program to the compiler, letting the compiler find errors in the program, correcting lines of code that are in error, and resubmitting the corrected source program as input to the compiler. Usually, this process must be repeated a half dozen or more times using a **batch** compiler. The length of time required to debug a program can be shortened considerably by the use of an **interactive** compiler. The interactive compiler checks the source program and displays any errors on a CRT or prints them on a typewriter terminal. The programmer corrects the indicated errors and initiates the interactive compiler as often as necessary until all errors are corrected. The interactive compiler allows the programmer to debug a program in a few hours, as opposed to a few days using a batch compiler.

Test program A careful and thorough testing of each program is imperative to the successful installation of any system. Approximately 30 to 50 percent of the resources required for program development should be allocated to program testing. The programmer should plan the testing to be performed, including testing all possible exceptions. The test plan should require the execution of all standard processing logic. Lastly, the program test plan should be discussed with the project manager and/or system users.

A log of test results and all conditions successfully tested should be kept. The log will prove invaluable in answering the inevitable question, "Did you ever test for this condition?"

Software packages are available that allow interactive testing of batch processing programs. They can greatly reduce the length of time required for testing. Interactive testing allows the programmer to monitor each step required to process a program input. If a problem in program logic flow is discovered, the programmer can stop the execution of the program, correct the problem, and have the program resume processing at that point just prior to the interruption.

Review program code for adherence to standards One of the preparatory steps of the system implementation phase is to set standards for program coding. It is necessary to review each program to ensure the standards are being met; ideally, someone who is not directly connected to the implementation team should review it so that the quality of the system and adherence to standards are evaluated objectively without concern for project schedule and budget. It is recommended that this evaluation occur long before the programmer has a clean compile. Because less than half of the total program development effort is complete at this stage, the programmer is much more receptive to constructive criticism and more open to changes. A second review should be scheduled sometime during the program testing stage to ensure that suggestions have been followed.

Document program The best program documentation is the program itself. Each program should contain an initial paragraph of comments describing the basic input, processing, and output of the program, as well as any assumptions or limitations concerning the program. The program itself should be heavily annotated. For modular programming, each module should have an introductory paragraph of comments. The program itself should be written in a clean and concise manner, with particular emphasis on implementing future program changes as deemed necessary. Overall, there must be an insistence on the development of programs that will be easy to maintain.

Equipment Acquisition and Installation

Hardware required to support the system was selected prior to the implementation phase. The necessary hardware should have been ordered in time to allow for installation and testing of equipment during the implementation phase. An installation checklist should be developed at this time utilizing advice from the vendor and DP personnel experienced in the installation of the same or similar equipment. Adequate time should be scheduled to allow completion of the following activities:

1. **Site preparation** An appropriate location must be found to provide an operating environment for the equipment that will meet the vendor's temperature, humidity, and dust control specifications. A reliable source of power for operating the equipment must be found. The human factor must also be considered: lighting, sound insulation, and safety measures must all be adequate.
2. **Equipment installation** The equipment must be physically installed by the manufacturer, connected to the power source, and wired to communication lines if required.
3. **Equipment check-out** The equipment must be turned on and tested under normal operating conditions. Not only should the routine "diagnostic tests" be run by the vendor, but also the implementation team should devise and run extensive tests of its own.

System Level Testing

System level testing must be conducted prior to installation of any system. It involves: (1) preparation of realistic test data in accordance with the system test plan, (2) processing the test data using the new equipment, (3) thorough checking of the results of all system tests, and (4) reviewing the results with the future users, operators, and support personnel. System level testing is an excellent time for training employees in the operation of the new system as well as maintaining it. Typically, it requires 25 to 35 percent of the total implementation effort.

One of the most effective ways to perform system level testing is to perform parallel operations with the existing system. **Parallel operations** consist of feeding both systems the same input data and comparing data files and output results. Despite the fact that the individual programs were tested, related conditions and combinations of conditions are likely to occur that were not envisioned. Last minute changes to computer programs are necessary to accommodate these new conditions.

For an interactive MIS project, the process of running dual operations for both new and old systems is more difficult than it is for a batch processing system, because the new system has no true counterpart in the old system. One procedure for testing the new interactive system is to have several remote input/output terminals connected on line which are operated by supervisory personnel, backed up by other personnel operating on the old system. The outputs are checked for compatibility and appropriate corrections are made to the on-line computer programs. Once this segment of the new system has proven satisfactory, the entire terminal network can be placed into operation for this one area. Additional sections of the system can be added by testing in this manner until all programs are operational.

During parallel operations, the mistakes detected are often not those of the new system, but of the old. These differences should be reconciled as far as it is feasible economically. Those responsible for comparing the two systems should clearly establish that the remaining deficiencies are caused by the old system. A poor checking job at this point can result in complaints later from customers, top management, salespersons, and others. Again, it is the responsibility of the MIS manager and assistants to satisfy themselves that adequate time for dual operations has been undertaken for each functional area changed.

The system implementation group must keep the entire organization posted on parallel operations and conversion activities. This can be accomplished via a series of bulletins which start at the inception of the feasibility study. Departmental personnel should be informed when they are to start on system implementation and what specific activities will be required of them. Department heads should be informed before the actual date of conversion activities so that anticipated problems can be worked out. The time spent instructing personnel on parallel operations or conversion activities will avoid the wasted motion and time that leads to costs greatly exceeding the original estimates. Activities must be organized, directed, and controlled around the original plan of the system project.

System Installation and Conversion

Installation and conversion includes all those activities which must be completed to successfully convert from the old system to the new one. Fundamentally, these activities can be classified as follows:
1. procedure conversion
2. file conversion
3. system conversion
4. scheduling personnel and equipment
5. alternative plans in case of equipment failure

Procedure conversion Operating procedures should be completely documented for the new system. This applies to both computer operations and functional area operations. Before any parallel or conversion activities can start, operating procedures must be clearly spelled out for personnel in the functional areas undergoing changes. Information on input, data files, methods, procedures, output, and internal control must be set forth in clear, concise, and understandable terms for the average reader. Written operating procedures must be supplemented by oral communication during the many training sessions on the system change. Despite the many hours of training, many questions will have to be answered during conversion activities. Brief meetings where changes are taking place must be held in order to inform all operating employees of a change that has been initiated. Qualified DP personnel must be in the conversion area to communicate and coordinate new developments as they occur. Likewise, revisions to operating procedures should be issued as quickly as possible. These efforts enhance the chances of a successful conversion.

Once the new system is completely operational, the system implementation group should spend several days checking with all supervisory personnel about their respective areas. As with every new installation, minor adjustments should be expected. Channels of communication should be open between the DP members and all supervisory personnel so that necessary changes can be initiated as conditions change. There is no need to get locked into a rigid system when it would be beneficial for the organization to make necessary changes. Thus, the proper machinery for making changes must be implemented.

File conversion   Because large files of information must be converted from one medium to another, this phase should be started long before programming and testing is completed. The cost and related problems of file conversion are significant whether they involve on-line files (common data base) or off-line files. Present manual as well as punched card files are likely to be inaccurate and incomplete where deviations from the accepted format are common. Both files suffer from the shortcomings of inexperienced and, at times, indifferent personnel whose jobs are to maintain them. Computer-generated files tend to be more accurate and consistent. The formats of the present computer files are generally unacceptable for the new system.

Besides the need to provide a compatible format, there are several other reasons for file conversion. The files may require character translation that is acceptable to the character set of the new computer system. Data from punched cards, magnetic tapes, and comparable media will have to be placed on magnetic disk, magnetic drum, and/or mass storage files in order to construct an on-line common data base. Also, the rearrangement of certain data fields for more efficient programming may be desired. A new format using packed decimal fields may be necessary for conversion.

In order for the conversion to be as accurate as possible, file conversion programs must be thoroughly tested. Adequate controls, such as record counts and control totals, should be required output of the conversion program. The existing computer files should be kept for a period of time until sufficient files are accumulated for back-up. This is necessary in case the files must be reconstructed from scratch when a "bug" is discovered at a later date in the conversion routine.

System conversion   After on- and off-line files have been converted and the new system's reliability has been proven for a functional area, daily processing can be shifted from the existing system to the new one. A **cut-off point** is established so that the data base and other data requirements can be updated to the cut-off point. All transactions initiated after this time are processed on the new system. DP members should be present to assist and answer any questions that might develop. Considerations should be given to operating the old system a short time longer to permit checking and balancing the total results of both systems. All differences must be reconciled. If necessary, appropriate changes are made to the new system and its computer programs. The old system can be dropped as soon as the data processing group is satisfied with the new system's performance. It is impossible to return to the old system if significant errors appear later in the new system; the operation of the existing system provides an alternate route in case of system failure during conversion.

Scheduling personnel and equipment Scheduling data processing operations of a new system for the first time is a difficult task for the MIS manager. As the individual becomes more familiar with the new system, the job becomes more routine. The objectives of scheduling both personnel and equipment are depicted in figure 17.4.

Maximize utilization of personnel and equipment, in particular, I/O terminal devices to further organization objectives.

Produce timely reports and meet deadlines for output desired.

Increase productivity of personnel by including time for training and on-the-job training.

Facilitate the planning of proposed new applications or modifications of existing applications for new and/or existing equipment.

Reduce conflicts of several jobs waiting for a specific piece of equipment which may result in delays of important outputs or unnecessary overtime.

---

figure 17.4 Objectives of scheduling personnel and equipment.

Before the system project is complete, it is often necessary to schedule the new equipment. Some programs will be operational while others will be in various stages of compiling and testing. Since production runs tend to push aside new program testing, the MIS manager must assign ample time for all individuals involved. This generally means second shift for those working on programs. Once all programs are "on the air," scheduling becomes more exacting.

Schedules should be set up by the MIS manager in conjunction with the departmental managers of the operational units serviced by the equipment. The master schedule for next month should provide sufficient computer time to handle all required processing. Daily schedules should be prepared in accordance with the master schedule and should include time necessary for reruns, program testing, special nonrecurring reports, and other necessary runs. In all cases, the schedules should be as realistic as possible since scheduling an interactive system is more difficult than a batch processing system.

The time to assign remote batch programs under normal operating conditions in real time is a problem, since the number of interruptions that will occur is generally unknown. Often, the solution is to assign a block of each day for the operation of remote input/output devices. If this arrangement is not feasible, the MIS manager must look to past experience. When total random and sequential demands are not high, the machine will have sufficient capacity to complete all scheduled work even though batch processing runs will be extended by random system inquiries.

The practice of attaching recording clocks to keep track of the machine time in executing instructions and awaiting instructions is quite common. These allow the data processing section to study the efficiency of each program and identify problem areas, and are helpful in determining how the equipment's cost is to be charged to an organization's functional areas. For a real-time system, however, recording clocks are of no real value since executive programs are running continuously whether or not demands are being made for service; the total time for a real-time system has no real meaning. However, information can be accumulated internally by determining the input source and allocating this cost to the respective areas on the monthly departmental statements.

Just as the equipment must be scheduled for its maximum utilization, so must the personnel who operate the equipment. It is also imperative that personnel who enter input data and handle output data be included in the data processing schedule. Otherwise, data will not be available when the equipment needs it for processing. It is essential that each person follow the methods and procedures set forth by the DP group; noncompliance with established norms will have an adverse effect on the entire system.

Alternative plans in case of equipment failure Alternative processing plans must be employed in case of equipment failure. Who or what caused the failure is not important; the fact is that the system is down. Priorities must be given to those jobs critical to an organization, such as billing, payroll, and inventory. Critical jobs can be performed manually until the equipment is functioning again.

Documentation of alternative plans is the responsibility of the DP section and should be a part of an organization's systems and procedures manual. It should state explicitly what the critical jobs are, how they are to be handled in case of equipment failure, where compatible equipment is located, who will be responsible for each area during downtime, and what deadlines must be met during the emergency. A written manual of procedures concerning what steps must be undertaken will help expedite the unfavorable situation. Otherwise, panic will result in the least efficient methods when time is of the essence.

OPERATIONAL REVIEW OF SYSTEM

Just after the system is installed, the MIS manager and DP staff should review the tangible and intangible system benefits set forth in the exploratory survey report. The purpose of such a review is to verify that these benefits are, in fact, being achieved. Discussions with managers of operating areas being serviced will determine how well the new system is performing. Tangible benefits, such as clerical reduction and lower inventory, and intangible benefits, like customer service and more managerial information, are open for constructive criticism. Typical comments will be: certain areas have been improved significantly, some are about the same, and others are not as good as before. The task of the DP section, then, is to make the necessary adjustments to accomplish the quantitative and qualitative goals of the feasibility study. It may take from several months up to one year to effect the changes, which include reprogramming the most frequently used programs, for greater efficiency.

As times passes, the workload of the present system increases. Factors which were not previously problems can become significant. Can the equipment run longer hours or should additional equipment be obtained? Can modification of methods and procedures be made to reduce processing time and cost? Can noncritical processing be shifted to another time? How does the time differential affect the system that is operating within the continental limits of the United States? Answers to these questions must be evaluated by the DP section through a periodic review of the existing system. It may be necessary to undertake a feasibility study periodically in order to evolve an optimum system for an organization's continually changing conditions.

SUMMARY: System Implementation

Preparatory Work for System Implementation Represents the work that must be undertaken before the daily operations of the new system can begin. It consists of:
1. Implementation planning: the selection of qualified personnel, a system test plan, a personnel training plan, an equipment installation plan, a system conversion plan, and an overall implementation plan.
2. team building activities: providing a system overview, expectation sharing, setting system implementation standards, and completing implementation planning.
3. development of standard subroutines: developing standard program modules that can be utilized by the entire programming staff.

Program Development Is the process of taking a set of program specifications and converting it into a tested and documented computer program. It includes:
1. finalize program design: reviewing program specifications for clarity and completeness with project manager and system user.
2. develop program code and debug: utilizes **modular programming,** in which the logical units of a computer program are divided into a series of individual routines so that each routine may be programmed independently.
3. test program: checking the accuracy of all computer processing logic for routine items and exceptions.
4. review program code for adherence to standards: determines whether or not the programmer has followed predetermined standards in computer program development.
5. document program: using the program itself as a form of documentation by employing comment paragraphs throughout the program.

Equipment Acquisition and Installation Centers on site preparation, equipment installation, and equipment check out.

System Level Testing Consists of running parallel operations, that is, feeding the same input data into both the old and new systems for comparison of data files and outputs.

System Installation and Conversion Focuses on converting to daily operations of the new system. Its major activities are:
1. procedure conversion: documenting detailed methods and procedures that will be used for day-to-day operations.
2. file conversion: converting data files from one processable medium to another for the new system.
3. system conversion: establishes a cutoff point for conversion from the present system to the new system.
4. scheduling personnel and equipment: setting up daily equipment schedules that involve organization personnel to operate the equipment efficiently.
5. alternative plans in case of equipment failure: giving preference to those jobs that are critical in case of equipment failure.

Examination of New System Approaches

Examination of new system approaches that lead to operational improvements must be explored by the DP section. The reason should be obvious: systems like an interactive management information system are so complex that it is almost impossible to complete the project in the most efficient manner for all functional areas the first time around. The efficiency of any large computer program is always open to question. Programming should be closely examined, along with the inputs, data base, off-line files, methods, procedures, and outputs that are related to it. Questions about new approaches eventually affect programming: Can the input cut-off be set earlier by rescheduling personnel? Is there a need to redesign certain inputs, forms, and documents? Can the regular work be separated initially from the exception items so that each can be expedited more quickly and accurately? Are present reports and outputs adequate in view of the organization's growth, technological improvements, and the increasing complexities of business? Numerous questions related (directly or indirectly) to programming should also be raised.

New system approaches should not be restricted to current applications, but also should include new areas reflecting changed conditions. The areas that previously processed a low volume of data may have experienced a phenomenal growth, making them suitable for interactive processing. A promising area for review is data stored on line; it may be that meaningful managerial reports can be prepared at a small additional cost, since data are available at all times for processing. Also, mathematical models for improving an organization's quantitative basis for decision making may be capable of using data stored on line. The ability to access large volumes of current data as well as to compile statistics on the data changes occurring within the system provide new approaches. An examination of these approaches may necessitate a new feasibility study when numerous changes are contemplated. This permits the implementation of a better system versus "patching" an existing system.

Evaluation of New Equipment

Periodic review of the existing system includes keeping abreast of the latest developments in equipment. Even though an organization may utilize a computer with interactive capabilities, the constant parade of newer equipment can make obsolete the present computer. The newer equipment may be lower in price. New design concepts and circuitry permit the introduction of minicomputers, microcomputers, intelligent terminals, OCR equipment as input to a DP system, faster data communication equipment, and many other developments.

Benefit and Cost Analysis of System

New equipment developments are fast changing the cost of data processing. This is caused by the addition of many newer models whose capabilities and performance are better than previous models. Later computer models of the same generation and different generations of computers have proven this to be true. Even though internal hardware speeds are faster with increased miniaturization of circuitry and components, many computer programs are input- and/or output-bound, which means the new speed of the central processor may be of no assistance in their operations. Thus, the real criterion of lower data processing costs is determined by throughput performance and turnaround time.

Estimated savings less estimated one-time costs and additional operating costs were computed for the estimated life of the new system in the exploratory survey report. Periodically, the MIS manager needs to compare these estimated figures with the actual amounts and evaluate the results with the vice-president in charge of finance. The study's figures could be different from the actual results. Unfavorable deviations need assessment for possible remedies of the situation. It should be noted that fewer large differences are likely to occur when the MIS manager knows there will be a series of reviews—that is, comparisons of the budgeted figures to actual amounts. Those data processing personnel who experience unfavorable results must be held accountable. For the most part, original estimates are stated realistically, but unforeseen factors can distort final results.

SUMMARY: Operational Review of System

Examination of New System Approaches   Involves the review of new approaches to the present system that may lead to operational improvements.

Evaluation of New Equipment   Relates not only to the review of computers, but also to all related equipment, particularly that with interactive capabilities.

Benefit and Cost Analysis of System   Centers on comparing estimated figures for the new system with the actual amounts and evaluating the results.

CHAPTER SUMMARY:

System implementation begins after the formal signing of the equipment contract. While the MIS feasibility study consists of four steps—the introductory investigations, systems analysis, systems design, and equipment selection—system implementation involves basically two steps, preparatory work and operation of new system. One-third of typical project time is spent on the feasibility study, and the remaining two-thirds on implementing the system. The number of operating personnel outside the DP group is substantially increased in the latter phase. Also, an organization will experience high costs during the system conversion phase.

Programming development, the mainstay of the preparatory work phase, is time-consuming and costly. Advanced programming techniques and languages should assist in developing computer programs and in reducing programming costs. Once a program is operational, system level testing should be initiated to check for conditions that might not have been present during program testing. This phase culminates in parallel operations and, finally, the conversion of activities for daily operations. Consideration must be given to alternative plans in case of unforeseen difficulties during conversion or later. Otherwise, an organization could be temporarily out of business.

Periodic operational reviews should follow the installation; they involve examination of new system approaches, new equipment, and benefit/cost factors. An evaluation of the existing system may signal the need for a new feasibility study; in essence, the system project cycle must start again. This permits making system changes necessary to accommodate external and internal factors and introducing more advanced hardware and software.

Questions

1. What type of implementation plans must be developed during the preparatory work phase of a new system?
2. How important are team-building activities to the success of a new system?
3. What is the preferred approach to program development and why?
4. a. What programming difficulties can be experienced by the DP section during the preparatory work phase?
   b. Explain how each can be resolved.
5. When installing a new computer system, what physical facilities are generally necessary?
6. Distinguish among program testing, system level testing, and parallel operations.
7. When should file conversion activities commence for a new system?
8. How important are parallel operations in getting a new system "on the air?"
9. Why should an organization periodically review its system for improvements? Explain.
10. How important is documentation for system implementation and operational review of a system?

Self-Study Exercise

True-False:

1. _____ Preparatory work for a new system requires very little time.
2. _____ An important part of implementation planning is team-building activities.
3. _____ Expectation sharing involves both the implementation team and the system users.
4. _____ Program development precedes system level testing.
5. _____ There is need to review program code for adherence to standards.
6. _____ Programming and testing of computer programs require little time for a system project.
7. _____ Scheduling the installation is the least important of all preparatory work activities.
8. _____ Documentation is the least important step in programming and testing.
9. _____ Generally, parallel operations constitute the next step after system level testing.
10. _____ Most organizations find that parallel operations are a waste of time and money.

Fill-In:

1. The job of _____ _____ is generally a major undertaking if it cuts across the entire organization for a system project.
2. The first major activity of the preparatory work to system implementation focuses on some type of _____.
3. The _____ _____ plan is established in order to verify that the system will produce the expected results.
4. The implementation of a complex system requires a coordinated team effort that can be effected through _____ _____ activities.
5. Programmers should review the computer program's input, processing, and output with the implementation project manager and/or _____ _____.
6. A preferred method of program development is to develop a program code that utilizes _____ _____.
7. Even though computer programs have been _____ _____, there is no way of duplicating the actual flow of work except with daily operations.
8. _____ _____ consist of feeding the present and new systems with the same input data, thereby allowing a comparison of files and outputs.
9. The scheduling of an _____ computer system is generally more difficult than it is for a batch-oriented computer system.
10. It is highly recommended that a _____ _____ be undertaken in order to maintain an optimum system for an organization's changing operating conditions.

appendix: answers to self-study exercises

CHAPTER 1

True-False:

1. F
2. F
3. T
4. F
5. T
6. T
7. T
8. T
9. F
10. T

Fill-In:

1. systems planning
2. detailed investigation
3. systems design
4. design principles
5. real time
6. sequential
7. interactive processing
8. distributed processing
9. building block
10. computer programming

CHAPTER 2

True-False:

1. T
2. F
3. F
4. F
5. F
6. T
7. T
8. T
9. F
10. F

Fill-In:

1. communicator and record
2. pictorial
3. document
4. decision
5. process
6. connector
7. compact
8. analyzing and designing
9. bar
10. events and activities

Problems:

1. a. completed shipping order forms for processing
   b. check customer credit by displaying past history
   c. separate customer orders that fail the credit check
   d. deduct items from inventory
   e. store production order data
   f. obtain shipping status of customer order
2. a. job schedule cards
   b. manual dispatching by production scheduling equipment
   c. on-line data collection (system) for recording time on jobs
   d. production evaluation program

Appendix: Answers to Self-study Exercises **477**

3.

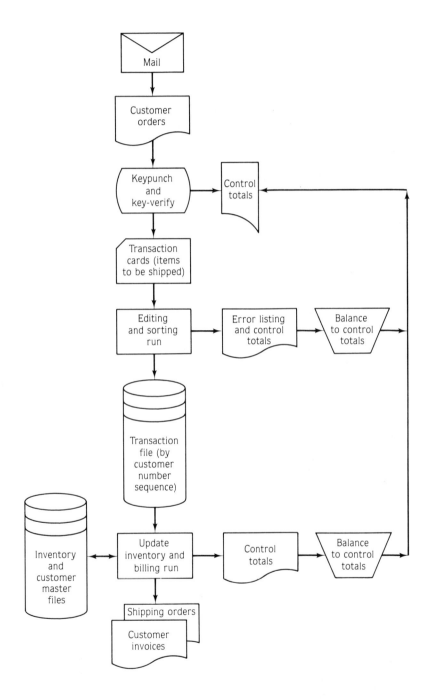

4.

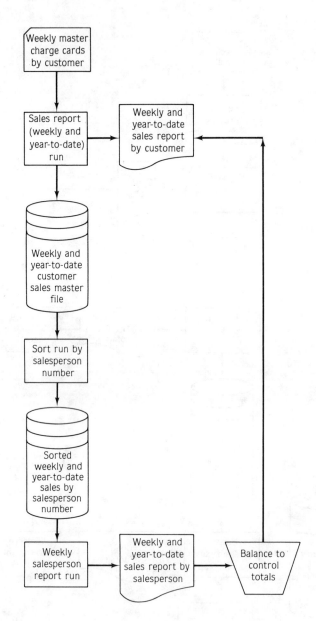

5. a. < (less than)
    b. write new master finished goods inventory file
    c. > (greater than)
    d. = (equal to)
    e. = (equal to)
    f. = (equal to)
    g. ≠ (does not equal)
    h. yes
    i. no

6.

Customer Billing Decision Table							
	Rule Number						
	1	2	3	4	5	6	7
Condition							
End of file shipment card	Y	N	N	N	N	N	N
First card	N	Y	N	N	N	N	N
Present card # = last card #	N	N	Y	N	N	N	N
Present card # ≠ last card #	N	N	N	Y	N	N	N
End of master file	N	N	N	N	Y	N	N
Master file # > card #	N	N	N	N	Y	N	N
Master file # < card #	N	Y	N	N	N	Y	N
Master file # = card #	N	N	N	N	N	N	Y
Action							
Print invoice total(s)	X	—	—	X	—	—	—
Read next record from master file	—	X	—	X	—	X	—
Calculate units × quantity; print line	—	—	X	—	—	—	X
Branch to read	—	—	X	—	—	—	X
Enter master file error routine	—	—	—	—	X	—	—
Advance **n** spaces, print name/address	—	—	—	—	—	—	X

7. a. **Gross Pay Program Flowchart**

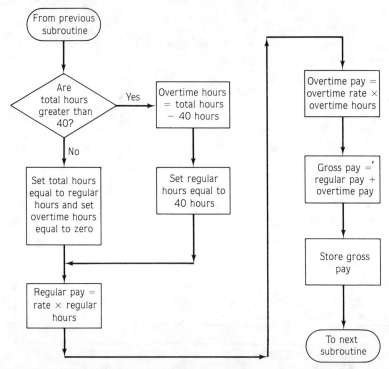

Gross Pay Decision Table							
	\multicolumn{7}{c}{Rule Number}						
	1	2	3	4	5	6	7
Condition							
Total hours > 40 hours	N	Y					
Action							
Set total hours = regular hours	X	—					
Set overtime hours = 0	X	—					
Compute overtime hours = total hours − 40 hours	—	X					
Set regular hours = 40 hours	—	X					
Compute regular pay = rate × regular hours	X	X					
Compute overtime pay = overtime rate × overtime hours	X	X					
Compute gross pay = regular pay + overtime pay; store gross pay	X	X					
Go to next subroutine	X	X					

7. b.

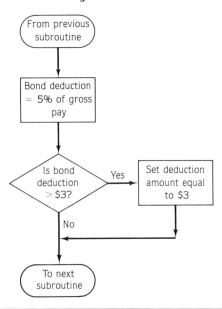

**Savings Bond Deduction
Program Flowchart**

Savings Bonds Deduction Decision Table							
	Rule Number						
	1	2	3	4	5	6	7
Condition							
(0.05 × gross pay) > $3.00	Y	N					
Action							
Set bond deduction = $3.00	X	—					
Set bond deduction amount = (0.05 × gross pay)	—	X					
Go to next subroutine	X	X					

7. c.

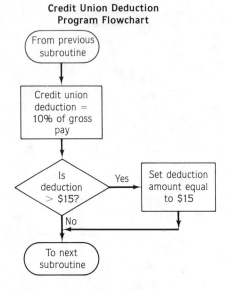

Credit Union Deduction Program Flowchart

Credit Union Deduction Decision Table							
	Rule Number						
	1	2	3	4	5	6	7
Condition							
(0.10 of gross pay) > $15.00	N	Y					
Action							
Set credit union deduction = (0.10 of gross pay)	X	—					
Set credit union deduction = $15.00	—	X					
Go to next subroutine	X	X					

CHAPTER 3

True-False:

1. F
2. T
3. T
4. F
5. F
6. F
7. F
8. T
9. F
10. T

Fill-In:

1. industrial and service
2. industrial organization
3. programming supervisor
4. materials flow
5. manufacturing and finance
6. marketing
7. engineering blueprints
8. purchase order
9. stock control
10. interactive

Problems:

1. marketing, accounting and finance, purchasing (1st level)
    R & D and engineering, information system (2nd level)
    manufacturing, personnel (3rd level)
2. a. 1) accounts receivable file 5) cost accounting file
        2) accounts payable file 6) budget file
        3) inventory file 7) general ledger file
        4) payroll file
    b. 1) aging of accounts receivable 5) work-in-progress status
        2) accounts payable status 6) budget adjustments
        3) inventory status 7) postings to general ledger
        4) payroll distribution
3. a. 1) customer billing data 5) cost accounting data
        2) accounts receivable data 6) budget data
        3) accounts payable data 7) general ledger data
        4) payroll data
    b. 1) amount of current orders 5) cost analysis by product
        2) overdue accounts receivable 6) expense accounts over budget
        3) total accounts payable 7) current general ledger balances
        4) employee year-to-date pay

CHAPTER 4

True-False:

1. F
2. T
3. T
4. F
5. T
6. F
7. F
8. T
9. F
10. T

Fill-In:

1. subsystems
2. human element
3. feasibility study
4. computer programs
5. introductory investigation
6. executive steering committee
7. functional
8. top management
9. scope
10. objectives

CHAPTER 5

True-False:

1. T
2. F
3. F
4. F
5. T
6. T
7. F
8. F
9. F
10. T

Fill-In:

1. systems and procedures
2. detailed investigation
3. source documents
4. data files
5. internal control
6. straight forward
7. intangible
8. outputs
9. decision table
10. exploratory survey

Problem

1. The Estimated Net Savings Report for system alternative 6 indicates that net savings before federal income taxes over five years are $201,170. After considering the federal income tax rate of 48 percent, the net savings after federal income taxes, discounted at 20 percent to the present time, results in a negative present value of $23,681. Thus, it is necessary to consider intangible benefits before a final evaluation of systems alternative 6 can be completed.

The American Company: Feasibility Study—SYSTEM ALTERNATIVE 6
Estimated net savings five-year period—198_)

	Years					
	1	2	3	4	5	TOTAL
Estimated savings	$100,000	$310,000	$325,500	$341,775	$358,864	$1,436,139
Estimated one-time costs	$505,000					505,000
Estimated additional operating costs	105,000	145,000	152,250	159,863	167,856	729,969
TOTAL estimated costs	$610,000	$145,000	$152,250	$159,863	$167,856	$1,234,969
NET SAVINGS (losses) before federal income taxes	($510,000)	$165,000	$173,250	$181,912	$191,008	$ 201,170

The American Company: Feasibility Study—SYSTEM ALTERNATIVE 6
Discounted cash flow—20 percent return after federal income taxes five-year period—198_)

Year	Net savings (losses) before federal income taxes	Federal income tax at 48 percent rate	Net savings (losses) after federal income taxes	At 20 percent Present value of $1.00	At 20 percent Present value of net savings (losses)
1	($510,000)	($244,800)	($265,200)	.833	($220,912)
2	165,000	79,200	85,800	.694	59,546
3	173,250	83,160	90,090	.579	52,162
4	181,912	87,318	94,594	.482	45,594
5	191,008	91,684	99,324	.402	39,929
TOTALS	$201,170	$ 96,562	$104,608		($ 23,681)

Note: All values are rounded to nearest dollar.

CHAPTER 6

True-False:

1. F
2. F
3. T
4. F
5. F
6. T
7. T
8. F
9. T
10. F

Fill-In:

1. 84
2. 12
3. 204
4. 205
5. sales order
6. shipment
7. customer order
8. computer processors
9. $12.63
10. WATS

CHAPTER 7

True-False:

1. F
2. F
3. F
4. T
5. F
6. T
7. F
8. T
9. F
10. F

Fill-In:

1. interactive
2. flexibility
3. simplicity
4. systems design
5. data
6. policies
7. data base
8. system performance
9. coding
10. bid invitations

Problems:

1.

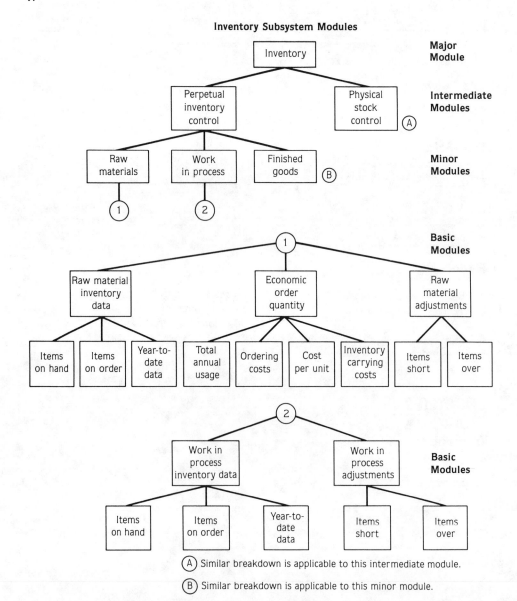

Appendix: Answers to Self-study Exercises **487**

2.

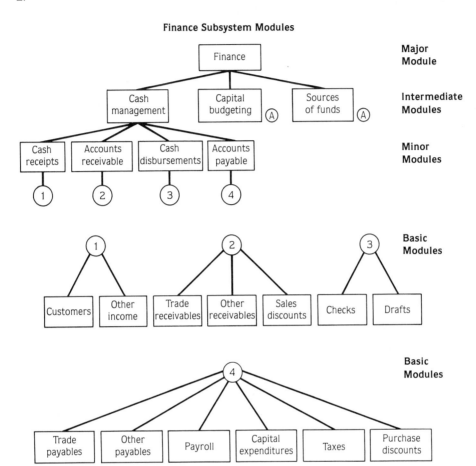

CHAPTER 8

True-False:

1. F
2. F
3. F
4. F
5. T
6. T
7. T
8. T
9. F
10. T

Fill-In:

1. output
2. visual
3. microfilm
4. acceptability
5. forward
6. duplication
7. human
8. requirements
9. detailed
10. editing

Problems:

1.

SYSTEM OUTPUT

Date: July 25, 198_
Preparer: R. J. Thierauf
Type of output: MONTHLY FINISHED GOODS INVENTORY SUMMARY

Informational elements	Type of characters	Number of characters	Comments
Inventory number	Numeric	6	
Description	Alphabetic	20	
Beginning balance	Numeric	5	Edit with comma
Receipts	Numeric	5	Edit with comma
Disbursements	Numeric	5	Edit with comma
Adjustments—Inc. (Dec.)	Numeric	5	Edit with comma
Ending balance	Numeric	5	Edit with comma
Unit price	Numeric	4	Edit with decimal point
Total finished goods amount by inventory number	Numeric	9	Edit with decimal point and commas
Total monthly finished goods inventory amount		11	Edit with decimal point and commas

Comments:
The computer program is to generate the report and column headings. The month, year, and page numbers are to be printed at the top of each page.

2.

150/10/6 PRINT CHART PROG. ID _____ PAGE _1 OF 1_
(SPACING: 150 POSITION SPAN, AT 10 CHARACTERS PER INCH, 6 LINES PER VERTICAL INCH) DATE _____
PROGRAM TITLE _MONTHLY FINISHED GOODS INVENTORY SUMMARY_
PROGRAMMER OR DOCUMENTALIST: _THIERAUF, ROBERT J._
CHART TITLE _MONTHLY FINISHED GOODS INVENTORY SUMMARY — FORMAT_

Line	Content
2	DATE XX/XX/XX MONTHLY FINISHED GOODS INVENTORY SUMMARY
4	INVENTORY BEGINNING DISBURSE- ADJUST
5	NUMBER DESCRIPTION BALANCE RECEIPTS MENTS MENTS
7	XXXXXX XXXXXXXX XXXXXXXXXXXXX XX,XXX XX,XXX XXX,XXX XX,XX
9	XXXXXX XXXXXX XXXXXXXXXXXXXX XX,XXX XX,XXX XXX,XXX XX,XX
11	XXXXXX XXXXXX XXXXXXX XXXXXX XX,XXX XX,XXX XXX,XXX XX,XX
21	XXXXXX XXXXXXXXX XXXXXXXXXXXX XX,XXX XX,XXX XXX,XXX XX,XX
24	TOTAL MONTHLY FINISHED GOODS INVENTORY

	PAGE XX
ENDING	UNIT TOTAL
BALANCE	PRICE AMOUNT

- X XX,XXX   XX.XX   X,XXX,XXX.XX
- X XX,XXX   XX.XX   X,XXX,XXX.XX
- X XX,XXX   XX.XX   X,XXX,XXX.XX

- X XX,XXX   XX.XX   X,XXX,XXX.XX

                      XXX,XXX,XXX.XX

CHAPTER 9

True-False:

1. F
2. T
3. T
4. T
5. T
6. T
7. T
8. F
9. T
10. T

Fill-In:

1. distributed
2. blocking
3. sequential
4. drum
5. mass storage
6. sequential file
7. direct file
8. index
9. capacity
10. data base

Problems:

1.

SYSTEM DATA FILE			
Date: February 15, 198_			Type of record: EMPLOYEE MASTER PERSONNEL FILE
Preparer: R. J. Thierauf			
Record elements	Type of characters	Number of characters	Comments
Employee number	Numeric (packed)	5	3 bytes (packed)
Employee name	Alphabetic	21	Last, first, M. I.
Employee address	Alphanumeric	19	
City and state	Alphanumeric	16	
Zip code	Numeric (packed)	5	3 bytes (packed)
Social security number	Numeric (packed)	9	5 bytes (packed); edit with hyphens
Starting date	Numeric (packed)	6	4 bytes (packed); edit with hyphens
Job code	Alphanumeric	2	
Pay rate	Numeric (packed)	4	3 bytes (packed); edit with decimal point
Number of exemptions	Numeric (packed)	2	2 bytes (packed)
Union code	Alphanumeric	2	

Comments:
The data base administrator has approved the record elements, types of characters, and number of characters for direct-access storage (magnetic disk) that utilizes the direct file organization method.

2.

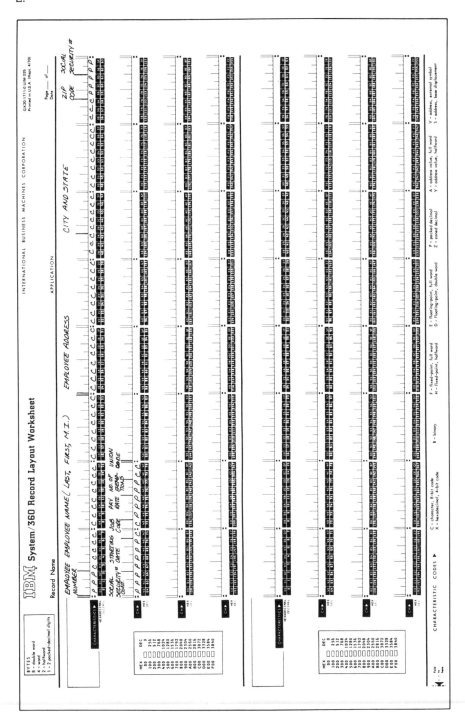

CHAPTER 10

True-False:

1. F
2. F
3. F
4. T
5. T
6. F
7. F
8. F
9. F
10. T

Fill-In:

1. shipment processing
2. easy
3. monthly
4. inventory
5. informational
6. quantity
7. distribution
8. table
9. output
10. volatility

CHAPTER 11

True-False:

1. F
2. T
3. T
4. T
5. F
6. T
7. F
8. F
9. T
10. F

Fill-In:

1. input
2. punched card
3. merging
4. data representation
5. human factor
6. duplication
7. coding
8. cooperation
9. input
10. method(s)

Problems:

1.

SYSTEM INPUT
Date: March 1, 198_
Type of input: **ACCOUNTS RECEIVABLE PAYMENTS**
Preparer: R. J. Thierauf

Information elements	Type of characters	Number of characters	Source	Comments
Account number	Numeric	5	Customer file	Must be numeric
Payment code	Numeric	1	Customer file	Must be numeric
Date of payment	Numeric	8	Deposit slip	Includes two hyphens
Payment received	Numeric	10	Deposit slip	Includes decimal point and commas

Comments:
Customer payments (checks, money orders, cash, and bank drafts) are received via mail, salespeople, or bank.

496 Appendix: Answers to Self-study Exercises

2.

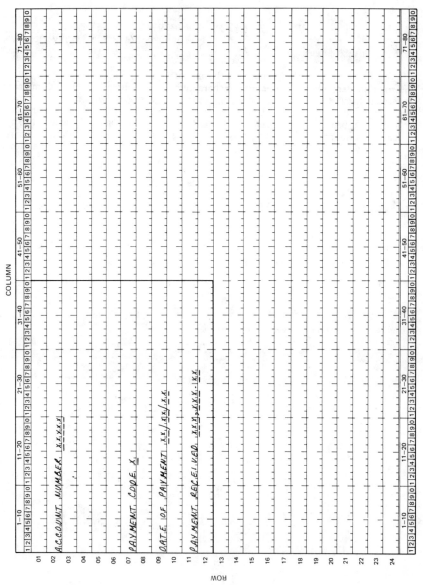

CHAPTER 12

True-False:

1. F
2. F
3. T
4. T
5. F
6. F
7. T
8. F
9. T
10. F

Fill-In:

1. classifying
2. updating
3. transaction
4. communicating
5. automated
6. economical
7. fast
8. corrections
9. data processing
10. system flowcharts

Problems:

1. Customer billing system flowchart: during the processing run, also update the master finished goods inventory file for items shipped.

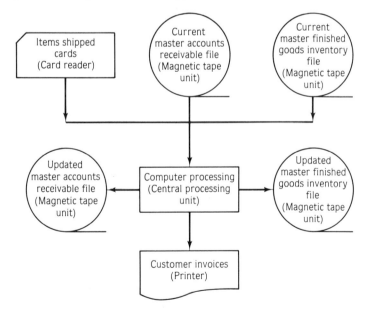

2. Weekly finished goods inventory system flowchart: for efficient processing, the two separate runs are combined. Also, inventory deductions will be handled by the customer billing program as set forth in problem 1 above.

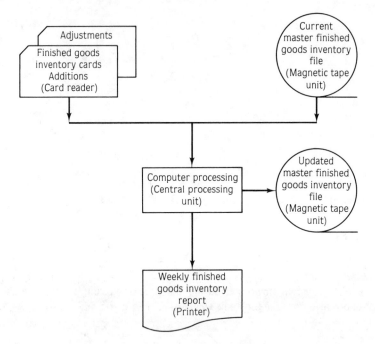

CHAPTER 13

True-False:

1. F
2. T
3. F
4. T
5. F
6. T
7. T
8. T
9. T
10. F

Fill-In:

1. human aspect
2. shipment processing
3. intelligent
4. field warehouse
5. predefined
6. total
7. product
8. modularity
9. specification
10. procedures

CHAPTER 14

True-False:

1. F
2. F
3. T
4. T
5. F
6. F
7. T
8. F
9. T
10. F

Fill-In:

1. accounting
2. control points
3. accounting control
4. management audit
5. suitability
6. reliability
7. key
8. batch
9. programmed
10. data base

Problems:

1. Several system controls are available to control inventory in an interactive information system environment. They are:
   - Use lockwords or passwords. These are matched by the sender before access to the inventory file is granted. They constitute several characters of a data file and should be changed every few days.
   - Use security lists. They indicate the type of inventory data the sender is permitted to receive.
   - Use limit checks or reasonable tests. These are magnitude tests designed to determine whether or not inventory data fall within predetermined limits. When data are less than or exceed the established limits, the on-line terminal operator follows instructions for exception items.
   - Check inventory data visually before executing the desired inventory program. The on-line operator should check the source document against the computer input values before executing changes (additions or deductions) to the on-line inventory file.
   - Use batch controls. Batch the inventory source documents in groups of twenty-five and total the quantity of additions. Compare these totals to those values accumulated on-line for the same batch.
2. Several system controls are available to control accounts receivable in an interactive information system environment. They are:
   - Use lockwords or passwords. These are matched by the sender before access to the accounts receivable file is granted. They constitute several characters of a data file and should be changed every few days.
   - Use security files. They indicate the type of accounts receivable data the sender is permitted to receive.
   - Use self-checking digit tests. This type of test is used extensively to check account numbers that are initially recorded by hand and later converted to the machine's lan-

guage. Accounts receivable programs should include provisions for reporting incorrect account numbers as soon as they are detected.
- Check accounts receivable number and corresponding data before executing the desired program. The computer operator should check the source document against the input values before executing changes (additions or deductions) in the on-line accounts receivable file.
- Use batch control totals. Batch the accounts receivable source documents in groups of twenty-five and total the values (debits and credits). Compare these totals to those values accumulated on-line for the same batch.

CHAPTER 15

True-False:

1. T
2. F
3. T
4. T
5. F
6. F
7. T
8. F
9. F
10. F

Fill-In:

1. control accountability
2. management by exception
3. check
4. edit
5. shipment processing
6. shipping terms
7. data base
8. record count
9. interactive controls
10. source listings

CHAPTER 16

True-False:

1. T
2. T
3. F
4. F
5. F
6. T
7. F
8. T
9. F
10. T

Fill-In:

1. equipment selection
2. system specifications
3. general information
4. system flowcharts
5. information
6. throughput performance
7. automation
8. delivery
9. decision table
10. one-third

Appendix: Answers to Self-study Exercises **501**

Problem:

1. Evaluation of the following decision table indicates that the third computer manufacturer (rule 3) should receive the contract.

Decision Table	Table name: CRITERIA TO SELECT EQUIPMENT MANUFACTURER												Page 1 of 1	
	Chart no.: FS–SEM–1	Prepared by: R. J. Thierauf										Date: July 25, 198_		
								Rule Number						
		1	2	3	4	5	6	7	8	9	10	11	12
Condition													
Major Criteria:													
High degree of automation proposed		Y	Y	Y	N								
Low-cost throughput performance		Y	N	Y	Y								
Low monthly rental cost		Y	N	Y	Y								
Programming assistance availability		N	Y	Y	Y								
Dependable and efficient software		Y	Y	Y	Y								
Backup equipment availability		N	Y	Y	Y								
Good reputation for meeting commitments		Y	Y	Y	Y								
Compliance with terms of bid invitation		Y	Y	Y	N								
Minor Criteria:													
Availability of equipment		N	Y	N	Y								
Acceptable quality of training		Y	Y	Y	Y								
Adequate equipment maintenance		Y	Y	Y	Y								
Free programming assistance		N	N	N	Y								
Action													
Add 10 points for each major criteria—yes (Y) answer		X	X	X	X								
Add 5 points for each minor criteria—yes (Y) answer		X	X	X	X								

Other Information:
Total points = 100 (8 major criteria × 10 points + 4 minor criteria × 5 points = 100)
Competitors' total points: **1**—70, **2**—75, **3**—90, **4**—80.

CHAPTER 17

True-False:

1. F
2. F
3. T
4. T
5. T
6. F
7. F
8. F
9. T
10. F

Fill-In:

1. system implementation
2. planning
3. system test
4. team building
5. system user
6. modular programming
7. system tested
8. parallel operations
9. interactive
10. periodic review

bibliography

Alexander, M. J. **Information Systems Analysis.** Palo Alto, California: SRA, Inc., 1974.
Brightman, Richard W. **Information Systems for Modern Management.** New York: The Macmillan Company, 1971.
Burch, John G., and Strater, Felix R., Jr. **Information Systems: Theory and Practice.** Santa Barbara, California: Hamilton Publishing Company, 1974.
Business Systems. Cleveland, Ohio: Systems and Procedures Association, 1966.
Cleland, David I., and King, William R. **Systems Analysis and Project Management.** New York: McGraw-Hill Book Company, 1975.
Clifton, H. D. **Data Processing Systems Design.** Princeton, New Jersey: Auerbach Publishers, 1971.
Condon, Robert, J. **Data Processing Systems Analysis and Design.** Reston, Virginia: Reston Publishing Company, Inc., 1975.
Couger, J. Daniel, and Knapp, Robert W. **Systems Analysis Techniques.** New York: John Wiley & Sons, Inc., 1974.
Fitzgerald, John M., and Fitzgerald, A. **Fundamentals of Systems Analysis.** New York: John Wiley & Sons, Inc., 1973.
Glans, Thomas B., et al. **Management Systems.** New York: Holt, Rinehart, and Winston, 1968.
Gore, Marvin, and Stubbe, John. **Elements of Systems Analysis.** Dubuque, Iowa: Wm. C. Brown Company Publishers, 1975.
Kindred, Alton R. **Data Systems and Management.** Englewood Cliffs, New Jersey: Prentice-Hall, Inc., 1973.
Lott, Richard W. **Basic Systems Analysis.** San Francisco: Canfield Press, 1971.
Lucas, Henry C., Jr. **The Analysis, Design, and Implementation of Information Systems.** New York: McGraw-Hill Book Company, 1976.

Mathews, Don O. **The Design of Management Information Systems.** Princeton, New Jersey: Auerbach Publishers, 1971.

Ramsgard, William C. **Making Systems Work: The Psychology of Business Systems.** New York: Wiley-Interscience, 1977.

Semprevivo, Philip C. **Systems Analysis: Definition, Process, and Design.** Palo Alto, California: SRA, Inc., 1976.

Shelly, Gary B., and Cashman, Thomas J. **Business Systems Analysis and Design.** Anaheim, California: Anaheim Publishing Company, 1975.

Silver, Gerald A., and Silver, Joan B. **Introduction to Systems Analysis.** Englewood Cliffs, New Jersey: Prentice-Hall, Inc., 1976.

Thierauf, Robert J. **Data Processing for Business and Management.** New York: John Wiley & Sons, Inc., 1973.

———. **Distributed Processing Systems.** Englewood Cliffs, New Jersey: Prentice-Hall, Inc., 1978.

———. **Systems Analysis and Design of Real-Time Management Information Systems.** Englewood Cliffs, New Jersey: Prentice-Hall, Inc., 1975.

Thierauf, Robert J., Klekamp, Robert C., and Geeding, Daniel W. **Management Principles and Practices: A Contingency and Questionnaire Approach.** Santa Barbara, California: Wiley/Hamilton, 1977.

Voich, Dan, Jr., Mottice, Homer J., and Shrode, William A. **Information Systems for Operations and Management.** Cincinnati, Ohio: South-Western Publishing Company, 1975.

Yourdon, Edward. **Design of On-Line Computer Systems.** Englewood Cliffs, New Jersey: Prentice-Hall, Inc., 1972.

index

ABC Company
   accounting and finance subsystem, 73-75
   accounts receivable, 162-65, 299-301, 389-90, 436
   concluding investigation, 146-53
   data processing organizational structure, 66-69
   detailed investigation, 134-46
   exploratory survey report to top management, 153-55
   finished product inventory, 158-61, 292-98, 385-88, 435
   introductory investigation, 127-34
   manufacturing subsystem, 71-72, 74
   marketing subsystem, 70-71, 74
   master case study, 64-66
   modular design of major systems, 267-68
   order-entry system, 126-55, 267-89, 359-83, 421-33
   overview of major subsystems, 68-75
   personnel subsystem, 74-75
   physical distribution subsystem, 73, 75
   purchasing and inventory subsystem, 72-73, 75
   research and development and engineering subsystem, 71, 74

Accuracy, 394-96
APOLLO, 206
Audio response, 198, 202

Batch controls, 404-405
Batch order-entry system, 126-55
Batch processing mode, 11-12, 17
   local batch, 12
   remote job entry, 12
Benchmark test, 446
Benefits and costs of MIS alternatives, 109-10, 112-19, 148-53
Bid invitations to manufacturers, 443-46
Block diagram, 41-43
Brainstorming, 173
Building block concept, 176, 250
Business organizations
   industrial, 63-64, 70-75
   service, 63-64, 75

Calling program, 461-63
Card punches, 306-309, 315
Charting techniques, 31
Classifying, 336-38, 343
Common date files, 19-21
Communicating, 336, 342-43

505

Computer programmers, 6
Computer scientists, 6
Concluding investigation, 110–19, 146–53
Consultant's role, 111
Continuous forms, 215–20
Conversational mode, 15
Cost displacement, 112
Cost of accuracy, 394–96
Creativity, 172–73

Data base, 228, 249
Data base administrator, 252
Data base management system, 257–59
Data code, 325
Data collection systems, 314
Data files maintained, 104, 106, 108, 134, 137–39, 146
Data management languages, 257
Data recorder, 307–309
Debugging, 463
Decision tables, 31, 43–48
  components, 44–45, 48
  evaluate MIS alternatives, 116–19
  value of, 43–44
Dedicated terminals, 16
Detailed investigation, 103–10, 134–46
Display tools, 31
Distributed data base, 228
Distributed data-entry systems, 311–12, 316
Distributed processing system concept, 18–23
  defined, 22–23, 25
  focus on local processing, 21
  hierarchical or tree network, 22
  order-entry system alternatives, 146–53
  relationship with centralized processing, 21
  three-level system approach, 21–22
Document flowchart, 31, 37, 43
Documenting the present system, 107–108

EBCDIC, 254
Editing, 215
Equipment selection, 7, 441–51
Equipment selection process
  approaches to, 442–43
  determine manufacturers, 443
  evaluate proposals, 446–48
  select manufacturers, 448–50
  submit bid invitations, 443–46
Equipment specifications, 441–42
Estimated savings, 112–19
Executive steering committee, 89–91
Exploratory survey report to top management, 119–20, 153–55

Feasible MIS alternatives, 110–19, 147–55
Feasibility study, 7, 87–88
Feedback, 20
File organization
  direct, 239–41, 243
  indexed sequential, 239, 241–43
  sequential, 239–40, 242
Files
  activity, 247–48, 281–82
  capacity, 248
  cost, 248
  direct (random) access, 11, 15
  magnetic disk, 232–36, 239
  magnetic drum, 236–37, 239
  magnetic tape, 230–32
  mass storage, 238–39
  master, 224
  processing speed, 248
  punched card, 228–30, 239
  sequential access, 11
  transaction, 227
  volatility, 281–82
Flowcharts
  block diagram, 41–43
  computer-prepared, 42
  defined, 32
  document, 31, 37, 43
  flow of symbols, 33
  flow process, 31, 37, 43
  procedural, 36–37, 43
  program, 31, 34–36, 41–43
  standard symbols, 33–36, 42
  system, 31, 34–37, 43
  type, 34–43
  value of, 32–33
Flow process chart, 31, 37, 43

Gantt, Henry, L., 49
Gantt charts, 48–51, 130–32
Geeding, Daniel W., 399

Index **507**

Historical aspects, 104–105, 108, 134, 136, 146
Human factor, 9
Human factors in new system, 180, 206–207, 246–47, 268–69

Incremental costs, 112–19
Infeasibility of new MIS alternatives, 120
Information
  constant, 215
  value, 394–95
  variable, 215
Input terminals, 312–14, 316
Inputs, 104–105, 108, 134–35, 146
Intangible benefits, 113–16, 148–50
Integrated data processing system, 18
Integrated management information system, 18
Intelligent terminal, 313, 316
Interactive processing mode, 15–17
  real-time, 16–17
  time-sharing, 15–17
Internal accounting control, 396, 398–99
Internal administrative control, 396, 399–400
Internal check, 396–99
Internal control, 104, 107–108, 134, 143–44, 146, 396–400
International Standards Organization, 33
Interview, 102–103
Introductory investigation, 7, 88–96, 127–34

Key-to-disk system, 306, 310–11, 315
Key-to-floppy-disk system, 306, 311, 316
Key-to-tape system, 306, 309–10, 315
Key verifier, 306–307, 315
Klekamp, Robert C., 399

Logical record, 231

Mainline program, 460–62
Management by exception, 204–205, 210, 401
Management by perception, 204–205
Management information system, 9–21
  batch, 10–11
  defined, 18–21, 25
  interactive, 12–15
  underlying concepts, 18–25
Manipulating, 336, 338–43
Master record, 324
Materials flow concept, 68
Method, defined, 106, 336
Methods and procedures, 104, 106, 108, 134, 139–42, 146
Microfilm or microfiche output, 198, 201
MIS task force, 89–92
Modular programming, 460–63
Modular system concept
  defined, 23, 25
  modular design, 231
  modular design method, 176–79
  modular programs, 23, 460–63

Off-line files, 251
On-line, 9–10
Operational review of system, 469–72
  benefit and cost analysis, 471–72
  evaluation of new equipment, 471
  examination of new system approaches, 471
Optical character readers, 314
Originating, 336–37, 343
Other analyses and considerations, 104, 107–108, 134, 146
Output analysis form, 209–10
Outputs, 104, 106–108, 134, 142–46, 196–221

Parallel operations, 465–66
Periodic review, 7
Personnel requirements, 109
PERT chart, 48–51
Physical block or record, 232
Plotter output, 198
Point-of-sale devices, 314
Printed output, 198, 202–203
Procedure, defined, 106, 336
Procedural flowchart, 36–37, 43
Program development, 460–65
Program flowchart, 31, 34–36, 41–43
Project description, 132–33

Real-time, 10
Real-time management information system, 16, 18

Recording, 336–38, 343
Records
  fixed length, 227–28
  variable length, 227–28
Remote job entry, 12
  off-line mode, 12
  on-line mode, 12
Report analysis sheet, 209–10
Reports
  batch processing, 210–12
  detailed exception, 210–11
  detailed printed, 210–11
  interactive processing, 212–14
  summary, 211–12
Responsibility reporting system, 18

Sampling techniques
  document, 47–48
  file, 47–48
  statistical, 45–48
Secondary storage output, 198, 200–201
Self-checking digit, 404, 425–26
Single record concept, 185
Sorting procedures, 339
Source document, 324–25
Standard flowchart symbols, 33–36, 42
Storage
  off-line, 228
  on-line, 228
Strategic control points, 400–401
Summarizing, 336, 342–43
System controls
  computer-programmed, 405–408, 424–28
  data base, 408–409, 428–30
  input, 403–405, 423–24
  interactive, 410–12, 431
  output, 409–10, 430–31
  overview, 431–32
  questionnaire, 413–15
  security, 412–13, 431
System data files (data base)
  analysis form, 252
  common types, 228–39
  design of, 246–59, 280–89
  determine content, 251–52, 256, 283–85
  human factor, 246–47, 256
  overview, 287–89
  review information, 247–48, 256
  specify requirements, 248–51, 256, 280–83
System flowchart, 31, 34–37, 43
System implementation, 7, 87–88, 457–69
  equipment selection and installation, 465
  preparatory work, 458–65
  system installation and conversion, 466–69
  system level testing, 465–66
System input
  analysis form, 323
  codes, 325
  common methods, 306–16
  design of, 318–27, 366–73
  determine input, 323–24, 363–66
  human factor, 319–20, 359–61
  overview, 373–74
  review information, 320
  specify requirements, 320–23, 361–63
System methods and procedures
  basic DP functions, 336–43
  design of, 345–52, 374–83
  determine the steps, 349–51, 378–80
  devise back-up procedures, 351–52, 380–81
  human factor, 346–48, 375
  introduction to design of, 335–36
  overview, 381–82
  review information, 348
  specify requirements, 348–49, 376–77
System output
  common types of, 198–203
  design of, 205–20, 275–77
  determine the content, 209, 221, 271–75
  human factor, 206–207, 220, 268–70
  introduction to design of, 197
  overview, 279–80
  review information, 207–208, 219
  specify requirements, 208–209, 221, 270–71
  use and distribution, 220–21, 277–79
System project affects on . . .
  communication process, 86–87
  entire organization, 85–87
  financial resources, 86–87
  major steps, 87–88
  organizational structure, 86–87, 128
  personnel, 85–87
  schedule, 95–96, 130–32

System project affects on . . . (cont.)
  scope, 92–96, 128–30
  subsystems (business functions), 86–87
Systems analyses, 7–9, 103–19, 134–46
  concluding investigation, 7, 110–19, 146–55
  detailed investigation, 7, 103–10, 134–46
  introductory investigation, 7, 88–96, 127–34
  methods of, 102–103
  objectives of, 101–102
Systems analysts, 5–6
Systems and procedures manuals, 102–103
Systems design, 7–9
  approached to, 171–72
  design the new system, 8–9
  determine requirements for the new system, 8
  imagination in, 172–73
  review appropriate data, 8
  steps in, 178–90
Systems design principles
  data files, 243–46
  input, 316–18, 360–61
  methods and procedures, 343–45, 375–76
  output, 203–205
  system controls, 400–403, 422–23
Systems planning, 6–7

Tangible benefits, 112–19, 148–53
Tape rings, 409
Time-charting techniques, 48–51
  Gantt, 48–51, 130–32
  PERT, 48–51
Thierauf, Robert J., 16, 21, 399, 404, 415, 442
Throughput performance, 446
Transaction record, 324
Turnaround document, 198–200

United Airlines, 206
United States of America Standards Institute (USASI), 33
Updating procedure, 339–42

Visual output, 198–99

Well-designed system
  characteristics of, 173–76
  modular design method, 176
Work volume, 108–109